Sharp Focusing of Laser Light

Sharp Focusing of Laser Light

Victor V. Kotlyar, Sergey S. Stafeev, and
Anton G. Nalimov

CRC Press
Taylor & Francis Group
Boca Raton London New York

CRC Press is an imprint of the
Taylor & Francis Group, an **informa** business

CRC Press
Taylor & Francis Group
6000 Broken Sound Parkway NW, Suite 300
Boca Raton, FL 33487-2742

First issued in paperback 2023

ISBN-13: 978-0-367-36444-1 (hbk)
ISBN-13: 978-1-03-265417-1 (pbk)
ISBN-13: 978-0-429-34607-1 (ebk)

DOI: 10.1201/9780429346071

Publisher's Note
The publisher has gone to great lengths to ensure the quality of this reprint but points out that some imperfections in the original copies may be apparent.

Visit the Taylor & Francis Web site at
http://www.taylorandfrancis.com

and the CRC Press Web site at
http://www.crcpress.com

Contents

Preface

In optical instrumentation and optical information systems, one of the limitations on the amount of information transmitted, on the resolution of optical memory devices, and on the minimum size of electronic microcircuits on microchips, is the diffraction limit. The diffraction limit was discovered in 1873 by Ernst Abbe and it means that light, including laser light, cannot be focused to a point. The size of the minimum focal spot in the half-maximum of the light intensity is equal to half the wavelength in the medium under consideration. This property of light follows from the general Heisenberg uncertainty relations, which for optics can be formulated as follows: the product of the width of a light beam by the width of its spatial spectrum cannot be less than a certain constant value. Therefore, to compress light into a region shorter than half the wavelength is an important task of photonics. The smaller the size of the focal spot, the smaller the line size in photolithography, the more information can be recorded on the optical disc, the greater the resolution that can be achieved in microscopy, and the easier it is to manipulate microparticles with the help of light. The diffraction limit can be reduced by choosing radiation with a shorter wavelength and by choosing a material for immersion with a large refractive index. It turns out that with sharp focusing of laser light with inhomogeneous polarization (radial or azimuthal), the light can be focused into a spot with a size 1.5 times smaller than the diffraction limit, only due to polarization effects in focus. Focusing light near the media interface, due to the constructive interference of transmitted light and surface waves and due to solid-state immersion, also makes it possible to reduce the size of the focal spot by several times. The diffraction limit can be overcome with the help of diffraction (nanostructured) optical elements focusing light near its surface. Due to the fact that the light is focused near the surface, firstly, the magnitude of the focal spot can be influenced by inhomogeneous surface (evanescent) waves that propagate along the interface, and, secondly, the focusing optical element can have small dimensions (tens and hundreds of micrometers). In this case, it is a component of microoptics. These include microaxicons (refraction and diffraction), zone plates, photonic crystal microlenses, gradient microlenses, Mikaelian lens, micropolarizers, and metalens. This book is devoted to such components of microoptics, which make it possible to overcome the diffraction limit and focus the laser light with super-resolution. The book contains research results financially supported by the Russian Science Foundation grant No. 18-19-00595.

Acknowledgment

The authors are grateful to Dr A.A. Kovalev for modeling the focusing of the logarithmic axicon and the propagation of hyper-geometric beams in a gradient-index waveguide. The authors are also grateful to Dr E.S. Kozlova for modeling the focusing of a femtosecond pulse by a microcylinder. The authors are especially grateful to Dr Liam O'Faolain and Dr M.V. Kotlyar for assistance in the manufacture of photonics components.

Authors

Victor V. Kotlyar is Head of the Laboratory at the Image Processing Systems Institute of the Russian Academy of Sciences, a branch of the Federal Scientific Research Center, "Crystallography and Photonics," and Professor of Computer Science at Samara National Research University. He received his MS, PhD and DrSc degrees in physics and mathematics from Samara State University (1979), Saratov State University (1988), and Moscow Central Design Institute of Unique Instrumentation, the Russian Academy of Sciences (1992). He is co-author of 300 scientific papers, 6 books, and 7 inventions. His current interests are diffractive optics, gradient optics, nanophotonics, and optical vortices.

Sergey S. Stafeev (b. 1985) received his master's degree in applied mathematics and physics from Samara State Aerospace University (2009). He received his PhD from Samara State Aerospace University (2012). He is researcher at the Laser Measurements Laboratory at the Image Processing Systems Institute of the Russian Academy of Sciences, a branch of the Federal Scientific Research Center "Crystallography and Photonics." His scientific interests are: diffractive optics, FDTD methods, near-field optics.

Anton G. Nalimov (b. 1980), graduated from Samara State Aerospace University in February, 2003. He entered into postgraduate study in 2003 with specialty "Mathematical Modeling and Program Complexes," completing it in 2006 with specialty "Optics." He works at the Technical Cybernetics Department at Samara National Research University as an associate professor, and as a scientist at the Image Processing Systems Institute of the Russian Academy of Sciences, a branch of the Federal Scientific Research Center "Crystallography and Photonics." He is a candidate in physics and mathematics, and co-author of 130 papers and 3 inventions.

Introduction

In their fundamental work *Principles of Optics* [1], M. Born and E. Wolf gave a brief historic overview of the first studies dealing with diffraction of light by a circular pinhole and the description of the electromagnetic field in the focus vicinity. E. Lommel [2] was the first to represent the electromagnetic field in the focus vicinity as a converging Bessel series. In Reference [1], the relationship describing a three-dimensional distribution in the region of a paraxial focus was expressed via the Lommel functions. Almost at the same time, H. Struve published his work [3] concerned with an approximate calculation of the intensity distribution near the shadow boundary due to diffraction of light by a circular pinhole. Later, P. Debye [4] established general patterns of the behavior of light near the focus, offering a familiar Debye integral to describe the expansion of the near-focus field into plane waves. However, when describing the light field, the above works relied on a scalar approach. V.S. Ignatowsky was the first [5] to derive formulae for the electric and magnetic vectors of the electromagnetic field in the image plane of an aplanatic system. The modern description of a vector light field near the focus of an aplanatic system was proposed in a classic work by B. Richards and E. Wolf [6], with all earlier derived formulae [2–5] being particular cases of the relations presented in Reference [6]. It is by using the Richards–Wolf formulae that the generation of focal spots was analyzed in a well-known monograph on nano-optics by L. Novotny and B. Hecht [7]. A different approach to analyzing the near-focus light field was taken in a well-known book by J.J. Stamnes [8], where asymptotic techniques for calculating rapidly oscillating integrals were treated using a scalar diffraction theory. However, considering that in References [1, 8] the description of the light field in the focal spot was conducted in terms of a scalar theory, this approach is unsuited when dealing with tightly focused light. When light is focused in an optical system with high numerical aperture (NA $= \sin\theta$, where θ is half of the maximum angle of light rays in the focus), the focal spot structure essentially depends on the polarization state of the incident wave. By way of illustration, a sharply focused linearly polarized beam has been known to produce an elliptic focal spot [6].

In a paraxial focus (for a low-NA system), the intensity in the focal plane is described by an Airy function $2J_1(x)/x$ [9], where $J_1(x)$ is the Bessel function of the first kind and first order. Based on this formula, the diffraction limit can be defined as the Airy spot diameter at full-width half-maximum intensity (FWHM), which is equal to FWHM $= 0.5\lambda/\mathrm{NA}$, where λ is the wavelength of light in free space. This half-wavelength diffraction limit defines the minimal resolution of an optical microscope, which was established by E. Abbe in 1873 [10].

In this book, we look into numerous examples of how to go beyond the diffraction limit by sharply focusing light near the surface of microoptic components. Why specifically microoptic components and not conventional optical elements? The fact is that the propagation of light through and its tight focusing by an optical element (a spherical lens or a zone plate) can adequately be described only by solving a complete set of Maxwell's equations, e.g. using a popular finite-difference time-domain

(FDTD) method [11]. Because of this, the propagation of light through a medium (optical element) is described in terms of linear Maxwell's theory, with the dielectric permittivity of the medium defined as a function of three Cartesian coordinates. Nonlinear effects as well as magnetic and anisotropic media are beyond the scope of this book. The majority of the numerical examples in this book have been implemented with an FDTD method. A number of illustrative examples have been obtained using vector Richards–Wolf formulae [6], but this approach only applies to focal lengths much larger than the incident wavelength. In the meantime, in this book we discuss only microoptic components that focus light at a wavelength distance. The limited computational capabilities of modern desktop computers mean that within a reasonable time and using the FDTD method it is possible to design optical elements and simulate the focusing of light within a three-dimensional space measuring dozens (up to a hundred) of wavelengths. This is the first reason why in this book we treat the tight focusing of light using microoptic components. The second consideration behind the choice of microoptic components is that the diffractive optical elements (DOE) – be that an axicon, a zone plate, or a metalens – are fabricated by electron beam lithography and reactive ion etching in a transparent substrate. Hence, the smaller the size, the less it takes to fabricate the element and the lower is the cost. Besides, if the focal length is several wavelengths, the outermost zone of the diffraction lens, whose radius is three to four times the focal length, appears to be smaller than the incident wavelength, meaning that light is not deflected to the first diffraction order. In this book, we discuss a binary axicon and a Fresnel zone plate, both 30 μm in diameter, fabricated by electron beam lithography in a quartz plate for the visible spectrum.

Other graded-index microoptic components treated in this book include a parabolic lens in the form of a fragment of a graded-index fiber, a planar graded-index Mikaelian microlens, Luneberg lenses, and a Maxwell fisheye lens. With light rays steeply converging within a graded-index focusing element, light is focused near the surface and the NA can be efficiently increased without increasing the diameter. Using the graded-index lenses as an example, it is demonstrated that a subwavelength focal spot smaller in size than the diffraction limit can be generated. A photonic crystal analog of the planar Mikaelian lens, whose refractive index is described by a hyperbolic secant, has been fabricated in a silicon waveguide.

Optical metasurfaces are also given consideration at some length, with the examples including metalenses synthesized in a thin-film amorphous silicon. The material has been chosen due to its high refractive index, allowing a phase shift of a half-wavelength to be realized in a nanostructured film as thin as just 120 nm. The metasurface enables the polarization, amplitude, and phase of incident light to be controlled simultaneously. The polarization plane tilt depends on the tilt of space-variant diffraction gratings. The phase shift depends on the angle between the grooves of the adjacent local gratings, so that the mutually perpendicular grooves result in a phase shift of π between of the adjacent gratings. Using fabricated metalenses with near-unity NA, subwavelength focal spots smaller in size than the diffraction limit are generated.

There is an interesting optical phenomenon that is briefly looked at in this book. When light is sharply focused on the optical axis, under certain conditions, there occurs a reverse flow of light energy. Mathematically speaking, the longitudinal component of the Poynting vector, which defines the magnitude and direction of the on-axis energy flow at each near-focus point, can take negative values. Here, it is shown that the reverse near-focus energy flow can be comparable in magnitude with the energy of incident light.

For our readers, we need to point out other books dealing with the description of light behavior in a subwavelength volume [12–14]. Topics of design, fabrications, and operation of nanophotonics devices are treated in Reference [12]. However, no consideration has been given to nanophotonics devices intended to focus light. The propagation of light in nanostructured media, including photonic crystals, is treated in detail in Reference [13], where quantum optical effects and optical properties of quantum dots are also discussed. In Reference [14], questions of attaining the resolution limit in microscopy are treated in detail using statistical analysis. Thus, the well-known books on nanophotonics are very interesting, being concerned with fundamental issues of light behavior on a subwavelength-scale level. At the same time, the description of light behavior in the vicinity of a sharp focus has not been given sufficient consideration in the mentioned books. In this book, we intend to "focus" on issues of sharp focusing of light.

With this brief introduction, we have tried to answer three questions that a potential reader may ask after browsing through the Table of Contents; who was historically the first to describe the behavior of light in the focus, what research methods were proposed, and what is the meaning of the diffraction limit? In which way the interaction of light with a microoptic element that generates a near-surface subwavelength focal spot can be rigorously described? What was the authors' motivation when choosing the topics to be included in the book?

1 Focusing Laser Light by Axicon and Zone Plate

1.1 MODELING THE SHARP FOCUS OF A RADIALLY POLARIZED LASER MODE USING A CONICAL AND A BINARY MICROAXICON

When focusing the electromagnetic field in free space in a region removed by more than a wavelength from objects, the focal spot cannot be made smaller than the diffraction limit, which is generally defined by the indeterminacy relation [15]: $k_x d \geq 2\pi$, where d is the focal spot diameter and k_x is the wave vector projection onto an axis perpendicular to the optical axis (the z-axis of beam propagation). Since the maximum value of k_x is $k_0 = 2\pi/\lambda$ (wave number), λ is the wavelength, the equation can be replaced with $d \geq \lambda$. For example, in Reference [7] the indeterminacy relation [15] is given in a different form: $\Delta x \geq \lambda/2\pi$, where Δx is a minimal width (diameter) of the focal spot. It stands to reason that in the formulae λ stands for the wavelength of light in vacuum. When focusing light in a homogeneous dielectric medium, the magnitude λ needs to be replaced with λ/n, where n is the refractive index of the medium.

However, what we describe above does not take place when focusing light in the proximity of the media interface or imaging (observing) the object in the near field where there are inhomogeneous surface (evanescent) waves. It was proposed [16] that an ideal image of source in the near field should be formed by means of a medium with the negative refractive index (the so-called superlenses, $n < 0$). A superresolution image was first modeled and then experimentally obtained in the near-field diffraction zone using a superlens [17, 18]. The superlens was made of a thin silver layer of thickness 50 nm. In References [19, 20] it was shown that for an ideal superlens the focal length f was in direct proportion to the resolution Δ (or the 2D focal spot diameter): $f \approx \Delta$, if $0 < f < 0.3\lambda$, and $\Delta = \lambda/2$, if $f > \lambda$. For a superresolution image to be optically resolved, the superlens needs to form a magnified image. By way of illustration, a cylindrical superlens ($n = -1$) of annular cross-section was proposed in Reference [21]. There are various types of practically implemented superlenses for the visible and IR ranges [22–28]. The evanescent light fields capable of producing a near-field focal spot of a diameter much smaller than the wavelength of light used were theoretically dealt with in Reference [29]. It stands to reason that the evanescent electromagnetic field modulation cannot be associated with the radiation propagation, most likely being linked with electrostatic effects taking place in the medium interface vicinity. In Reference [30] it was numerically shown that a hyperbolic lens to form a near-field subwavelength magnified image could be implemented not only as a cylindrical lens but as a plane-parallel layer as well.

There have been publications reporting the focusing of light by means of 2D photonic crystals (PhC) [31–34]. The superfocusing of a Gaussian beam of waist full-width at half-maximum (FWHM) = 3.2λ using a 4-layer 2D PhC slab consisting of a four-row lattice of cross-shaped holes was modeled in Reference [31]. A focal spot of diameter FWHM = 0.25λ was shown in Reference [32] to be formed inside the PhC slab. For transverse electric polarization (TE-polarization), such a PhC slab was shown to have a negative refraction ($n = -1$), producing a focal spot of size FWHM = 0.4λ. The focusing properties of a 2D PhC with negative refraction composed of an array of magnetic and dielectric nanorods were modeled in Reference [33]. The PhC superlens focused a weakly diverging Gaussian beam within itself into a focal spot of diameter FWHM = 1.4λ [34].

If the periodic holes in a PhC are varying in diameter, the PhC is referred to as a gradient PhC. The gradient 2D PhC lenses focus light outside the PhC surface [35, 36]. The minimal focal spot of size FWHM = 0.54λ was larger than the diffraction limit [35]. In Reference [36] a similar PhC lens with a triangular hole lattice in Si ($n = 3.46$) was modeled. The resulting focal spot diameter was FWHM = 1.27λ.

Reaching the subwavelength focusing or superresolution is possible using the solid immersion lens (SIL) [7], the numerical aperture increasing lens (NAIL) [37], the solid immersion axicon (SIAX) [38], and the SIL of the wavelength scale (nSIL) [39]. By immersing the object in a high refractive index medium, the diffraction limit is thus rescaled to FWHM = $\lambda/2\,n$ [7]. In Reference [37] it was shown that using NAIL on silicon one can theoretically obtain the resolution limit FWHM = 0.14λ. The experimentally achieved resolution was FWHM = 0.23λ [37]. A scheme proposed in Reference [38] used the SIL axicon to generate a Bessel beam. It is possible to form a focal spot with a diameter equal to FWHM = 0.36λ/nNA, NA = $\sin\varphi$, φ is equal to the convergence angle of the Bessel beam. In Reference [39] it was demonstrated that the use of an objective-lens-nanoSIL system reduced the FWHM of the focal spot by more than 25% compared to a conventional macroscopic SIL. The size of the focal spot was FWHM = 0.23λ [39], but the sidelobe intensity was 0.3 of the maximum intensity in the spot center. Using an axicon that forms the Bessel beam the minimal focal spot diameter is FWHM = 0.36λ/NA, where NA is the numerical aperture of the axicon [40–42].

In this section, we analyze the sharp focus of radially polarized laser light by means of microoptics. Thus, it would be appropriate to briefly discuss methods for generating radially polarized laser beams. Why is it just the radially polarized light that is used? The reason is that when focusing the linearly and elliptically polarized light the radial symmetry of the focused beam is disturbed: the focal spot has an elliptic shape and, besides, the axial component of the electric vector in the focal region is small compared to the transverse components. The other types of polarization (azimuthal and radial) result in light beams of annular cross-section, which produce either annular (azimuthal polarization) or circular (radial polarization) focal spots. Besides, the azimuthal polarization is devoid of the axial component of the electric vector even in the sharp focus region, whereas the radial polarization can result in a much greater axial component of the electric vector compared to the transverse component in the focal region. Note that it is the axial component of the electric vector of the electromagnetic wave which is responsible for the sharp focus of the radially polarized beam.

It has been known that radially polarized laser beams can be generated by use of modified laser cavities [43–46] or subwavelength optical elements [44, 47, 48]. The beam was experimentally shown to present an R-TEM$_{01}$ mode, retaining its annular structure upon propagation. The radial component of such a beam is given by

$$\vec{E}_r\left(r\right) = \vec{e}_r\left(\frac{r}{w}\right)\exp\left\{\frac{-r^2}{w^2}\right\}, \tag{1.1}$$

where r is the radial coordinate at the beam section, w is the Gaussian beam waist radius, and $\vec{e}_r$ is a unit vector on the radial coordinate.

Thus, it can be seen from the aforesaid that there have been few publications dealing with focusing the laser light into a near-interface subwavelength region beyond the diffraction limit by means of conventional refractive and diffractive optics and an axicon. In the SIL, NAIL, nSIL, and SIAX schemes multiple optical elements were used; however, we obtain a subwavelength focal spot using only a microaxicon. In this section, based on the Radial Finite-Difference Time-Domain (R-FDTD) method [49] we show that when a conventional glass ($n = 1.5$) microaxicon (NA = 0.6) is illuminated by a radially polarized laser mode R-TEM$_{01}$ in the direct vicinity of the surface, it becomes possible to generate a focal spot of diameter FWHM = 0.39λ (for a binary axicon) and FWHM = 0.30λ (for a refraction axicon). This value is smaller than the previously achieved value of FWHM = 0.40λ in Reference [49] or that reported in References [32, 34–36], being far smaller than the diffraction limit of FWHM = 0.51λ. It is also smaller than the diameter of a focal spot formed by a lens with NA = 0.6 in an immersion medium $n = 1.5$: FWHM = 0.51λ/(nNA) = 0.55λ, smaller than the diffraction limit in the medium ($n = 1.5$, NA = 1) FWHM = 0.51λ/n = 0.33λ and smaller than the focusing limit obtained for the zero-ordered Bessel beams 0.36λ [40–42].

This sharp focal spot is formed near the top (longitudinal focus width is FWHM$_z$ = 0.12λ) of the glass axicon (full angle at the vertex of the axicon $2\alpha = 100°$) because of a surface wave with an amplitude $J_0(krn\cos\theta')$ (where $J_0(x)$ is the Bessel function, $\theta' = 90° - \alpha = 40°$ is the angle of light incidence onto the conical axicon surface) propagating along the axicon surface and focusing on its top. Therefore the full width at half-maximum of the focal spot can be estimated from the condition $J_0^2(1.1) \approx 0.5$, being equal to FWHM = 0.31λ.

Using a surface plasmon [50] one can reduce the diameter of the focal spot near the surface of an axicon to λ/50. But in this case because the surface plasmon exists only near the surface, they can be used in a near-field optical microscope only in the reflection mode. In our case we use a glass axicon, thus forming a light field with diameter λ/3.3 near its top, but it can be used in the near-field transmission microscope.

A specific feature of this focusing is that converging near the vertex of the axicon is not only the surface wave that propagates on the surface of the axicon, but also a conical wave of amplitude $J_0(kr\sin\varphi)$, where φ is the half angle at the vertex of the conical wave (in our case $\sin\varphi = 0.6$). The laser beam diameter formed by the conical wave is FWHM = 0.35λ/$\sin\varphi$ = 0.58λ. The surface wave remains near the axicon surface, and the conical wave continues its propagation along the optical axis. The diameter of the focal spot on the optical axis is gradually increasing.

1.1.1 RADIALLY POLARIZED LASER MODE R-TEM$_{01}$

In Reference [43] the central section of the R-TEM$_{01}$ mode was shown to coincide within a high accuracy with the Hermite–Gauss (HG) mode (0.1). Therefore, it becomes possible to put down the propagation of such a mode analytically and then model how it can be focused by microoptics. The Hermite–Gauss mode takes the form

$$E_{mn}(x,y,z) = \left(\frac{\sigma_0}{\sigma(z)}\right)\exp\left\{i(m+n+1)\eta(z) - \frac{ik(x^2+y^2)}{2R(z)} - \frac{(x^2+y^2)}{\sigma^2(z)}\right\}$$
$$\times H_m\left(\frac{\sqrt{2}x}{\sigma(z)}\right)H_n\left(\frac{\sqrt{2}y}{\sigma(z)}\right),$$
(1.2)

where $\eta(z) = \arctan(z/z_0)$, $R(z) = z(1+z_0^2/z^2)$, $\sigma(z) = \sigma_0(1+z^2/z_0^2)^{1/2}$, $z_0 = k(\sigma_0)^2/2$, $H_n(x)$ is the Hermite polynomial, $H_0(x) = 1$, $H_1(x) = 2x$, and $\sigma_0 = w$ is the Gaussian beam waist radius. The radially polarized laser mode (1.1) can be represented as a sum of two linearly polarized modes $E_{1,0}$ and $E_{0,1}$, being polarized along the x- and y-axis, respectively:

$$E_r(x,y,z) = \left(\frac{2\sqrt{2}\sigma_0}{\sigma^2(z)}\right)\exp\left\{i2\eta(z) - \frac{ik(x^2+y^2)}{2R(z)} - \frac{(x^2+y^2)}{\sigma^2(z)}\right\}(x\vec{e}_x + y\vec{e}_y),\quad (1.3)$$

where the Cartesian unit vectors are put in the parenthesis.

Introducing a unit vector $\vec{e}_r$ on the polar system radius, Equation (1.3) can be replaced by the final relations for the electric field strength of the radially polarized laser mode R-TEM$_{01}$

$$\vec{E}_r(x.y.z) = \left(\frac{2\sqrt{2}\sigma_0}{\sigma^2(z)}\right)\exp\left\{i2\eta(z) - \frac{ikr^2}{2R(z)} \frac{r^2}{\sigma^2(z)}\right\}r\vec{e}_r.$$
(1.4)

1.1.2 FOCUSING THE LASER MODE R-TEM$_{01}$ WITH A MICROAXICON

At $z=0$ (at the waist plane where the wavefront is plane) Equation (1.4) changes to Equation (1.1). Figure 1.1 shows an absolute value of the amplitude of the mode R-TEM$_{01}$ for the Gaussian beam waist radius of $w=3$ μm.

From Equation (1.1) it follows that the field amplitude takes a maximal value when $r = w/\sqrt{2} = 2.14$ μm. When the above-described mode of wavelength $\lambda = 1$ μm is incident on a glass ($n=1.5$) microaxicon of height $h=6$ μm and radius $R=7$ μm, with the radial section as shown in Figure 1.2, a sharp intensity peak is formed near the on-axis axicon vertex: $|E|^2 = |E_r|^2 + |E_z|^2$.

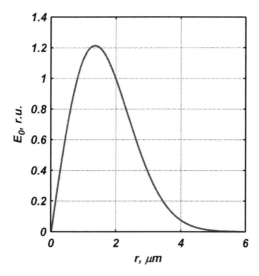

FIGURE 1.1 The (absolute value of) radial component of the electric field strength of the mode R-TEM$_{01}$.

Numerical aperture of the axicon (Figure 1.2) is given by

$$\text{NA} = \frac{nh/R - \sqrt{(h/R)^2 - n^2 + 1}}{(h/R)^2 + 1}, \tag{1.5}$$

in our case NA = 0.6.

Figure 1.3(a) depicts the axial (longitudinal) intensity distribution $|E|^2$ within and outside the axicon (Figure 1.2). Shown in Figure 1.3(b) show a scale-up pattern: the intensity distribution region near the axicon vertex. The vertical lines in Figure 1.3 designate the axicon boundaries.

In our case, only three components of the electromagnetic field are non-zero: $E_{r,0}$, $E_{z,0}$, $H_{\varphi,0}$. These are the radial and longitudinal components of the electric field and the azimuthal component of the magnetic field. The zero indices mean that all three components are independent on the azimuthal angle. Therefore, six Maxwell's equations for radially polarized light reduce to three equations [49]:

$$-\frac{\partial H_{\varphi,0}}{\partial z} = \varepsilon\varepsilon_0 \frac{\partial E_{r,0}}{\partial t} + \sigma E_{r,0}, \tag{1.6}$$

$$\frac{1}{r}\frac{\partial(rH_{\varphi,0})}{\partial r} = \varepsilon\varepsilon_0 \frac{\partial E_{z,0}}{\partial t} + \sigma E_{z,0}, \tag{1.7}$$

$$\frac{\partial E_{r,0}}{\partial z} - \frac{\partial E_{z,0}}{\partial r} = -\mu\mu_0 \frac{\partial H_{\varphi,0}}{\partial t}, \tag{1.8}$$

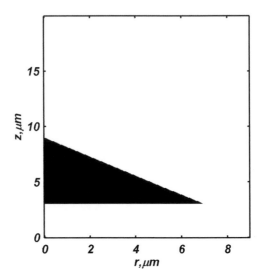

FIGURE 1.2   Radial section of a conical microaxicon of radius $R = 7$ μm and height $h = 6$ μm.

where ε and μ are the relative electric permittivity and magnetic permeability of the optical element material (further, $\mu = 1$), ε_0 and μ_0 are the electric permittivity and magnetic permeability of vacuum, σ is the relative conductivity (further, $\sigma = 0$).

The calculation was conducted using the R-FDTD method [49] with space sampling interval $\lambda/50$ and time sampling $T/100$, where T is the electromagnetic oscillation period. It can be seen from Figure 1.3(b) that the subwavelength focusing of light occurs at a distance of 0.02 μm from the axicon surface, with the axial (longitudinal) focal spot size at half-intensity being equal to $\text{FWHM}_z = 0.08\lambda$.

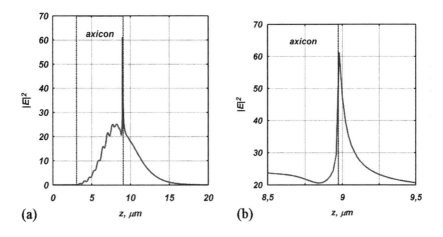

FIGURE 1.3   (a) Intensity distribution $|E|^2$ on the optical axis when focusing the R-TEM$_{01}$ mode (Figure 1.1) with the microaxicon of Figure 1.2; and (b) a scaled-up fragment of the curve in (a).

Shown in Figure 1.4 is the radial intensity distribution $|E_z|^2$ (curve 3), $|E_r|^2$ (curve 2), and $|E|^2=|E_r|^2+|E_z|^2$ (curve 1) in the focal plane at a 20-nm distance from the axicon apex (Figure 1.3).

When the axicon is illuminated by a wave of maximal amplitude of 1.2 arbitrary units (Figure 1.1), the maximal intensity $|E_z|^2$ in the focus (Figure 1.4) is about 50 arbitrary units. Thus, the focus intensity is 50 times larger than the illuminating field intensity. The focal spot diameter (Figure 1.4) at half-intensity is FWHM=0.30λ, with the focal spot area at half-intensity being HMA=0.071λ^2. Note, for comparison, that the said diameter is 1.7 times smaller than the diameter of the Airy spot (FWHM=0.51λ), whereas the spot area is 2.87 times smaller than the Airy spot area (HMA=0.204λ^2). Figure 1.5 shows the FWHM (curve 1) and intensity (curve 2) in the focal spot as a function of the axicon height.

The curves in Figure 1.5 confirm the optimality of the selected parameters. In Figure 1.5 it is shown that with other parameters of the axicon (different from the height $h=6$ μm and radius $R=7$ μm) the diameter of the focal spot at the vertex of the axicon is greater and the intensity is smaller.

The size of the near-apex focal spot (Figure 1.4) can be estimated theoretically using the following reasoning. The incidence of light onto the conical surface is nearly at the total internal reflection angle $\sin\theta = 1/n = 0.67 \geq \sin\theta' = h^2/(R^2+h^2)=0.65$, where θ is the angle of total internal reflection and θ' is the angle of light incidence onto the conical axicon surface.

Note that the surface wave propagates along the conical surface toward the axicon apex, producing a focus both inside and outside the axicon apex vicinity (Figure 1.3(b)), with the focus amplitude described by the zero-order Bessel function

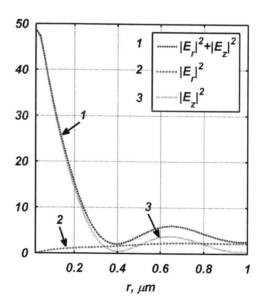

FIGURE 1.4 Radial intensity distributions: $|E_z|^2$ (curve 3), $|E_r|^2$ (curve 2), and $|E|^2=|E_r|^2+|E_z|^2$ (curve 1) in the focal plane 20-nm distance from the axicon apex (Figure 1.3).

$J_0(krn\cos\theta')$. Since $J_0^2(1.1)\approx 0.5$, then $krn\cos\theta' = \dfrac{2\pi rnR}{\lambda\sqrt{R^2+h^2}} = 0.35\pi$, whence it follows that the focal spot diameter at half-intensity can be estimated to be

$$\text{FWHM} = 2r = \frac{0.35\sqrt{R^2+h^2}}{nR}\lambda \approx 0.31\lambda. \tag{1.9}$$

This value is close to that following from Figure 1.4: FWHM=0.30λ.

The fact that in the resulting focal spot (Figure 1.3 and Figure 1.4) the transverse size (FWHM=0.30λ) is larger than the longitudinal size (FWHM=0.08λ) corroborates that the wave propagation has mainly been not along the z-axis but along the inside and outside surface of the axicon cone at angle arctan(R/h)=arctan(1.2)>$\pi/4$ to the optical axis. The exponential distribution of the longitudinal intensity near the surface of the axicon (Figure 1.3(b)) is given by exp(–2$kzn\sin\theta'$). From this expression the longitudinal width of the intensity evanescent Bessel beams is:

$$\text{FWHM} = \frac{\ln 2}{4\pi n\sin\theta'}\lambda \approx 0.07\lambda. \tag{1.10}$$

This value is close to that following from Figure 1.3(b): FWHM=0.08λ.

Serving to prove that nearly all radiation incident on the axicon contributes to the focal spot is the comparison between the maximal intensity in the focus $|E|^2 \approx 60$ (arbitrary units, see Figure 1.3) and the ratio of the incident beam energy to the wavelength squared (see Equation (1.4) for z=0):

$$W_0/\lambda^2 = \left(\frac{2\sqrt{2}}{w\lambda}\right)^2 \iint r^2 \exp\left\{-2\frac{r^2}{w^2}\right\} r\,dr\,d\varphi = 2\pi\left(\frac{w}{\lambda}\right)^2 \approx 56.5. \tag{1.11}$$

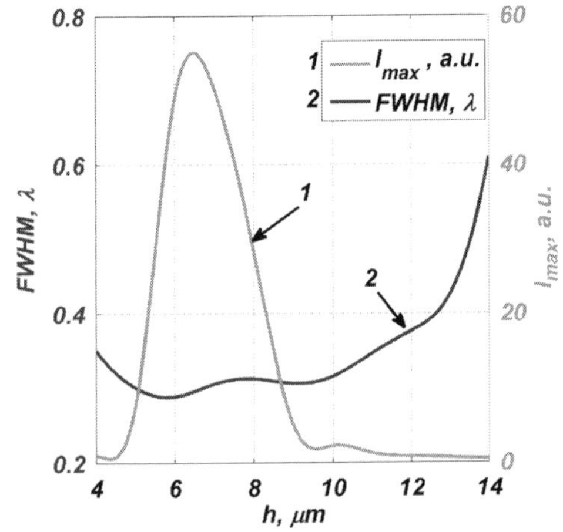

FIGURE 1.5   FWHM (curve 1) and intensity I_{max} in the focal spot (curve 2) as a function of the axicon height h.

1.1.3 FOCUSING THE R-TEM$_{01}$ MODE WITH A BINARY AXICON

A binary axicon is easier to manufacture compared to the conical axicon discussed in the previous section. The binary axicon can be manufactured by photolithography using a single binary amplitude mask composed of concentric bright and dark rings of equal width. Shown in Figure 1.6 is the radial profile of a binary microaxicon corresponding to the conical axicon in Figure 1.2.

The axicon in Figure 1.6 has a step height of $H = \lambda/2(n-1) \approx 633$ nm for wavelength $\lambda = 633$ nm and refractive index $n = 1.5$. The step width $d = 0.74$ µm equals the groove depth $D - d = 0.74$ µm and the binary axicon period is $D = 2d = 1.48$ µm. The total height of the axicon (on the z-axis) is $2H = 1.266$ µm. The axicon radius equals three periods $R = 3D = 4.44$ µm. In this section, we discuss the simulation of sharply focusing the laser mode R-TEM$_{01}$ with the binary axicon of Figure 1.6(a). Figure 1.6(b) depicts the amplitude distribution of the mode R-TEM$_{01}$ of waist radius $w = 1.9$ µm. The amplitude is found to be maximal at $r = 1.36$ µm.

Table 1.1 below gives the dependence of the axicon focal length (f) and the focal spot diameter at half-intensity (FWHM) on the illuminating wavelength λ.

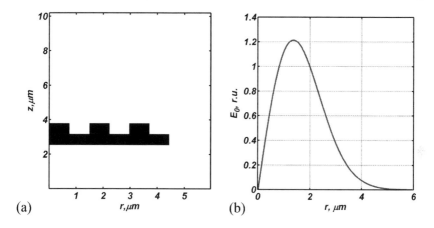

FIGURE 1.6   (a) The radial binary axicon profile and its position in the calculation window; and (b) the amplitude distribution of the mode R-TEM$_{01}$ of waist radius $w = 1.9$ µm.

TABLE 1.1

The Focal Spot Diameter at FWHM and Focal Length f vs. the Wavelength λ

λ, µm	FWHM, λ	f, µm
0.600	0.69	0.64
0.630	0.56	0.53
0.700	0.45	0.19
0.750	0.46	0.19
0.850	0.39	0.08

Table 1.1 suggests that the chromatic aberration of the binary microaxicon coincides (in sign) with that of a conventional diffraction grating: longer waves are diffracted at larger angles to the optical axis. Therefore, it follows from Table 1.1 that with increasing wavelength λ from 0.600 μm to 0.850 μm, the focal spot is moving toward the microaxicon apex, whereas its diameter (in wavelengths) is decreasing. The minimal diameter of FWHM=0.39λ was formed in the close proximity to the microaxicon apex (f=0.08 μm). Figure 1.7 shows (a) the intensity distribution $|E|^2$ on the optical axis and (b) the radial intensity profiles $|E_z|^2$ (curve 3), $|E_r|^2$ (curve 2), and $|E|^2=|E_r|^2+|E_z|^2$ (curve 1) in the focal plane for wavelength 850 nm.

It is seen from Figure 1.7 that a sharp focus is formed near the central annular step of the binary axicon (Figure 1.6), with the intensity peak being equal to 7.5 relative units (r.u.), the diameter at half-intensity FWHM=0.39λ, and the spot area at half-intensity HMA=$0.119\lambda^2$. Being somewhat smaller than in Reference [49], this is larger than the focal spot diameter produced by the conical axicon of the previous section. Because the total energy of the laser mode (Figure 1.6(b)) R-TEM$_{01}$ equals (by analogy with Equation (1.11)) $W_0/\lambda^2 \approx 32$, given that w=1.9 μm, only 25% of this light energy contributes to the focal spot generation (Figure 1.7). In other words, with the mode illuminating three periods of the binary axicon (Figure 1.6(a)), approximately one-and-a-half central periods contribute to the near-axicon axial focal spot.

Figure 1.8 shows the instantaneous intensity distributions of (a) amplitude E_r and (b) amplitude E_z resulting from the diffraction of the R-TEM$_{01}$ mode of wavelength 850 nm (Figure 1.6(b)) by the binary axicon of Figure 1.6(a) in the region under study. We can see from Figure 1.8 that each step of the axicon profile tends to form an individual lobe of the diffraction pattern: the central near-axis step produces the major first lobe of the diffraction pattern, the second annular step is responsible for the second annular lobe, and the third is for the third diffraction lobe. Thus, being easier to manufacture than a conical axicon, the binary axicon is less efficient in

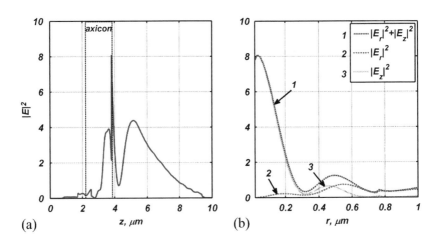

(a)   (b)

FIGURE 1.7 (a) Intensity distribution $|E|^2$ on the optical axis; and (b) radial intensity profiles $|E_z|^2$ (curve 3), $|E_r|^2$ (curve 2), and $|E|^2=|E_r|^2+|E_z|^2$ (curve 1) in the focal plane for wavelength 850 nm.

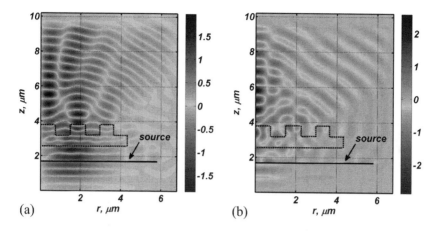

FIGURE 1.8 Instantaneous amplitude distributions of (a) E_r and (b) E_z resulting from the diffraction of the R-TEM$_{01}$ mode of wavelength 850 nm (Figure 1.6(b)) by the binary axicon of Figure 1.6(a) in the region under study.

sharply focusing, though producing the focal spot (FWHM = 0.39λ) beyond the diffraction limit (FWHM = 0.51λ).

Summing up, note that it is possible to focus the R-TEM$_{01}$ mode of waist radius $w = 3$ μm (Figure 1.1) or $w = 1.9$ μm (Figure 1.6(b)) by use of conventional focusing optics and an outgoing R-TEM$_{01}$ mode of several mm waist radius. This becomes possible because when propagating in space and being focused, the mode structure remains unchanged (up to scale).

In this section, by modeling in the R-FDTD method [49] we have shown that when illuminating a conical glass microaxicon of base radius 7 μm and height 6 μm (NA = 0.6) by a radially polarized annular laser mode R-TEM$_{01}$ of wavelength $\lambda = 1$ μm, in close proximity (20 nm apart) to the cone apex, we obtain a sharp focus of transverse diameter at half-intensity FWHM = 0.30λ and axial depth of focus (DOF) at half-intensity DOF = 0.08λ. The focal spot area at half-intensity equals HMA = 0.071λ^2. Note, for comparison, that the diameter of the resulting focal spot is 1.7 times smaller than the diameter of the minimal Airy spot (FWHM = 0.51λ), with its area being 2.87 times smaller than that of the Airy disk (HMA = 0.204λ^2). The focal spot reported here is smaller than earlier reported in References [32, 34–36, 44]. It is also smaller than the diameter of a focal spot formed by a lens with NA = 0.6 in an immersion medium $n = 1.5$ FWHM = 0.51λ/(nNA) = 0.55λ, smaller than the diffraction limit in the medium ($n = 1.5$, NA = 1) FWHM = 0.51λ/n = 0.33λ, and smaller than the focusing limit for the Bessel beam (NA = 1) FWHM = 0.36λ [41–43]. In Reference [39] using an nSIL the focal spot size FWHM = 0.23λ was obtained, but the sidelobe intensity was only 0.3 of the maximum intensity. In our case (Figure 1.7(b)), the sidelobe intensity equals 0.1.

This sharp focal spot is formed near the top (longitudinal focus width is FWHM$_z$ = 0.08λ) of the glass axicon (full angle at the vertex of the axicon $2\alpha = 100°$) because of the surface wave with an amplitude $J_0(krn\cos\theta')$, which is propagating on the surface of axicon and focusing on its top. Therefore, the full width at half-maximum of the focal spot can be estimated from the condition $J_0^2(1.1) \approx 0.5$ being equal to FWHM = 0.31λ. If we disregard the influence of the surface wave and assume that near the top of the axicon the focal spot is formed only by the conical wave of amplitude $J_0(kr\sin\varphi)$, where φ is half angle at the vertex of the conical wave (in our case, NA = $\sin\varphi = 0.6$), the diameter of the laser beam formed by such a conical wave is equal to FWHM = $0.35\lambda/\sin\varphi = 0.58\lambda$. But this value is larger than the one we have obtained (FWHM = 0.30λ).

1.2   NEAR-FIELD DIFFRACTION FROM A BINARY MICROAXICON

Axicons are known to be suitable for generating a diffraction-free laser Bessel beam in a definite range of the optical axis. Such beams continue to attract researchers' interest. In Reference [51] a coreless silica fiber, of diameter 30 μm and thickness 3 μm, combined with a lens of radius 70 μm, was used for generating a Bessel beam of diameter 20 μm maintained over 500-μm distance at wavelength $\lambda = 1.55$ μm. In Reference [52] the finite-difference time-domain (FDTD) method was used to model a 2D photonic crystal composed of an axicon-shaped rectangular array of dielectric rods: axicon base, $20a$; height, $10a$; refractive index of the rods, $n = 3.13$; radius of the rods, $0.22a$; and wavelength, $\lambda = a/0.36$, where a is the period of the rod array. A diverging Bessel beam of diameter FWHM = 1.5λ at half-maximum intensity was shown to be generated at a distance of $z < 30a$. A surface plasmon wave in the form of concentric rings described by the first-order Bessel function was reported in Reference [53] in experiments using a radially polarized laser beam ($\lambda = 532$ nm), a conical axicon, and an immersion microlens with numerical aperture NA = 1.25 found in a silver film of thickness 50 nm (permittivity $\varepsilon = -10.1786 - i0.8238$). The central axial ring diameter was 278 nm, the thickness being 250 nm $\approx 0.5\lambda$. The surface plasmon pattern was observed with a near-field microscope Veeco Aurora 3 with a 50–100-nm resolution. In a similar work [54], a scheme including a radially polarized beam of an He–Ne laser ($\lambda = 632.8$ nm), an axicon, and an immersion lens with NA = 1.4 in a 44-nm thick Au film ($\varepsilon = 0.3 + i3.089$) was used to form a surface plasmon with a central focal spot of diameter FWHM = 0.22 μm = 0.35λ. The plasmon was observed with the aid of a latex ball 175 nm in diameter.

Focusing the laser light in the neighborhood of an annular structure on metal was discussed in References [55, 56]. In Reference [55], the focusing of light with a zone plate of the ring radii $r_n^2 = 2nf\lambda + n^2\lambda^2$, $f = 1$ μm, $\lambda = 633$ nm, was simulated by the FDTD method. The plate was realized in thin silver (50 nm) and golden (50 nm) films deposited on silica. The diameter of the ring array was 13 μm. A focal spot of half-intensity diameter FWHM = 0.3λ (spot's full width being 0.7λ) was shown to occur at a distance of $z = 1.5$ μm from the plate. Similar ring arrays (8 μm in diameter) in a 100-nm thick golden film were experimentally obtained in Reference [56]. Using a near-field microscope NTEGRA (NT-MDT) with 100-nm resolution, a focal spot of diameter FWHM = 0.5 μm at half-intensity (full diameter 5λ) was observed at distance $z = 1.6$ μm for $\lambda = 633$ nm. Note that the theoretical estimation for the focal spot gives FWHM = 0.5λ. In Reference [57], eight nanoholes of diameter 200 nm arranged symmetrically in a circle of diameter 1 μm in the polymethyl methacrylate (PMMA) resist on glass were illuminated with incoherent light of wavelength $\lambda = 650$ nm, producing a focal spot of size FWHM = 0.4λ (full diameter 1.2λ) at distance 500 nm from the surface. A Fresnel lens of focal length $f = 5$ μm and diameter 50 μm in the 120-nm thick amorphous silicon film was realized in Reference [58] for wavelength $\lambda = 575$ nm, offering a 26% transmittance. Upon immersion, the lens numerical aperture was NA = 1.55, producing a focal spot of diameter FWHM = 0.9λ. It is noteworthy that the spot was measured using a fluorescent sphere of diameter 0.5 μm.

Other theoretical [59] and experimental [60] works reported on the studies into the near-field focusing of light with a binary diffraction axicon. An approximate theory that adequately described a diffraction axicon of ring period $T < 5\lambda$ was proposed in Reference [59]. In the said theory, with its central part shielded with an opaque disk, the binary axicon can be treated as a diffraction grating. The axicon of period $T = 5\lambda$ and radius 40λ was shown to produce a focal spot of diameter FWHM = 0.88λ at distance 40λ from the surface. Experimental studies of a binary axicon (30 mm in diameter) of period $T = 33$ μm (which is equivalent to a conical glass axicon with the apex angle of 88°) fabricated on the ZEP520A resist (refractive index $n = 1.46$) was reported in Reference [60]. The laser beam diameter was shown to be independent of the wavelength, with the Bessel beam radius increasing from 1.2 μm to 12.5 μm ($\lambda = 532$ nm) over a distance range from 0 to $z = 50$ mm, retaining its radius upon further propagation from $z = 50$ mm to $z = 100$ mm.

In this section, we study binary axicons of period 4, 6, and 8 μm fabricated by photolithography with a 1 μm resolution, of depth 500 nm and diameter 4 mm. We experimentally show that the near-field diffraction produces focal spots within a 40 μm distance from the axicon, varying in diameter from 3.5λ to 4.5λ (for the axicon with $T = 4$ μm) and from 5λ to 8λ (for $T = 8$ μm), λ is the incident wavelength ($\lambda = 0.532$ μm). Note that with the first focus appearing at distance 2 μm ($T = 4$ μm), the focal spots recur with a period of 2 μm (for $T = 4$ μm) and 4 μm (for $T = 8$ μm). We have also modeled the diffraction of plane and diverging linearly polarized light waves using FullWAVE (RSoft) and a proprietary program R-FDTD that use finite-difference schemes to solve Maxwell's equations in the Cartesian and cylindrical coordinates. The numerical values of paraxial focal spot diameters for the near-field diffraction derived for the axicon with period $T = 4$ μm are in good agreement with the experimental results.

1.2.1 Scalar Non-Paraxial Diffraction by a Binary Axicon

The electromagnetic theory [59] is not suited for the analysis of the near-axicon field, because the central part of the axicon cannot be treated as a diffraction grating.

With a plane linearly polarized wave incident on the axicon, the electric field of the binary axicon in the initial plane $z = 0$, which is coincident with the axicon output surface, is given in the transparency approximation by

$$E_{y0}(r) = \left(1 - e^{i\phi}\right) \sum_{n=0}^{N-1} (-1)^n \operatorname{circl}\left(\frac{r}{r_n}\right), \qquad (1.12)$$

where $r_n=(n+1)r_0$, n is an integer, r_n are radii of the axicon binary relief steps along the radial coordinate r, φ is the phase delay due to the axicon relief steps, and N is the number of jumps of the axicon relief. The amplitudes of the spectrum of plane waves given by the initial field in Equation (1.12) are derived from the expression

$$A(\rho) = \frac{k^2}{2\pi}\left(1-e^{i\phi}\right)\sum_{n=0}^{N-1}(-1)^n\int_0^{\infty}\text{cicrl}\left(\frac{r}{r_n}\right)J_0(kr\rho)rdr$$

$$= \left(1-e^{i\phi}\right)k^2r_0^2\sum_{n=0}^{N-1}(-1)^n(n+1)^2\frac{J_1\left[k\rho(n+1)r_0\right]}{\left[k\rho(n+1)r_0\right]},$$

(1.13)

where ρ is a dimensionless variable. Then, the y-component of the electric field in any plane z is derived from

$$E_y(r,z) = \left(1-e^{i\phi}\right)k^2r_0^2\sum_{n=0}^{N-1}(-1)^n(n+1)^2$$

$$\times \int_0^{\infty}\frac{J_1\left[kr_0(n+1)\rho\right]}{\left[kr_0(n+1)\rho\right]}J_0(k\rho r)e^{ikz\sqrt{1-\rho^2}}\rho d\rho.$$

(1.14)

The problem of finding the light field amplitude in the neighborhood of the binary axicon (1.14) reduces to taking the integral:

$$I = \int_0^{\infty}J_1(ax)J_0(\beta x)e^{i\gamma\sqrt{1-x^2}}dx,$$

(1.15)

where α, β, γ are constant numbers. Since we have failed to find the integral (1.4) in the reference literature, we will calculate it in the limiting cases. Assuming $\gamma=kz\ll 1$, we will find the near-axicon field. Then, by expanding the exponent into the Taylor series and retaining two terms, we obtain, instead of Equation (1.15):

$$I \approx \int_0^{\infty}J_1(ax)J_0(\beta x)dx + i\gamma\int_0^{\infty}\sqrt{1-x^2}J_1(ax)J_0(\beta x)\,dx.$$

(1.16)

The first integral can be found in the reference literature, Reference [61]:

$$\int_0^{\infty}J_1(ax)J_0(\beta x)dx = \begin{cases}1/a, & a > \beta, \\ 0, & a < \beta.\end{cases}$$

(1.17)

Note that in our case $\alpha=kr_0(n+1)$, $\beta=kr$. The second integral in Equation (1.16) will be sought-for near the optical axis, on the assumption that the radial coordinate r is much smaller than the radius of the axicon relief first step r_0: $r\ll r_0$;

then the zero-order Bessel function can be replaced with the quadratic relationship:
$J_0(x) \approx 1 - \left(\dfrac{x}{2}\right)^2$ at $x \ll 1$. This done, the second integral in Equation (1.16) can be derived using the integral from Reference [61]:

$$\int_0^1 x^{a-1}\left(1-x^2\right)^{b-1} J_n(cx)dx = \frac{c^n}{2^{n+1}} \frac{\Gamma(b)\Gamma\left(\dfrac{n+a}{2}\right)}{\Gamma\left(b+\dfrac{n+a}{2}\right)\Gamma(n+1)}$$

$$\times {}_1F_2\left[\frac{n+a}{2},\frac{n+a}{2}+b,n+1,-\frac{c^2}{4}\right],$$

(1.18)

where ${}_1F_2(a,b,c,x)$ is a hyper-geometric function. Then, for the second integral in Equation (1.16) we can write the expression:

$$\int_0^1 \sqrt{1-x^2}\,J_1(ax)dx - \left(\frac{kr}{2}\right)^2 \int_0^1 x^2\sqrt{1-x^2}\,J_1(ax)dx$$

$$= \left(\frac{kr_0}{6}\right)^2\left\{{}_1F_2\left(1,\frac{5}{2},1,-y^2\right) - \frac{(kr)^2}{10}\,{}_1F_2\left(2,\frac{7}{2},2,-y^2\right)\right\},$$

(1.19)

where $y = kr_0/2$. Replacing the infinite limits of integration in Equation (1.16) by finite limits of integration in Equation (1.19) is a routine procedure in which the contribution of evanescent inhomogeneous waves to the amplitude is disregarded.

In view of Equations (1.7) and (1.9), the paraxial and near-axicon field ($kz \ll 1$, $r \ll r_0$) is given by

$$E_y(r,z) \approx \frac{1}{kr_0} + ikz\frac{(kr_0)}{6}\left[{}_1F_2\left(1,\frac{5}{2},2,-y^2\right) - \frac{(kr)^2}{10}\,{}_1F_2\left(2,\frac{7}{2},2,-y^2\right)\right].$$

(1.20)

From Equation (1.20), the diameter of the central peak of the paraxial light field is given by the relation

$$2r = k^{-1}\left[\frac{10\,{}_1F_2\left(1,\dfrac{5}{2},2,-y^2\right)}{{}_1F_2\left(2,\dfrac{7}{2},2,-y^2\right)}\right]^{1/2},$$

(1.21)

whence follows the numerical estimate of the focus diameter:

$$2r \approx 0.6\lambda.$$

(1.22)

The relation in Equation (1.22) suggests that the central peak diameter of the near-axicon field is independent of its period $T = 2r_0$, being nearly equal to the diffraction limit (FWHM=0.51λ). To verify the expression (1.11), we can evaluate the central peak diameter of the field intensity based on different considerations.

1.2.2 PARAXIAL ESTIMATE OF THE AXIAL BEAM DIAMETER

In Reference [60] the scalar paraxial theory was employed to show that since the axicon generates a Bessel beam, the Bessel beam diameter can be deduced from the expression:

$$J_0^2(k \sin \theta \cdot r) = 0. \tag{1.23}$$

Then, we obtain

$$2r = \frac{2.4\lambda}{\pi \sin \theta}, \tag{1.24}$$

where θ is half angle at the apex of a conical wave generated by the axicon.

With the binary axicon treated as a diffraction grating [59], the conical wave angle θ can simultaneously be interpreted as the angle of a diffraction grating with period T:

$$\sin \theta_m = \frac{\lambda m}{T}, \tag{1.25}$$

where m is the diffraction order number.

In view of Equations (1.24) and (1.25), the final estimate of the axicon field diameter on the optical axis is

$$2r = \frac{2.4}{\pi} \frac{T}{m} \approx 0.774 \frac{T}{m}. \tag{1.26}$$

From the relation (1.26) it can be seen that the Bessel beam diameter produced by the binary axicon is independent of the wavelength [60], only depending on the axicon period and the diffraction order number. The relation (1.26) suggests that different diffraction orders reach the near-axicon axial point, and thus the light field formed in the axicon vicinity at $z < z_0$ (where $z_0 = RT/2\lambda$ is the distance beyond which a single diffraction order of the axicon field is produced and R is the axicon radius) will have the diameter varying in a complex manner along the optical axis at $0 < z < z_0$.

From Equation (1.26) it follows that at $z \geq z_0$ the axial beam diameter is equal to $2r = 6.2\lambda$ for the axicon with period $T = 4$ μm $= 8\lambda$.

1.2.3 ESTIMATING THE DIAMETER OF THE AXIAL BEAM AS A WAVEGUIDE MODE

In the axicon vicinity, when $r < r_0 = T/2$ and $z < T$, we can estimate the axial light field diameter using the waveguide theory, because the axicon microrelief depth is $h = \lambda/(2(n-1)) = \lambda$ at $n = 1.5$ (refractive index), whereas the central axicon part can be considered as a fragment of a circular fiber with the core radius of $r_0 = T/2$.

The number of modes in the circular stepped-index fiber is derived from the dispersion equation [62]:

$$\frac{uJ_0(u)}{J_1(u)} = -\frac{wI_0(w)}{I_1(w)}, \tag{1.27}$$

where J_0, J_1, I_0, I_1 are the standard and modified Bessel functions of zero- and first-orders, $u^2 + w^2 = V^2$, $V = kr_0\sqrt{n_1^2 - n_2^2}$, where n_1 and n_2 are the refractive indices of the fiber core and cladding, and r_0 is the core radius. Because the maximal root of Equation (1.27) for a waveguide mode is less than the cut-off number, $u_{max} < V$, we may infer that the waveguide mode with a minimal diameter has the amplitude proportional to the Bessel function: $J_0(Vr/r_0)$, $r < r_0$. Hence, the diameter of a mode with the maximal number is ($n = 1.5$):

$$2r = 2\frac{2.4r_0}{V} = \frac{2.4\lambda}{\pi\sqrt{n^2 - 1}} \approx 0.7\lambda. \tag{1.28}$$

It is noteworthy that in a similar way to the estimate (1.22), the diameter estimate (1.28) is also independent of the axicon period T.

1.2.4 NON-PARAXIAL RELATIONS FOR THE LONGITUDINAL INTENSITY

The relation between the central peak diameter of the light field and the distance to the axicon should be similar to that between the axial intensity and the distance to the axicon. Actually, if there is a bright ring on the axis, its diameter is supposed to be larger than the diameter at focus, and, on the opposite, the axial intensity has a local maximum at focus and a minimum in the ring center. Thus, we will arrive at the relation for the axial intensity of light produced by the binary axicon. The axial amplitude of the scalar non-paraxial field from a circular aperture of radius R as a function of the longitudinal coordinate is given by Reference [63]:

$$E_y(z) = e^{ikz} - \frac{ze^{ik\sqrt{R^2+z^2}}}{\sqrt{R^2+z^2}}. \tag{1.29}$$

For a binary axicon of transmittance in Equation (1.12), we can derive a similar relation for the electric field amplitude on the optical axis:

$$E_y(z) = e^{i\phi}\left[e^{ikz} - \frac{ze^{ik\sqrt{r_{2N+1}^2+z^2}}}{\sqrt{r_{2N+1}^2+z^2}}\right] + \left(e^{i\phi} - 1\right)z\sum_{m=1}^{2N}\frac{e^{ik\sqrt{r_m^2+z^2}}}{\sqrt{r_m^2+z^2}}(-1)^m, \tag{1.30}$$

where $r_m = mr_0$ is the radius of the axicon binary relief step, and $r_0 = T/2$ is half the axicon period.

From Equation (1.30) it follows that at $\phi = \pi$, secondary spherical waves scattered from any point of relief jump (i.e. with half-period intervals) will be contributing to any point on the z-axis. Because the number of the said points equals $2N + 1$, it would be difficult to predict the result of such multi-ray interference. It can only be said that for $z \ll R$, Equation (1.30) reduces to Equation (1.29), suggesting that in the vicinity of the axicon surface the minimal period of the axial intensity variations is as small as λ.

1.2.5 FABRICATION OF BINARY AXICONS

Three diffractive binary axicons of periods 4 μm, 6 μm, and 8 μm and diameter 4 mm were fabricated. The process involved the deposition of a 100-nm thick chromium layer onto a silica substrate ($n = 1.46$) of thickness 1 mm and diameter 30 mm using the unit UBM2M1. On this chromium-plated substrate was written a photomask on a circular laser writing system CLWS–200 of minimal laser spot diameter 0.8 μm and a 20-nm positioning accuracy. In the process of the photomask writing, a protective oxide film was produced on the chromium surface through its exposure to a focused light of the Ar laser of wavelength 500 nm. In the next step, the photomask underwent plasmo-chemical etching. The etching was conducted on the unit UTP PDE-125-009, producing the relief features of horizontal measures less than 100 nm. The silica was etched at the following parameters: power of high-frequency current, 800 W, interim vacuum $2.1*10^{-2}$ Pa, current of the plasma-trapping electromagnet, 0.8 A. The etching was conducted in the Freon-12 atmosphere for 21 minutes. The etch rate of the silica substrate was 20–25 nm per minute. Using the above-described procedure, three axicons 4 mm in diameter and with periods 4 μm, 6 μm, and 8 μm were fabricated.

Figure 1.9 depicts the SEM image (top view) of the binary axicon-on-silica of period 4 μm. The image was produced on the scanning electron microscope SUPRA-25-30-85 with 1000x magnification. Figure 1.10 shows the image (top view) of the binary axicon with period 8 μm obtained with an atomic force microscope Solver Pro.

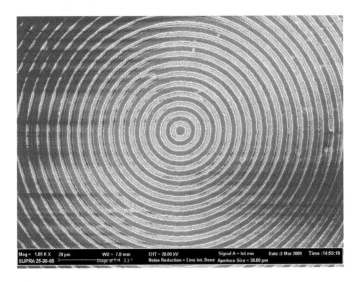

FIGURE 1.9 SEM image (top view) of the binary axicon of period 4 μm, produced with a scanning electron microscope Supra-25-30-85 with 1000x magnification.

FIGURE 1.10 Top view of the binary axicon with period 8 μm, produced with an atomic force microscope Solver Pro.

Figure 1.11 shows oblique images of the binary axicon reliefs obtained on the Solver Pro microscope: (a) a peripheral part of the axicon with period 6 μm; and (b) the central part of the axicon with period 8 μm.

Shown in Figure 1.12 is the radial section of the peripheral part of relief of the binary axicon of period 6 μm (Figure 1.11(a)), which suggests that the relief depth is 450–500 nm. Because all three axicons in question are designed for the wavelength 532 nm, the required relief depth is

$$h = \frac{\lambda}{2(n-1)} \approx 578 \text{ nm.} \tag{1.31}$$

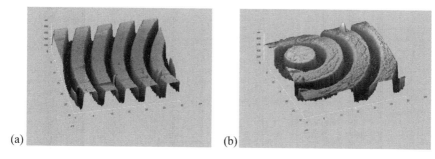

(a)   (b)

FIGURE 1.11 An oblique image of (a) the peripheral fragment of the binary axicon with period 6 μm; and (b) the central part of the binary axicon of period 8 μm, produced with the Solver Pro microscope.

Thus, the etch depth error (underpickling) is 20%. It is clearly seen from Figure 1.12 that the relief top features have a slope of about 1/150 rad, whereas the bottom has irregularities of height 30 nm. Note, also, that the binary relief has a trapezoid shape: the top-to-bottom ratio is 3:4. Besides, it is clearly seen from Figure 1.10 that the step width is larger than the step span, with the width-to-span ratio being 2:1.

1.2.6 EXPERIMENTAL RESULTS

The experiment aimed to look into the relation between the central spot diameter and the distance on the optical axis. The binary axicons were sequentially placed into an optical setup shown in Figure 1.13, where the near-field diffraction pattern was measured with a charge-coupled device (CCD)-camera at different distances upon illumination by a laser light of wavelength 532 nm.

The beam of a solid-state laser of wavelength $\lambda = 532$ nm and diameter 1.4 mm was focused with a microlens L_1 into a point diaphragm of diameter 15 μm. The central part of a uniform light spot produced by the diaphragm was collimated with a lens L_2, producing a light beam in the form of a bounded near-plane wave. The central portion of the resulting beam was coupled into an optical system of the microscope Biolam-M and then focused by a microlens L_3, so that the laser beam diameter should be equal to the axicon diameter D_3 (the matching of the beam and axicon diameters was intended to reduce the energy losses). The resulting diffraction pattern was recorded on a CCD-camera with the aid of a microlens L_4. By displacing the axicon D_3 we can obtain the diffraction patterns at different distances from the element. The origin of coordinates was placed on

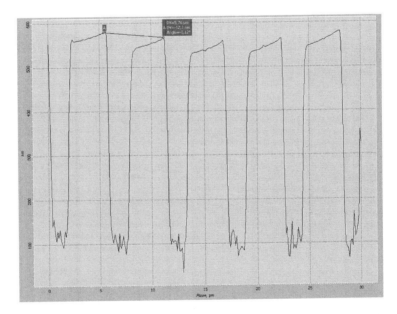

FIGURE 1.12 Profile of the binary axicon with period 6 μm of Figure 1.11(a), produced on the Solver Pro microscope.

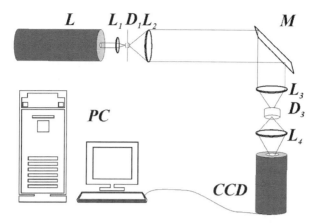

FIGURE 1.13 An optical arrangement for measuring the near-field diffraction pattern from the binary axicons: L – laser, L_1 – microlens (20x, NA=0,4), D_1 – point diaphragm (diameter=15 µm), L_2 – collimating lens (f=100 mm), M – deflecting mirror, L_3 – focusing microlens (8x, NA=0.2), L_4 – imaging microlens (20x, NA=0.4), D_3 – binary axicon.

the microrelief valley surface, which was brought to sharp focus in white light. The axicon D_3 was displaced with a micrometric screw with a 1-µm division value. The CCD-camera resolution was 2048*1536 pixels, the pixel size being 6.9 µm.

Figure 1.14 shows the central maximum diameter as a function of distance from the axicon surface (with respect to a transverse Cartesian coordinate) for period (a) 4 µm and (b) 8 µm.

It is seen from Figure 1.14 that in the interval from 0 to 40 µm, the central maximum diameter is changing quasiperiodically with a period of about 2 µm (Figure 14(a)) and 4 µm (Figure 1.14(b)). Note that the diameter values larger than 5λ (Figure 1.14(a)) and 9λ (Figure 1.14(b)) correspond to the annular intensity distribution centered on

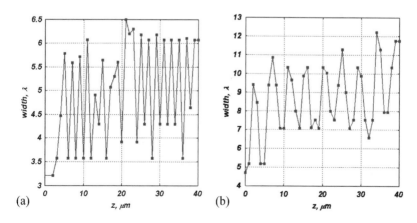

FIGURE 1.14 The diameter of the light spot on the axis (in wavelengths) as a function of distance from binary axicons with period (a) 4 µm and (b) 8 µm.

the axis (on which it has a local minimum), while the diameter values smaller than 4.5λ (Figure 1.14(a)) and 8λ (Figure 1.14(b)) correspond to the local maxima intensity (focal points). The analysis of Figure 1.14 suggests that the diameter of the near-field axial focal spots (at the distance ≤ 40 μm) is varied from 3.5λ to 4.5λ (Figure 1(a)) with a 0.4λ error and from 5λ to 8λ (Figure 1.14(b)) with a 0.5λ error, respectively, for the axicons with period 4 μm and 8 μm.

Figure 1.15 shows the recorded diffraction patterns of laser light produced by binary axicons with period (a), (b) 4 μm and (c), (d) 8 μm at distances of (a) 5 μm, (b) 8 μm, (c) 16 μm, and (d) 18 μm. From Figure 1.15, the annular intensity distribution on the optical axis (a, c) is seen to be alternated with a central focal spot (b, d) at 2–3-μm intervals over a distance less than 40 μm. Because the diffraction patterns in Figure 1.15 are shown in the same scale (75×55 μm), it can be seen that the corresponding diameters of the ring and the focus for the axicon of period 8 μm are 1.5 times larger than those for the axicon of period 4 μm.

Figure 1.16 shows (a) the diffraction pattern from the axicon of period 4 μm recorded 2-μm away from its surface and the pattern cross-sections on the (b) x- and (c) y-axes. It is seen from Figure 1.16 that at small distances of $z < 5$ μm the focal spot in the diffraction pattern features an ellipticity of eccentricity $\varepsilon = 0.63$ (which is not found in Figure 1.15) associated with the linear polarization of the incident laser light (the electric field of the incident beam directed along the y-axis in Figure 1.16).

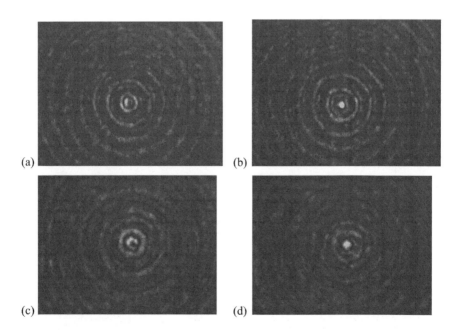

(a)   (b)

(c)   (d)

FIGURE 1.15   Diffraction patterns recorded with the CCD-camera from the axicons with period (a), (b) 4 μm and (c, d) 8 μm at different distances: (a) 5 μm, (b) 8 μm, (c) 16 μm, and (d) 18 μm.

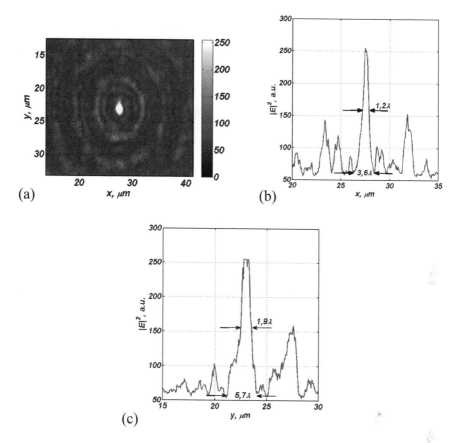

FIGURE 1.16 (a) The intensity distribution recorded at a 2-μm distance from the binary axicon of period 4 μm and the cross-sections thereof on the (b) x- and (c) y-axes.

1.2.7 SIMULATION RESULTS

The diffraction of the linearly polarized plane wave by the diffractive binary axicons was simulated using two similar but still different techniques: the R-FDTD method that realizes a finite-difference algorithm for solving Maxwell's equations in cylindrical coordinates using MATLAB® [49] and the FDTD method implemented in the FullWAVE (RSoft) program.

Figure 1.17 depicts the simulated distributions of the squared electric vector magnitude in the xz-plane of a linearly polarized plane wave having passed the binary axicon of period 4 μm, calculated using two approaches. Figure 1.17 suggests that both diffraction patterns show a qualitative similarity, which is evidenced by the fact that in both pictures each asperity of the axicon relief focuses light into local foci at a distance of 2 μm (Figure 1.17(a)) and 1.7 μm (Figure 1.17(b)) from the axicon surface. These values agree well with the estimate that follows from Equation (1.29). In the paraxial case, it follows from Equation (1.29) that the intensity along the optical axis will vary according to $\cos\left(kR^2/4z\right)$ with a period of $z_0 = R^2/4\lambda = 1.9$ μm when $T = 4$ μm.

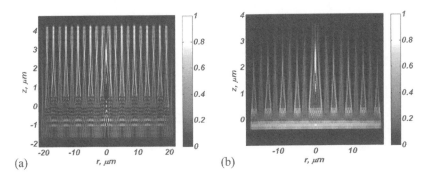

FIGURE 1.17 Simulated intensity distribution in the xz-plane when a linearly polarized plane wave (the electric vector directed along the y-axis) is diffracted by a binary axicon of period 4 µm: (a) Matlab2008a, (b) FullWAVE (RSoft).

Shown in Figure 1.18 are (a) the diffraction patterns in the xy-plane at distance $z = 1.7$ µm (in the local focus plane), (b) its radial section, and, for comparison, the intensity distribution in the xy-plane at a distance of $z = 0.3$ µm at which a light ring is formed on the optical axis.

From Figure 1.18(a, b), the local maximum diameter can be estimated at 1.25λ ($\lambda = 532$ nm). Comparing this value with the experimentally measured diameter of

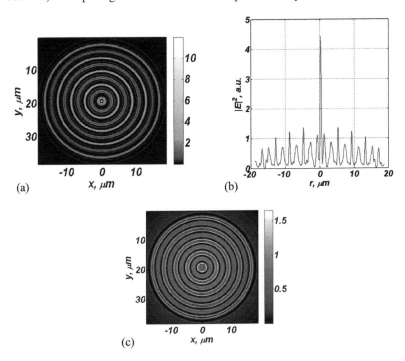

FIGURE 1.18 The intensity distribution in the transverse xy-plane at distance (a) $z = 1.7$ µm, (c) $z = 0.3$ µm; and (b) the radial section of the distribution in (a) for the binary axicon of period $T = 4$ µm, simulated using FullWAVE.

the local focal spot in Figure 1.16 (of diameter 3.6λ on the x-axis), we can see that the former is nearly three times smaller.

Derived with the R-FDTD method, Figure 1.19 depicts the axial intensity distribution for an axicon of radius 28 μm and period 4 μ, which is illuminated by a linearly polarized plane wave.

From Figure 1.19, the local peaks are seen to alternate on the optical axis in a quasiperiodic manner, with the peak-to-peak spacing increasing from 2 μm to 8 μm over the distance $z < 60$ μm.

Figure 1.20 depicts the axial intensity distribution (in arbitrary units) for the binary axicon of period 4 μm and radius 28 μm. As distinct from Figure 1.19, the distribution of Figure 1.20 resulted from illuminating the axicon by a linearly polarized diverging Gaussian beam, with the electric vector amplitude given by

$$E_y(r, z = 0) = \exp\left(-\frac{r^2}{w^2} + i\frac{kr^2}{2f} \right),$$ (1.32)

where $w = 2$ mm is the waist radius of the Gaussian beam, $f = 2.5\, w$ is the focal length of a parabolic lens. The use of the diverging beam (in contrast with the previously studied plane beam) was prompted by the effort to match the simulation with the experimental results arrived at using the setup in Figure 1.13. In the arrangement of Figure 1.13, the binary axicon D_3 of diameter 4 mm was illuminated by a diverging laser light having passed the microlens L_3. The diverging light beam was used to match the diameter of the illuminating light spot with that of the axicon, D_3. In Figure 1.20, the peak-to-peak spacing is increasing from 2 μm to 4 μm at $z < 50$ μm. The comparison of the plots in Figures 1.19 and 1.12 shows that in the interval $0 < z < 30$ μm the periods of intensity oscillations are nearly the same, while in the interval $30 < z < 50$ μm the intensity oscillation period in Figure 1.20 is smaller.

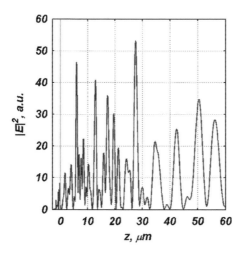

FIGURE 1.19   The axial intensity distribution for the axicon of period 4 μm illuminated by a linearly polarized plane wave, simulated with the R-FDTD method.

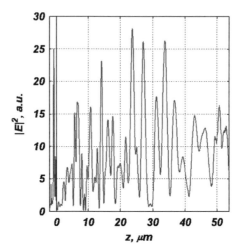

FIGURE 1.20 The axial intensity distribution calculated using the R-FDTD method for the axicon of period 4 µm and radius 28 µm illuminated by a linearly polarized diverging Gaussian beam.

Calculated with the R-FDTD method, Figure 1.21 depicts diameters of the central spots produced by the diverging wave of Equation (1.32) as a result of diffraction from the axicon of period $T = 4$ µm versus the distance to the axicon.

Figure 1.21 shows local maxima that correspond to the annular intensity distribution in the central spot of the diffraction pattern, whereas local minima correspond to the focal spots on the optical axis. From Figure 1.21, the diameter of the focal spots on the optical axis produced by the binary axicon of period 4 µm is also seen

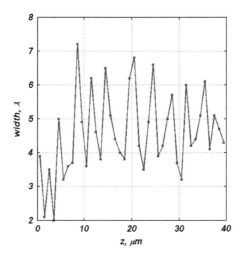

FIGURE 1.21 R-FDTD simulation: the profile of the full width (diameter on the x-axis in wavelengths) of the central spot intensity of the diverging laser beam diffracted by the binary axicon of period 4 µm.

to vary in the near field ($z < 40$ µm) from 2λ to 4.3λ with a longitudinal period of about 3 µm (with 12 local minima found at length 40 µm in Figure 1.21). From the comparison of the experimental and simulated curves for the central spot diameters in the diffraction pattern (Figure 1.14(a) and Figure 1.21) it follows that the focal spot diameters conform well at a distance up to 40 µm: from 3.5λ to 4.5λ (Figure 1.14a) and from 2λ to 4.3λ (Figure 1.21). Note, however, that the longitudinal period of the focal spot variation was 3 µm in Figure 1.21, whereas in the experiment it was 2 µm.

Comparison of Figures 1.20 and 1.21 shows that there is agreement in the number of the local foci (intensity minima in Figure 1.21 and maxima in Figure 1.20): in the interval 30 µm $< z <$ 40 µm, there are 4 minima in Figure 1.21 and 4 maxima in Figure 1.20; meanwhile in the interval 20 µm $< z <$ 30 µm there are 2 minima in Figure 1.21 and two maxima in Figure 1.20.

From Equation (1.26) it follows that the focal spot diameter produced by the axicon of period 4 µm equals $2r = 0.774T = 3.1$ µm $= 6\lambda$ (for $m = 1$) and 3λ (for $m = 2$). Comparing the above values with the experimental (Figure 1.14(a)) and simulation data (Figure 1.21) we can infer that for the near-field diffraction the contribution to the foci on the optical axis comes from the first and second diffraction orders of the binary axicon.

We have obtained the following results. It has been shown that the calculation of the non-paraxial complex amplitude of the light field resulting from the diffraction of a plane wave by a binary axicon reduces to calculating the integral of the product of two Bessel functions of the zero- and first-orders. By approximate calculations, we have shown that at $z \ll \lambda$, the central maximum diameter of the light field is independent of the axicon period, being equal to the diffraction limit of 0.6λ. The said estimate of the diameter of the near-axicon focal spot conforms with the estimate of the minimal modulation diameter (0.7λ) of the light field composed of spatial modes of a stepped-index fiber whose diameter equals the axicon period. A relationship has been deduced of the field complex amplitude on the optical axis for a plane wave diffracted from a binary axicon, which suggests that, in the axicon vicinity, the minimal period of axial intensity oscillations is λ. Using a laser writing system CLWS-200 and plasmo-chemical etching of a fused-silica substrate ($n = 1.46$), we fabricated binary axicons of period 4 µm, 6 µm, and 8 µm, diameter 4 mm, and depth 500 nm. The analysis of the axicon surface with a Solver Pro microscope has shown that the axicon was fabricated with a 20% error. Top features of the relief have a slope of about 0.01 rad, the binary relief steps are a 3:4 trapezoid, and the relief bottom has 30-nm asperities. Experimental relationships for the variation of the focal spot diameter along the optical axis in the near field of the binary axicon ($z < 40$ µm) have been derived, showing that for the axicon of period 4 µm, the focal spot diameter varies from 3.5λ to 4.5λ with a 2-µm period, and for the axicon of period 8 µm – from 5λ to 8λ with a 4-µm period. This result agrees with Reference [60]. The minimal period of an elliptic focal spot equal to 3.6λ µm (FWHM $= 1.2\lambda$) was found at a 2-µm distance from the axicon of period 4 µm. This result agrees well with the findings reported in References [52, 56]. We have experimentally shown that in the near-axicon region ($z < 5$ µm), elliptical focal spots are formed (with eccentricity of 0.63), which are elongated along the linear polarization vector of the incident light wave. For comparison, the diffraction of a linearly polarized plane wave by the binary axicon of period 4 μm and diameter 40 μm was simulated using (1) the FullWAVE (by RSoft), which solves Maxwell's equation in the Cartesian coordinates using a 3D-FDTD method, and (2) the earlier developed proprietary R-FDTD software for solving Maxwell's equations in the cylindrical coordinates by a finite-difference scheme for radially symmetric microoptics elements. In both simulations, the diffraction patterns show that the first focus appears 1.7–2 μm away from the axicon surface. The numerically simulated relationship for the central spot diameter variation on the optical axis when a linearly polarized plane wave is diffracted from the binary axicon of period 4 μm has been shown to conform with the similar experimental curve (Figures 1.6(a) and Figures 1.13): in the interval $0 < z < 40$ μm, the focal spot diameter on the optical axis varies from 3.5λ to 4.5λ (experiment) and from 2λ to 4.3λ (simulation), with the local foci appearing on the axis with period 2 μm (experiment) and 3 μm (simulation).

1.3   TIGHT FOCUSING WITH A BINARY MICROAXICON

Axicons are known to be suitable for generating a diffraction-free laser Bessel beam at a certain optical axis segment. Despite being widely known in optics for quite some time, interest in axicons has not weakened. Surface plasmons with a sharp central peak were generated in metal films by use of an axicon in References [53, 54]. A binary axicon of period 33 μm was studied in Reference [60]. The minimal diameter of the laser beam was 1.2 μm (at wavelength $\lambda = 532$ nm). The laser beam diameter was shown to be independent of the wavelength. In Reference [64] an axicon made of glass with the vertex angle of 175 degrees was employed as a wavefront sensor, whereas Reference [65] reported on the generation of an annular collimated laser beam of diameter 8 mm with the aid of two axicon mirrors. A diffractive axicon of diameter 2 mm and period 20 μm was investigated using a Mach-Zehnder interferometer in Reference [66]. The fabrication of a microaxicave (inverted microaxicon) of period 20 μm to generate a ring of radius 1.5 mm at a distance of 50 mm was reported in Reference [67].

To the best of our knowledge, Reference [43] is the only publication where the use of the axicon for the tight focusing of laser light was reported. In Reference [43] a radially polarized Bessel beam formed by passing an Ar laser beam (wavelength $\lambda = 514$ nm) through an axicon with numerical aperture NA = 0.67 was registered in a PMMA-DR1 (polymer matrix PMMA with azobenzene molecules) layer, which selectively registered the longitudinal E-vector component. The Bessel beam diameter was experimentally measured to be FWHM = 0.62λ. In a similar way, a circularly polarized Bessel beam of diameter FWHM = 0.60λ was registered in the PMMA-DR1 medium. It is noteworthy that without the special registration medium PMMA-DR1, the beam diameter would have been FWHM = 0.89λ.

In a number of papers [68–71], the focus depth was reported to be increased using axicon-like binary phase diffractive optical elements (DOE) with their radius jumps fitted in a special manner. Thus, in References [68, 69] by modeling based on Richards and Wolf formulae, a four-ring DOE matched to a lens with NA = 0.95 was shown to form a focal segment of diameter FWHM = 0.43λ and depth of focus (DOF) = 4λ. The DOE was illuminated with a radially polarized Bessel–Gaussian

(BG) beam. In similar modeling reported in Reference [70], a lens with spherical aberration (lens axicon) was employed. In this way, it became possible to obtain the best characteristics of the focal line, namely, FWHM=0.395λ and DOF=6λ. In Reference [71], using a radially polarized Bessel–Gaussian beam bounded with a narrow annular aperture and a microlens (NA=0.9) a focal spot of diameter FWHM=0.61λ was experimentally obtained (although the simulated value was as small as FWHM=0.4λ). Note that all these articles [68–71] consider the far-field focusing (in the lens focus).

In this section, we studied the tight focusing by a binary microaxicon of diameter 14 µm, period 800 nm, microrelief depth 465 nm, and numerical aperture NA=0.665. Using a near-field scanning microscope, with a 100-nm-aperture cantilever positioned 1 µm away from the axicon surface, we measured the intensity distribution in the focal spot produced by linearly polarized laser light ($\lambda = 532$ nm). The focal spot diameter was measured to be FWHM=0.58λ, with the focal depth being DOF=5.6λ. The root mean square (rms) deviation of the intensity distribution in the focal spot from the calculated value was 6%. The resulting focal intensity was five times higher than the illumination beam maximal intensity. Note that the measured value of diameter (0.58λ) was smaller than that reported in Reference [71] (0.61λ), whereas the axicon's numerical aperture (0.67) was smaller than that of the lens (0.9) in Reference [71].

A high-quality binary axicon was fabricated using electron beam lithography. A thin layer of ZEP520A resist coated on a glass substrate was then exposed to 180 C heat for 10 min in order to dry off the solvent. The resist thickness was carefully controlled so that it would be sufficient to ensure the required phase shift of the axicon. The pattern of concentric rings was written in the resist with an electron beam using the electron microscope ZEISS GEMINI with a lithographic device RAITH ELPHY PLUS at voltage 30 kV. The pixel size was 10 nm, the exposure energy being 45 mAc/cm². Following the exposure, the sample was developed in xylene at 23 C and then rinsed in isopropanol The remaining resist formed an axicon of period 800 nm (the resist refractive index $n = 1.5$). Figure 1.22 presents an electron image of the axicon under characterization, the profile height being 465 nm. From Figure 1.22, the binary microrelief height is seen to be approximately equal to one reading (500 nm) of the scale shown on the right bottom. For clarity, Figure 1.22(c) depicts an optical arrangement with the axicon and the incident Gaussian beam in the Cartesian coordinates.

Using the near-field scanning microscope NT-MDT, we studied the passage of a linearly polarized Gaussian beam (radius 7λ) of wavelength $\lambda = 0.532$ µm through the aforesaid binary axicon of period 800 nm. Figure 1.23(a) shows an enlarged image of a cantilever with a 100-nm aperture, which was utilized for the measurements. The aperture in Figure 1.23(a) is shown as a horizontal white segment. The near-field intensity distribution was measured in the following manner. Linearly polarized light from a solid-state laser L with wavelength 532 nm was focused by the lens $L1$ onto the surface of a glass substrate with a microaxicon $A1$ located in it. Passing through the substrate, the light was diffracted by the axicon. A cantilever with aperture $C1$, found immediately behind the axicon, conducted the parallel scanning of the axicon surface at different distances from it. Having passed through the cantilever aperture,

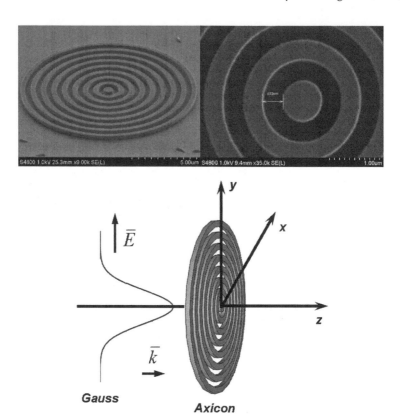

FIGURE 1.22 A SEM image of a binary axicon of period 800 nm: (a) tilted view; (b) top view (enlarged); and (c) an optical arrangement containing the axicon and the Gaussian beam.

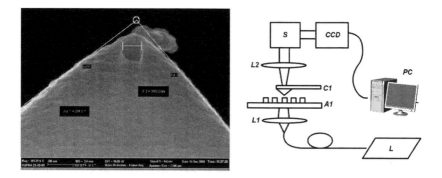

FIGURE 1.23 (a) Shape of a four-sided cantilever with a 100-nm aperture (is shown as a horizontal white segment) that was used in the near-field microscope NT-MDT; and (b) an experimental optical schematic: L – laser, $L1$, $L2$ – lenses, $A1$ – the axicon under study on the substrate, $C1$ – cantilever with an aperture, S – spectrometer, CCD-camera, and PC – computer.

the light was then focused by the lens $L2$, transmitted through the spectrometer S (to filter off the irrelevant light), and registered by the CCD-camera.

Figure 1.24 illustrates an example of the intensity distribution measured with the microscope NT-MDT (Figure 1.23(b)) 1 μm away from the axicon surface. The spot diameter was FWHM = 0.58λ. The intensity in Figure 1.24(b) (curve 2) is presented in relative units. Note that the focal intensity was found to be 5 times higher than the maximal intensity of the illumination Gaussian beam. This was attained by shifting the axicon on the substrate $A1$ (Figure 1.23(b)) across the optical axis by 20 μm, so that the incident beam should only pass through the substrate, and repeatedly measuring the intensity. The conversion efficiency was as low as 6%. This is because the contribution to the focal spot (Figure 1.24(a)) came only from the first circle and ring of the axicon relief (Figure 1.22(b)). This conversion efficiency is lower than that reported in References [68] (20%, DOF = 4λ) and [70] (15%, DOF = 6λ). Note, however, that in References [68, 70] it was only the theoretical efficiency that was considered.

A comparison between the experimental intensity distribution in the focus (formed 1 μm away from the axicon surface along the optical axis z) and the simulated intensity distribution is depicted in Figure 1.24(b). The simulation was conducted using an R-FDTD (radial finite-difference time-domain) method with the code written in MATLAB [49].

The root mean square deviation of the curves in Figure 1.24(b) is 6%. Note that major differences between the curves in Figure 1.24(b) are seen to be found in the side lobes of the diffraction pattern, whereas the deviation of the major intensity peak at $|x| \leq 1$ μm from the simulated value is as small as 1%. A noticeable increase in the side lobes in comparison with the theoretical estimation is due to the fact that, among other factors, the amplitude of the signal that comes from the cantilever depends on its distance from the surface (on the local relief shape). Figure 1.25 shows the simulated intensity distribution along the optical axis z. The vertical line in Figure 1.25 marks the axicon surface. Experimental values of the intensity in Figure 1.25 are marked as squares crossed with vertical lines, with the lines denoting

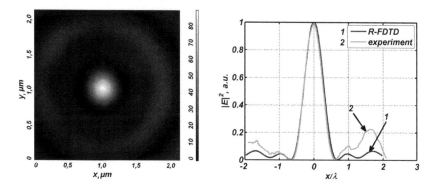

FIGURE 1.24 (a) An NSOM image of the intensity distribution in the spot; and (b) the intensity profile along the x-axis: simulation (curve 1) and experiment (curve 2).

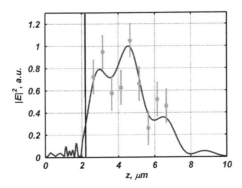

FIGURE 1.25 The longitudinal intensity profile along the optical axis: experiment (marked as squares with vertical lines) and simulation (curve).

the admissible magnitude of the intensity measurement error. The longitudinal depth of the axicon focus at half-maximum is DOF$=3$ µm$=5.6\lambda$ (Figure 1.25).

It is well known that an axicon forms a Bessel beam, with the Bessel beam diameter derived from the expression $J_0^2(kr\sin\theta)=0$. In Reference [72] the above formula was shown to define the smallest possible spot for any vector field. Although, strictly speaking, it should be noted that for the linearly polarized beam, the contribution to the focal spot comes from three Bessel functions, J_0, J_1, and J_2 [73]. Thus, we find that the beam diameter is $2r=2.4\lambda/(\pi\sin\theta)$, where θ is the half angle at the vertex of a conical wave formed by the axicon. If the binary axicon is interpreted as a diffraction grating [60], the angle θ of the conical wave simultaneously represents the diffraction order angle for a grating of period T: $\sin\theta_m=m\lambda/T$, where m is the diffraction order number. The final expression for evaluating the diameter of the axicon light field on the optical axis can be obtained in the form: $2r=2.4T/(m\pi)=0.774T/m$. This expression suggests that for the binary axicon, the Bessel beam diameter is independent of the illumination wavelength [60], being only determined by the axicon period and the diffraction order number. Let us evaluate the diameter of a focal spot produced by our binary microaxicon of period $T=800$ nm (NA$=\lambda/T=0.665$ at $m=1$): FWHM$=0.36\lambda/$NA$=0.54\lambda$ (according to Reference [72], this is the smallest possible value at a given NA). Thus, the experimentally derived value of the focal spot diameter (FWHM$=0.58\lambda$) differs from the theoretical estimate (FWHM$=0.54\lambda$) by as little as 8%. The larger value of the focal spot size (0.89λ) reported in Reference [43] for a similar axicon (NA$=0.67$) can be ascribed to the use of the near-field microscope with a bare fiber sharp tip. It can be conjectured that the dielectric tip increases the focus diameter to a greater extent than the metal cantilever with a pinhole, which was utilized in this work.

Summing up, the following results have been derived. By e-beam lithography on ZEP520A resist, a binary microaxicon of diameter 14 µm, period 800 nm, and relief depth 465 nm has been fabricated. Using a near-field scanning microscope NT-MDT with a 100-nm-aperture cantilever positioned 1 µm away from the axicon surface, we have registered a focal spot produced by linearly polarized laser light of wavelength 532 nm with diameter at half-maximum equal to FWHM$=308$ nm, which amounts to 0.58λ. The average deviation of the experimental intensity in focus from the simulated value is not larger than 6%. The experimentally derived value of the focal spot diameter differs from the theoretical estimate (FWHM$=0.54\lambda$) by as little as 8%. The focus depth at half-maximum is DOF$=5.6\lambda$. The intensity maximum on the optical axis has been found to be 5 times larger than the illumination beam intensity.

Such types of generated beams have many important potential applications in optical manipulations [74], optical microscopy [73, 75], and particle acceleration [76].

1.4 FOCUSING LASER LIGHT BY A LOGARITHMIC AXICON

The interest in focusing the laser light into on-axis lines, including those with subwavelength diameter, is unabated. In an experiment reported in Reference [77] a femtosecond laser pulse of wavelength $\lambda=800$ nm and an axicon were employed to generate a Bessel beam, which was used to machine a nanochannel of diameter 200 nm$=0.25\ \lambda$ and length 30 µm in the glass. In Reference [78], a tapered glass microtube (similar to a hollow axicon) was used to experimentally generate at distance 2.2 µm a focal spot of near-subwavelength size FWHM$=435$ nm$=0.65\lambda$, where $\lambda=671$ nm. In Reference [79], a logarithmic axicon (LA) of radius 6.5 mm with the phase function given by $S(r)=\gamma\ln(a+br^2)$, where γ, a, and b are constants, and r is the transverse radial coordinate in the cylindrical coordinate system, was fabricated from the plastic using a diamond cutter. The axicon was reported to focus the He–Ne laser light into an axial line of length 10 cm and diameter 10 µm. In Reference [80] it was numerically demonstrated that a subwavelength focal spot of size FWHM$=0.30\lambda$ occurred at certain parameters near the glass axicon vertex. Note that analytical expressions to describe the on-axis intensity resulting from the diffraction of a plane wave and a Gaussian beam by a conventional conical axicon were for the first time reported in Reference [81]. An LA with the quadratic dependence of phase on the radial coordinate, $S(r)=\gamma\ln(a+br^2)$, realized with the aid of a computer-generated hologram and intended to focus light into an axial line was, for the first time, proposed in Reference [82]. The simulation of the LA performance in Reference [83] showed the average axial intensity to be more uniformly distributed along the focusing line than that produced by a linear conical axicon [81], in which the average intensity increased along the axial line.

In this section, we have investigated the scalar paraxial diffraction of the Gaussian beam by a logarithmic axicon described by the phase function of References [79, 82, 83], given that $a=0$. This axicon produces an axial focal line that begins directly behind it. The phase function of the axicon is given by $S(r)=\gamma\ln(r/\sigma)$, having a singularity at $r=0$. At this point, the phase $S(r)$ tends to plus (or minus) infinity. Note, however, that this singularity only occurs in the original plane at $z=0$. In any other plane $(z>0)$, when illuminated by the Gaussian beam, the LA produces a finite-energy light field without singularities. An analytical relationship for the complex amplitude of the field for the Fresnel diffraction is derived. The diameter of the transverse focusing line intensity is found to be in the inverse proportion to the LA parameter, $|\gamma|^{-1/2}$. Thus, at sufficiently large $|\gamma|\gg 1$ it becomes possible to obtain a subwavelength diameter of the laser beam in the LA vicinity. This can be confirmed by the numerical simulation.

1.4.1 ANALYTICAL RELATIONSHIPS

Below, using the paraxial scalar approximation, we consider the diffraction of the Gaussian beam by a spiral logarithmic axicon (SLA) and by a conventional logarithmic axicon. The transmission function in the thin transparency approximation for the SLA in the polar coordinates (r, φ) is given by

$$T(r, \phi) = \exp\left[i\gamma \ln\left(\frac{r}{\sigma}\right) + in\,\phi\right], \tag{1.33}$$

where n is an integer (the topological charge of the optical vortex), γ is a real number (the axicon "force" parameter), and σ is the axicon scaling parameter. Then, directly behind the SLA, the complex amplitude of monochromatic light field takes the form:

$$E_0(r, \phi) = \exp\left[-\left(\frac{r}{w}\right)^2 + i\gamma \ln\left(\frac{r}{\sigma}\right) + in\,\phi\right], \tag{1.34}$$

where w is the Gaussian beam waist radius. The Fresnel transform of the function (1.34) takes the form:

$$\begin{aligned}
E(\rho, \theta, z) = \frac{(-i)^{n+1}}{n!} &\left(\frac{z_0}{z}\right)\left(\frac{w}{\sigma}\right)^{i\gamma}\left(\frac{kw\rho}{2z}\right)^n \\
&\times \Gamma\left(\frac{n+2+i\gamma}{2}\right)\left(1 - \frac{iz_0}{z}\right)^{-\frac{n+2+i\gamma}{2}} \exp\left(in\,\theta + \frac{ik\rho^2}{2z}\right) \\
&\times {}_1F_1\left[\frac{n+2+i\gamma}{2}; n+1; -\left(\frac{kw\rho}{2z}\right)^2\left(1 - \frac{iz_0}{z}\right)^{-1}\right],
\end{aligned} \tag{1.35}$$

where (ρ, θ) are the transverse polar coordinates in the observation plane, z is the coordinate on the optical axis, $k = 2\pi/\lambda$ is the wave number for wavelength λ, $z_0 = kw^2/2$ is the Rayleigh range, $\Gamma(x)$ is the gamma-function, and ${}_1F_1(a; c; x)$ is the confluent hyper-geometric function [84]. Note that the relationship (1.35) presents an exact solution of the paraxial wave equation (similar to the Schroedinger equation), being a particular case of the earlier derived solution for the family of hyper-geometric laser beams [85, 86]. The light field intensity $I(\rho, z) = |E(\rho, z)|^2$ derived from Equation (1.35) takes the form:

$$\begin{aligned}
I(\rho, z) = \frac{1}{(n!)^2} &\left(\frac{z_0}{z}\right)^2\left(\frac{kw\rho}{2z}\right)^{2n}\left|\Gamma\left(\frac{n+2+i\gamma}{2}\right)\right|^2 \\
&\times \left(1 + \frac{z_0^2}{z^2}\right)^{-\frac{n+2}{2}} \exp\left[-\gamma \arctg\left(\frac{z_0}{z}\right)\right] \\
&\times \left|{}_1F_1\left[\frac{n+2+i\gamma}{2}; n+1; -\left(\frac{kw\rho}{2z}\right)^2\left(1 - \frac{iz_0}{z}\right)^{-1}\right]\right|^2.
\end{aligned} \tag{1.36}$$

Note that the intensity (1.36) is independent of the axicon scaling parameter σ in Equation (1.33). Everywhere on the optical axis (except for the original plane $z=0$), the intensity (1.36) equals zero for $n\neq0$. Considering the focusing of light with the LA below we put $n=0$. Then, the light field intensity will be given by, instead of Equation (1.36)

$$I_0(\rho,z) = \frac{z_0^2}{z^2+z_0^2}\left|\Gamma\left(1+\frac{i\gamma}{2}\right)\right|^2\exp\left[-\gamma\,\mathrm{arctg}\left(\frac{z_0}{z}\right)\right]$$

$$\times\left|{}_1F_1\left[1+\frac{i\gamma}{2};1;-\left(\frac{kw\rho}{2z}\right)^2\left(1-\frac{iz_0}{z}\right)^{-1}\right]\right|^2.$$

(1.37)

Putting $\rho=0$ in Equation (1.37), the expression for the axial intensity takes the form:

$$I_0(z) = \left(\frac{\pi\gamma}{2}\right)\mathrm{sh}^{-1}\left(\frac{\pi\gamma}{2}\right)\frac{z_0^2}{z_0^2+z^2}\exp\left[-\frac{\pi\gamma}{2}+\gamma\,\mathrm{arctg}\left(\frac{z}{z_0}\right)\right].$$

(1.38)

In deriving Equation (1.38), we made use of the fact that ${}_1F_1(a;c;0)=1$, $|\Gamma(1+ix)|^2=(\pi x)$ $\mathrm{sh}^{-1}(\pi x)$ and $\arctan(x)=\pi/2-\arctan(1/x)$.

In Figure 1.26, the axial intensity at zero ($\rho=0$, $z=0$), as derived from Equation (1.38), is

$$I_0(0) = \pi\gamma\left[\exp(\pi\gamma)-1\right]^{-1} \geq 1.$$

(1.39)

The axial intensity in Figure 1.26 calculated using the non-paraxial Rayleigh–Sommerfeld integral has the aim of demonstrating that, first, the paraxial and

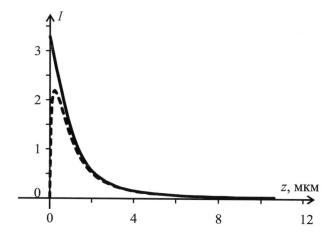

FIGURE 1.26   The axial intensity ($\rho=0$) produced by the focusing LA ($\gamma<0$) ($\gamma=-1$, $w=\sigma$, $\lambda=532$ nm) derived from Equation (1.38) (solid line) and from the Rayleigh–Sommerfeld integral (dashed line).

non-paraxial axial intensities are considerably different at $z < 2\lambda$ and, second, neither curve yields the correct result of $I_0(0) = 1$ at $z = 0$ and $\rho = 0$.

Figure 1.27 shows the intensity distribution in the transverse plane $z = 2\lambda$ derived using the Rayleigh–Sommerfeld integral (curve 1) and Fresnel integral (curve 2), the remaining parameters being the same as in Figure 1.26.

From Figure 1.27 the LA is seen to form a light beam nearly devoid of sidelobes, as distinct from the conventional axicon that forms a beam with the amplitude proportional to the Bessel function $J_0(\gamma\rho)$ and the sidelobes amounting to 0.4 of the axial intensity.

The value of $I_0(0)$ in Equation (1.39) is larger than or equal to 1 (at $\rho = 0$), which contradicts to Equation (1.34) from which it follows that the intensity at zero, $I_0(0)$, should be equal to 1. The fact is that a change from Equations (1.37) to (1.38) is possible at $\rho = 0$ for any z, except for when $z = 0$. Note that as z and ρ simultaneously tend to zero, the relationship resulting from the transformation of Equation (1.37) will depend on the relative rates at which the two variables z and ρ tend to zero. For instance, by tending z to zero and making $\rho \neq 0$ fixed, we derive from Equation (1.37), instead of (1.38):

$$I_0(\rho, z) = \frac{z_0^2}{z_0^2 + z^2} \exp\left[-\frac{2\rho^2}{w^2} + 2\gamma \operatorname{arctg}\left(\frac{z}{z_0}\right) \right]. \qquad (1.40)$$

In deriving Equation (1.40), the following asymptotic expression for the hyper-geometric function at $x \to \infty$ was used [84]:

$$ {}_1F_1(a; c; x) = \frac{\Gamma(c)\exp(x)}{\Gamma(a)x^{c-a}}. \qquad (1.41)$$

At $z = 0$, Equation (1.40) gives the correct expression for the original intensity of the field (1.34), $I_0(\rho, z = 0) = \exp[-2(\rho/w)^2]$. Equation (1.40) is true at $z \to 0$ and $\rho \neq 0$. Figure 1.28 shows the form of the function (1.40) for near-zero values of ρ.

Though at first sight Figures 1.26 and 1.28 seemingly contradict each other, this is not the case. At $z = 0$ and $\rho = 0$ there takes place the radial phase singularity for which

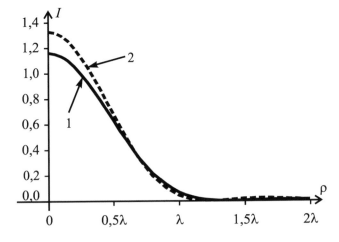

FIGURE 1.27 The intensity in the transverse plane $z = 2\lambda$ ($\gamma = -1$, $w = \sigma$, $\lambda = 532$ nm derived from the Rayleigh–Sommerfeld integral (curve 1) and the Fresnel integral (curve 2).

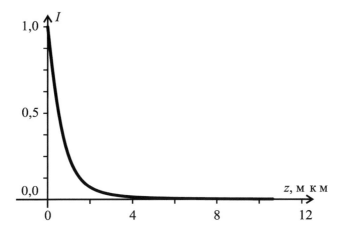

FIGURE 1.28 The z-dependence of the intensity in the vicinity of $z=0$ and at near-zero values of ρ ($\gamma=-1$, $w=\sigma$, $\lambda=532$ nm).

the phase in Equation (1.34) and the derivatives thereof at $n=0$ tend to infinity for $r \to 0$, which means that the intensity experiences a jump on the optical axis, directly behind the LA, for $z>0$: the LA "infinitely quickly" focuses the light into a point on the optical axis. Because of this, the original intensity of $I_0(0)=1$ cannot be arrived at by the continuous passage to the limit along the optical axis to point $z=0$. However, if the passage to the limit is conducted at $\rho \neq 0$ (in the optical axis vicinity), we obtain the correct intensity value at $z=0$: $I_0(0) \sim 1$ (Figure 1.28).

Below, we evaluate the width of the transverse intensity distribution near the optical axis. It has been known that the first guess of the coordinate x_1 of the first zero of the hyper-geometric function ${}_1F_1(a; c; x)$ is deduced from the relation [84]:

$$x_1 = \frac{\gamma_{c-1,1}^2}{2(c-2a)},\tag{1.42}$$

where $\gamma_{c-1,1}$ is the first root of the Bessel function of the $(c-1)$th-order: $J_{c-1}(\gamma_{c-1,1})=0$. Considering Equation (1.37), in Equation (1.42) one should substitute the values $a=1+i\gamma/2$, $c=1$, $\gamma_{0,1}=2.4$. Since in our case, the quantities a and x are complex, the relation in Equation (1.42) should be written for the modules of the complex numbers:

$$|x_1| = \frac{\gamma_{c-1,1}^2}{2|c-2a|}.\tag{1.43}$$

Then, the real coordinate of the first complex zero (local minimum) ρ_1 of the intensity (1.37) can be evaluated as

$$\rho_1 = 2.4w \left[\frac{1+\dfrac{z^2}{z_0^2}}{2(1+\gamma^2)} \right]^{1/4}.\tag{1.44}$$

From Equation (1.44) it follows that at large $|\gamma| \gg 1$ the effective radius of the transverse intensity distribution is given by ($z \ll z_0$):

$$\rho_1 \approx 2w|\gamma|^{-1/2}. \tag{1.45}$$

Besides, Equation (1.44) suggests that choosing a sufficiently large value of $|\gamma|$ ($\gamma < 0$) makes possible obtaining a focal spot of arbitrarily small subwavelength diameter near the plane $z = 0$ using the LA. For instance, at $w = \sigma$, $z_0 = kw^2/2 = \pi\lambda$, $z = \lambda$, and $\gamma = -400$, it is possible to obtain a focal spot of size FWHM $= \rho_1 = \lambda/10$, because the gradient of the phase in Equation (1.33) at $n = 0$ equals $\gamma\sigma/r$ and tends to infinity at $r \to 0$. Thus, not only does the LA focus the propagating waves on the optical axis but also excites evanescent waves with a large magnitude of the wave vector projection onto the transverse axis $k_r \gg k$. It is the existence of such surface waves near $z = 0$ that enables the subwavelength size of the focal spot. Let us conduct the simulation to corroborate the relationship (1.45).

Table 1.2 contains n, the order number; γ_n, the LA parameter; ρ_n, the radius of the first zero (or the first local minimum) in terms of wavelengths at $z = 10\lambda$ (the other parameters being $w = 2\lambda$, $z_0 = 4\pi\lambda$) that was numerically calculated using the Fresnel transform; ρ_n/ρ_{n+1}, the ratio of two adjacent radii; $(\gamma_{n+1}/\gamma_n)^{1/2}$, the square root from the ratio of two adjacent LA parameters. According to the relationship (1.45), the values in the third and fourth rows in each column of Table 1.2 should be the same, because $\rho_n/\rho_{n+1} = (\gamma_{n+1}/\gamma_n)^{1/2}$. The comparison of the third and the fourth rows of Table 1.2 shows that with increasing absolute value of the parameter γ the values of the rows are getting increasingly closer.

1.4.2 SIMULATION

Let us consider the propagation of a radially polarized mode R-TEM$_{01}$ through the diffractive logarithmic microaxicon. The simulation is conducted using the R-FDTD method [49]. The simulation parameters are as follows: the calculation domain, $20\lambda \times 20\lambda$, spatial resolution, $\lambda/20$, temporal resolution, $\lambda/40c$, where c is the speed of light in free space. The axicon parameters are: the maximum height is $h_{\max} = \lambda/(n' - 1)$, where n' is the refractive index of the glass ($n' = 1.5$), the current profile height is derived from $\mathrm{mod}\left(\gamma \ln(r/\sigma)/h_{\max}\right)$, where $\gamma = -2$ and $\sigma = 20$ μm, and $\mathrm{mod}(x)$ denotes the residue of division. The mode parameters are: the radius is $w = 6\lambda$ and

TABLE 1.2
Radii of the Transverse Intensity Distribution

n	1	2	3	4	5
$-\gamma_n$	1	3	5	7	9
ρ_n, λ	2.80	1.93	1.53	1.29	1.13
ρ_n/ρ_{n+1}	1.45	1.26	1.19	1.14	–
$(\gamma_{n+1}/\gamma_n)^{1/2}$	1.52	1.29	1.18	1.13	–

the wavelength is $\lambda = 532$ nm. Figure 1.29 shows the radial profile of the LA and Figure 1.30 shows the radial profile of the illumination beam amplitude. Figure 1.31 presents the axial intensity profile and Figure 1.32 shows the radial intensity profile in the focal plane (the one where the intensity is maximal).

The simulation has shown that the longitudinal focus width is $DOF = 2\lambda$ (Figure 1.31), whereas the focal spot size measured at a wavelength distance from the "sawtooth" axicon surface is $FWHM = 0.44\lambda$ (Figure 1.32). This value is smaller than the diffraction limit of $FWHM = 0.51\lambda$.

Let us consider another example. Let us consider a 2D LA, conducting the simulation using the FullWAVE software.

The relief of the 2D axicon is shown in Figure 1.33. The simulation domain boundary is $[-4\lambda, +4\lambda] \times [0, 3\lambda]$, the simulation time is $cT = 30\lambda$, the sampling on the

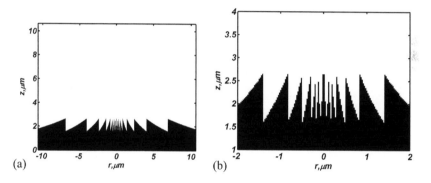

FIGURE 1.29 (a) The logarithmic microaxicon in the calculating domain and (b) a magnified near-axis fragment thereof.

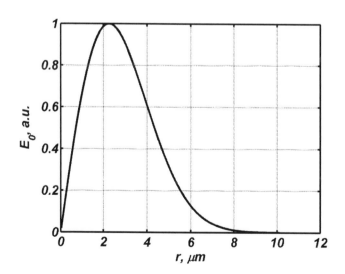

FIGURE 1.30 Radially polarized laser mod R-TEM$_{01}$.

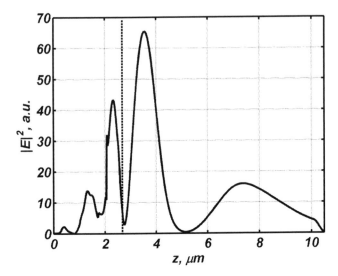

FIGURE 1.31 Intensity distribution along the z-axis (the vertical line is where the LA relief tops are found).

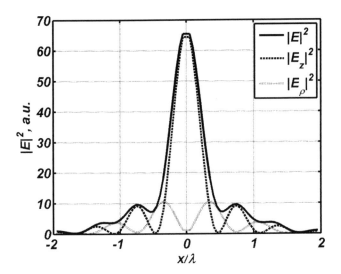

FIGURE 1.32 Intensity distribution in the focal spot.

x- (horizontal) and z-axes is $\lambda/50$, time resolution is $T/100$. The microrelief height is $\lambda/(n-1) = 2\lambda$, the refractive index is $n = 1.5$, the illumination wavelength is 532 nm, and the axicon parameter is $\gamma = -20$. The LA is illuminated by a Gaussian beam with TE-polarization and the waist radius of 3λ.

The focal spot size of the central spot in Figure 1.34 is FWHM = 108 nm = 0.20λ, which is below the diffraction limit of FWHM = $0.44\lambda/n' = 0.293\lambda$ (for the glass, $n' = 1.5$) in the 2D medium.

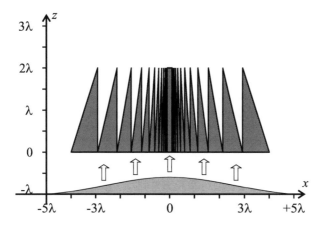

FIGURE 1.33 The appearance of the 2D LA with the radius 4λ, microrelief height $\lambda/(n-1)=2\lambda$, and $\gamma=-20$.

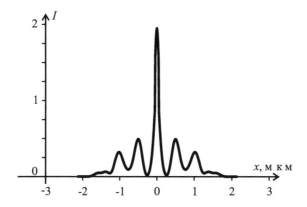

FIGURE 1.34 Averaged intensity distribution directly behind the axicon.

The results arrived at in this section are as follows. An explicit analytical relationship for the complex amplitude of light field that describes the Fresnel diffraction of a Gaussian beam by a spiral logarithmic axicon has been derived. An explicit relation of the axial intensity for the Fresnel diffraction of the Gaussian beam by the LA has been derived. The evaluation formula derived for the diffraction pattern radius shows the inverse proportion to the axicon parameter γ. The FDTD-aided simulation has shown that the output light beam diameter for the 3D LA with $\gamma=-2$ is FWHM$=0.44\lambda$ and for the 2D LA with $\gamma=-20$ is FWHM$=0.20\lambda$ (for the glass, $n'=1.5$).

1.5 TIGHT FOCUSING WITH A BINARY ZONE PLATE

The near-field focusing of light enables a deliberately small size of focal spot to be obtained with the aid of evanescent waves. However, in this case, there is no radiant flux from the surface toward the far-field observer. For far-field diffraction, with distances from the surface much larger than the wavelength, the light cannot be focused into a spot smaller than the scalar diffraction-limited size defined as $FWHM = 0.51\lambda$ (Airy disk diameter) or vector diffraction-limited size of $FWHM = 0.36\lambda$ (zero-order Bessel beam diameter). When the focal spot is formed at a wavelength distance from the surface, this case is on the boundary between the near- and far-field diffraction. For example, Richards–Wolf formulae are not suitable in this case (as we show in this work). Light can be focused at a wavelength distance from the surface using a conventional binary zone plate (ZP) with the near-unity numerical aperture (NA). There are numerous publications that study the focusing of light with the aid of the ZP with the focal length comparable with the incident wavelength. In Reference [87] it was numerically shown that the ZP could focus the radially polarized wave into a smaller-size spot when compared with focusing with the same-NA aplanatic objective (AO). A focal spot of size $FWHM = 0.42\lambda$ was reported in Reference [87] to be generated with a ZP of $NA = 0.98$. The focusing of the Bessel–Gaussian beam with the aid of a ZP and an AO was also simulated in Reference [88]. The focal spot size was reported to be decreased by combining the focusing elements with a three-zone plate: in the AO-aided scheme, the focal spot size was reduced from $FWHM = 0.584\lambda/NA$ till $FWHM = 0.413\lambda/NA$ and in the ZP-aided scheme – from $FWHM = 0.425\lambda/NA$ till $FWHM = 0.378\lambda/NA$. In References [87, 88], the simulation was conducted using Richards–Wolf (RW) formulae [6] that were adapted for the radially polarized light [89]. It should be noted that the Debye vector theory [4] that forms the basis of the RW formulae is only valid for focal distances much larger than the incident wavelength. When the focusing distance is equal to or smaller than the wavelength the simulation is conducted using the FDTD method. For instance, using the FDTD method and experimental measurements on a near-field scanning optical microscope (NSOM) [90], the diffraction-limited focal spot size was shown to be broken by use of a plasmonic ZP: the simulated focal spot size was $FWHM = 0.41\lambda$ and the experimental size was $FWHM = 0.48\lambda$. The focusing of light with a silver amplitude ZP fabricated on a silica substrate was reported in Reference [91], with the FDTD modeling having shown that such a ZP produced the focal spot size of 0.33λ. A nine-ring binary amplitude ZP with $f = 2\lambda$ was discussed in Reference [92]. Using the integral Rayleigh formulae it was shown that for the linearly and circularly polarized light the focal spot size (considering the transverse intensity $I_x + I_y$ and disregarding the axial intensity I_z) was equal to $FWHM = 0.45\lambda$. A plasmonic ZP in which the even-number zones were coated with silver and glass layers was modeled in Reference [93]. The use of the structure was shown to increase the ZP diffraction efficiency up to 40% for the focal spot size of $FWHM = 0.48\lambda$. The feasibility to perform the near-field focusing by means of a dielectric binary phase ZP was numerically demonstrated in Reference [94]. The resulting focal spot size was $FWHM = 0.52\lambda$. Focusing of the linearly polarized light of wavelength 633 nm using a binary ZP of focus 0.5 µm was numerically and experimentally studied in Reference [95]. The minor size of the experimentally observed elliptic focal spot was $FWHM = 0.63\lambda$. Note that the FDTD-based simulation predicted a smaller focal spot size of $FWHM = 0.36\lambda$. The comparative numerical study of focusing a plane linearly polarized wave and a radially polarized BG beam by means of a phase ZP with focus 0.5 µm and $NA = 0.996$ was reported in Reference [96]. The said phase ZP was shown to be able to focus the BG beam into a focal spot of size FWHM$=0.39\lambda$ and a plane linearly polarized wave into an elliptic spot of sizes FWHM$_x=0.87\lambda$ (polarization axis) and FWHM$_y=0.39\lambda$.

In this section, we studied a binary phase ZP whose focus is equal to the incident wavelength, $f=\lambda=0.532$ µm, radius 7 µm, and groove depth 510 nm. The propagation of a linearly polarized Gaussian beam through such a ZP was studied with the aid of the NSOM. The experimentally observed focal spot size was FWHM$=0.44\lambda$, which is smaller than the diffraction limit (FWHM$=0.51\lambda$) and the experimental results reported in References [90, 94, 95]. By the FDTD-based numerical simulation, it was also shown that the ZP under study was able to focus the light beam into a focal spot of size FWHM$=0.42\lambda$, with the root mean square deviation of the experimental curve from the calculated value being 5%.

We aim to find the smallest focal length for which the FDTD method and the RW formulae give similar results for the ZP-aided focusing. The focusing of the mode R-TEM$_{01}$ with the aid of a ZP with varying foci ($\lambda \leq f \leq 15\lambda$) was simulated. Figure 1.35(a) depicts the plots of the focal spot size FWHM as a function of the ZP focal length, which were calculated using two methods – the FDTD method (solid curve) and the RW formulae (dotted curve). In a similar way, Figure 1.35(b) depicts the plots of the depth of focus as a function of the focal length. The FDTD-based simulation parameters were the following: spatial sampling – $\lambda/50$, time sampling – $\lambda/100c$, where c is the speed of light in vacuum. The rms deviation δ of the focal spot size calculated using the FDTD method and the RW formulae was found not to exceed 6% (Figure 1.35(a)), whereas the rms deviation of the DOF was increasing with decreasing focal length, reaching 30% at $f=\lambda$ (Figure 1.35(b)). For $f\geq 4\lambda$ (NA ≤ 0.98), both deviations were not larger than 6%. Thus, the RW formulae may be suited for modeling the tight focusing of light when the ZP focus is larger than 4λ. Figure 1.35(a) suggests that the diffraction limit is broken for both simulation techniques. For the FDTD simulation, the diffraction limit is overcome at $f=4.7\lambda$ (NA $=0.97$), whereas for the RW formulae this occurs at $f=5.4\lambda$ (NA $=0.96$).

The minimal focal spot size amounted to FWHM$=0.37\lambda$ (for RW formulae–aided simulation) and FWHM$=0.39\lambda$ (for FDTD-aided simulation) for NA $=0.999$ ($f=\lambda$).

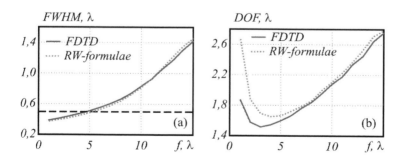

FIGURE 1.35   The focal spot (a) size and (b) DOF (both at FWHM) against the focal length f of a ZP of radius $R=20\lambda$ illuminated by a radially polarized mode R-TEM$_{01}$ of radius $\omega=10\lambda$: FDTD (solid curve) and RW formulae (dotted curve). The dashed line in Figure 1.35(a) denotes the diffraction limit.

Shown in Figure 1.36 is the intensity pattern in the focal plane of a ZP with focus $f=\lambda$. The diffraction limit is overcome due to the increase of the side lobes energy (i.e. the increase of DOF). At the same time, the focusing efficiency (the proportion of light energy contributing to the central part of the focal spot) is decreasing. From Figure 1.35(b), the DOF calculated with the FDTD method is seen to be always smaller than the DOF derived from the RW formulae. This is because the light beam is bounded in the axial direction by ZP surface, as clearly seen in Figure 1.36(b). Figure 1.35(b) shows that there is a minimum of the DOF, with the FDTD-based simulation producing the value of $DOF=1.51\lambda$ at $f=3\lambda$ ($NA=0.99$) and with the RW formulae–based simulation producing $DOF=1.65\lambda$ for $f=4\lambda$ ($NA=0.98$). From Figure 1.35(b) it is seen that as the ZP focal length is decreasing to $f=3\lambda$, the DOF is also decreasing; but the further decrease of the focal length below $f<3\lambda$ leads to an increase in the DOF. This serves to confirm the general concept that with the focal spot size decreasing below the diffraction limit, the DOF starts to be increasing in such a manner that the diffraction-limited focus volume remains unchanged.

The focusing efficiency can be estimated as $\eta=W1/W0$, where W0 is the input beam power, W1 is the light power in the focal spot. When focusing the mode of radius $\omega=10\lambda$ with a ZP of focus $f=\lambda$ and radius $R=20\lambda$, the focusing efficiency was found to be $\eta=5\%$. The FDTD-based simulation of the propagation of a linearly polarized Gaussian beam through a ZP of $f=\lambda$ has shown that for the superresolution to be achieved on the x-axis, it was sufficient to employ a 7-ring ZP, with the minor focal spot size being $FWHM_x=0.42\lambda$. The DOF was also found to be smaller than the diffraction limit, $DOF=0.86\lambda$; the major-axis focal spot size was nearly twice as large as the minor-axis spot size, $FWHM_y=0.84\lambda$. Note that a ZP with $f=\lambda$ (see Figure 1.35) illuminated by a radially polarized beam was found to form a focal spot of size $FWHMr=0.39\lambda$ and $DOF=1.9\lambda$. It is that the ZP with $f=\lambda$ forms a comparable focal volume with the linearly polarized incident beam when compared with the radially polarized incident beam:

$$V_{rad} = \pi\left(0.39\lambda\right)2\left(1.9\lambda\right) = 0.91\lambda^{3}, \qquad (1.46)$$

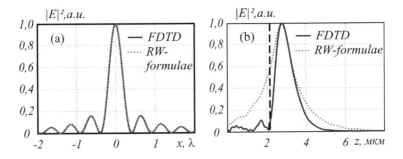

FIGURE 1.36 The intensity profile (a) in the focal plane along the x-axis and (b) along the z-axis of symmetry of the ZP with $R=20\lambda$ and $f=\lambda$ produced by the incident radially polarized mode R-TEM$_{01}$ with $\omega=10\lambda$: the FDTD method (solid curve) and the RW formulae (dotted curve). The dashed vertical line in Figure 1.36(b) denotes the edge of the ZP.

$$V_{\text{lin}} = \pi\left(0.42\lambda\right)\left(0.84\lambda\right)\left(0.86\lambda\right) = 0.95\lambda^3. \tag{1.47}$$

We can infer that a decrease in the focal spot transverse size when illuminating the ZP by the radially polarized light does not substantial result in the reduction of the total focal spot volume.

A high-quality ZP was fabricated by lithography in a ZEP resist with $n = 1.52$. In Figure 1.37 is a SEM image of the ZP: groove depth – 510 nm, ZP diameter – 14 μm, and the peripheral zone – $0.5\lambda = 266$ nm. The ZP has 12 rings and a central disk.

The ZP radii were derived from the well-known formula $r_m = (m\lambda f + m^2\lambda^2/4)^{\frac{1}{2}}$, where $f = \lambda = 532$ nm is the ZP focal length and m is the radius number. The experiment was conducted using a near-field scanning optical microscope NTEGRA Spectra (Figure 1.38(a)). The ZP was illuminated by a linearly polarized Gaussian beam and the intensity distributions at different distances from the ZP surface were then measured. The intensity pattern in the ZP focal plane in Figure 1.38(a) shows that the focal spot is elliptic, with the ellipse minor semi-axis located in the plane perpendicular to the polarization direction of the beam under focusing (the polarization is along the y-axis). The smallest focal spot size (along the x-axis) was measured to be FWHM $= 0.44\lambda$.

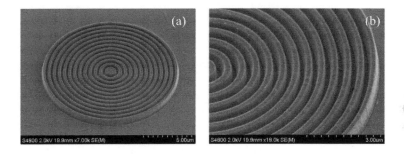

FIGURE 1.37 A SEM image of the ZP with (a) 7000× and (b) 18 000× magnification

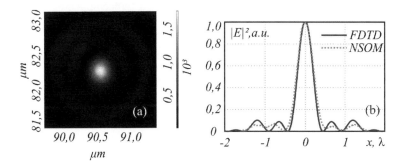

FIGURE 1.38 (a) An experimentally measured intensity pattern (relative units) in the focus; (b) the intensity profile in the focal spot along the x-axis: FDTD (solid curve) and NSOM (dotted curve).

To compare the experiment and calculation results, focusing a linearly polarized Gaussian beam of radius $\omega = 7\lambda$ and wavelength $\lambda = 532$ nm was simulated using the FDTD method. Then, the comparison of the focal spot sizes obtained by the simulation and the experiment was conducted, showing the focal spot size along the x-axis to be similar for the simulation (FWHM$_x$=0.42λ) and experiment (FWHM$_x$=0.44λ). Meanwhile, along the y-axis, the focal spots were different in size: FWHM$_y$=0.52λ (experiment) and FWHM$_y$=0.84λ (simulation). However, the experimentally measured spot size along the y-axis (FWHM$_y$=0.52λ) was equal to the transverse intensity component obtained by simulation (FWHM$_y$=0.52λ). Figure 1.38(b) shows the intensity profiles along the x-axis obtained by simulation (solid curve) and experimentally (dotted curve). From Figure 1.38(b), both curves are seen to be in good agreement. In a similar way, Figure 1.39(a) shows the intensity profiles in the focal spot along the y-axis. The curves are seen to be in a good coincidence in Figure 1.39(a) and different in Figure 1.39(b). Thus, it can be inferred that a metallic pyramid-shaped cantilever with a 100-nm hole in the tip that is used in the NSOM is only able to record the transverse component of the electric field strength.

Thus, considering that for the linearly polarized light incident on the ZP of focus $f = \lambda = 532$ nm, only the transverse E-field component contributes to the total light intensity in the xz-plane perpendicular to the polarization vector, the minor focal spot size measured by the NSOM cantilever (FWHM$_x$=0.44λ), which is coincident with the calculated size (FWHM$_x$=0.42λ) within 5% accuracy, is correct. This is the smallest focal spot yet experimentally obtained for the binary phase ZP.

1.6 THE SHAPE OF SUBWAVELENGTH FOCAL SPOT FOR THE LINEARLY POLARIZED LIGHT

The subwavelength focusing of light by means of microoptic elements has been a highly relevant problem because the reduced focal spot size not only produces a higher resolution in uses such as lithography, microscopy, and optical memory, but also increases the radiation power density, which is important for micro-manipulation. In recent years, significant advances have been reported in this area. Thus, the subwavelength focusing is performed using planar plasmonic structures [97, 98] or plasmonic lenses [99, 100]. The tight focusing of laser light can also be achieved in

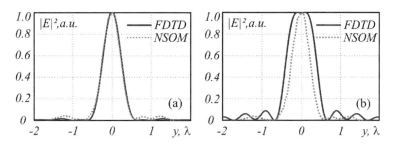

FIGURE 1.39   The radial intensity profiles in the focal spot along the y-axis: FDTD (solid curve) and NSOM (dotted curve) for (a) the transverse intensity component $I_x + I_y$; and (b) the total intensity $I_x + I_y + I_z$.

the near-surface regions of conventional optical elements like a microaxicon [101], a zone plate [95, 102], a microlens [103], a solid immersion lens (SIL) [104], and a conventional high NA lens [105–108]. The subwavelength focal spot can also be obtained on the tips of dielectric [80, 109] and metallic [110] microcones. The microoptics elements can not only perform the subwavelength focusing but also form the micro- and nano-object image with subwavelength resolution (or superresolution) [111].

Note, however, that the above-cited publications have not been concerned with processes whereby the near-field radiation is registered using a small-hole cantilever of the microscope. The following questions have remained unanswered so far. Which radiation component – energy (power) density or energy (power) flux – is being registered by the near-field microscope? Why for the linearly polarized light is the focal spot intensity (power density) in the form of an ellipse, whereas the Poynting vector's optical-axis projection (power flux) is focused into a circle? Also, the publications quoted above lack the comparative analysis of the experimental data on the subwavelength focal spot measurement and the results of the rigorous simulation based on Maxwell's equations.

In this section, by decomposing the linearly polarized light field in terms of plane waves we show that the elliptic shape of the intensity cross-section is determined by the E-vector's longitudinal component. Considering that the Poynting vector's projection on the optical axis (power flux) is independent of the said E-vector's longitudinal component, the power flux cross-section has a circular form. Using a near-field microscope with a small-hole metallic cantilever, we experimentally show that a glass binary zone plate with a wavelength focal length focuses the linearly polarized Gaussian beam into a weakly elliptical focal spot whose size in the Cartesian coordinates is $FWHM_x = (0.44 \pm 0.02)\lambda$ and $FWHM_y = (0.52 \pm 0.02)\lambda$, with the depth of focus being $DOF = (0.75 \pm 0.02)\lambda$, where λ denotes the incident wavelength. The comparison of the experimental results with the FDTD-based numerical simulation suggests an unambiguous conclusion that the near-field microscope measures the transverse intensity (power density) rather than the power flux or the total intensity. The fact that the small-hole metallic cantilever measures the transverse intensity is in compliance with the Bethe–Bouwkamp theory.

1.6.1 LINEARLY POLARIZED LIGHT: FOCAL SPOT INTENSITY AND THE POYNTING VECTOR'S PROJECTION

Assume that light is propagating along the optical axis z from a transverse plane P_1 (referred to as the initial plane) toward a transverse plane P_2, which is parallel to and offset by distance z from the first plane. We introduce the Cartesian coordinates (x, y) and (u, v) and the polar coordinates (r, φ) and (ρ, θ) for the said planes. Let there be a linearly polarized electromagnetic field with a radial symmetry in the initial plane:

$$\begin{cases} E_x(r,\varphi,0) \equiv E_x(r), \\ E_y(r,\varphi,0) \equiv 0, \\ E_z(r,\varphi,0) \equiv 0. \end{cases} \tag{1.48}$$

In the plane P_2, we will measure the intensity:

$$I = |\mathbf{E}|^2 = |E_x|^2 + |E_y|^2 + |E_z|^2 \tag{1.49}$$

and the power flux (the Poynting vector component, which is parallel to the optical axis z):

$$S_z = \frac{1}{2} \operatorname{Re} \left\{ (\mathbf{E} \times \mathbf{H}^*)_z \right\} = \frac{1}{2} \operatorname{Re} \left\{ E_x H_y^* - E_y H_x^* \right\}. \tag{1.50}$$

According to the Rayleigh–Zommerfeld diffraction integrals, the E_y-component is also equal to zero in the plane P_2. Thus, Equations (1.49) and (1.50) can be rearranged as

$$I = |E_x|^2 + |E_z|^2, \tag{1.51}$$

$$S_z = \frac{1}{2} \operatorname{Re} \left\{ E_x H_y^* \right\}. \tag{1.52}$$

Using Maxwell's equation for the monochromatic light of frequency ω:

$$\operatorname{rot} \mathbf{E} = -i\omega\mu_0\mu\mathbf{H}, \tag{1.53}$$

where μ is the permeability and μ_0 is vacuum permeability, we derive from Equation (1.52) that

$$S_z = \operatorname{Re} \left\{ \frac{-i}{2\omega\mu_0\mu} E_x \left(\frac{\partial E_x^*}{\partial z} - \frac{\partial E_z^*}{\partial u} \right) \right\}. \tag{1.54}$$

Let us decompose the E_x-component in terms of the angular spectrum of the plane waves. To these ends, we introduce the Cartesian coordinates (α, β) and the polar coordinates (ζ, ϕ) in the spectral plane. Then,

$$E_x(u,v,z) = \iint_{\mathbb{R}^2} A(\alpha,\beta) \exp\left\{ ik\left[\alpha u + \beta v + z\sqrt{1-\alpha^2-\beta^2} \right] \right\} d\alpha d\beta. \tag{1.55}$$

From the third Maxwell's equation

$$\frac{\partial E_x}{\partial u} + \frac{\partial E_y}{\partial v} + \frac{\partial E_z}{\partial z} = 0 \tag{1.56}$$

and considering that $E_y \equiv 0$, the E_z component is given by

$$E_z(u,v,z) = -\iint_{\mathbb{R}^2} \frac{\alpha}{\sqrt{1-\alpha^2-\beta^2}} A(\alpha,\beta)$$
$$\times \exp\left\{ ik\left[\alpha u + \beta v + z\sqrt{1-\alpha^2-\beta^2} \right] \right\} d\alpha d\beta + C(u,v). \tag{1.57}$$

The constant $C(u, v)$ that arises from the integration with respect to z, denotes the infinite (on the z-axis) constant field. From physical considerations, we may put it equal to zero. Also, the expression in the brackets in Equation (1.54) can be rearranged as

$$\frac{\partial E_x^*}{\partial z} - \frac{\partial E_z^*}{\partial u} = -ik \iint_{\mathbb{R}^2} \frac{1-\beta^2}{\sqrt{1-\alpha^2-\beta^2}} A^*(\alpha,\beta)$$

$$\times \exp\left\{-ik\left[\alpha u + \beta v + z\sqrt{1-\alpha^2-\beta^2}\right]\right\} d\alpha d\beta. \tag{1.58}$$

Considering that the E_x-component is radially symmetric in the initial plane, its angular spectrum is also symmetric, i.e. $A(\zeta, \phi) \equiv A(\zeta)$. Taking this into account, Equations (1.55), (1.57), and (1.58) can be rewritten in the polar coordinates. In this case, all the integrals with respect to ϕ will be expressed through the Bessel functions:

$$E_x(\rho,\theta,z) = 2\pi \int_0^\infty A(\zeta) \exp\left(ikz\sqrt{1-\zeta^2}\right) J_0(k\rho\zeta) \zeta d\zeta, \tag{1.59}$$

$$E_z(\rho,\theta,z) = -2\pi i \cos\theta \int_0^\infty A(\zeta) \exp\left(ikz\sqrt{1-\zeta^2}\right) J_1(k\rho\zeta) \frac{\zeta^2 d\zeta}{\sqrt{1-\zeta^2}}. \tag{1.60}$$

$$\frac{\partial E_x^*}{\partial z} - \frac{\partial E_z^*}{\partial u} = -2\pi i k \int_0^\infty A^*(\zeta) \exp\left(-ikz\sqrt{1-\zeta^2}\right)$$

$$\times \left[\left(1-\frac{\zeta^2}{2}\right) J_0(k\rho\zeta) - \frac{\zeta^2}{2} J_2(k\rho\zeta)\cos(2\theta)\right] \frac{\zeta d\zeta}{\sqrt{1-\zeta^2}} \tag{1.61}$$

Substituting (1.59)–(1.61) into (1.51) and (1.54) gives:

$$I = 4\pi^2 \left| \int_0^\infty A(\zeta) \exp\left(ikz\sqrt{1-\zeta^2}\right) J_0(k\rho\zeta) \zeta d\zeta \right|^2$$

$$+ 4\pi^2 \cos^2\theta \left| \int_0^\infty A(\zeta) \exp\left(ikz\sqrt{1-\zeta^2}\right) J_1(k\rho\zeta) \frac{\zeta^2 d\zeta}{\sqrt{1-\zeta^2}} \right|^2, \tag{1.62}$$

$$S_z = -\frac{2\pi^2 k}{\omega\mu_0\mu} \text{Re}\left(\left\{\int_0^\infty A(\zeta) \exp\left(ikz\sqrt{1-\zeta^2}\right) J_0(k\rho\zeta) \zeta d\zeta\right\}\right.$$

$$\times \left\{\int_0^\infty A^*(\zeta) \exp\left(-ikz\sqrt{1-\zeta^2}\right) \left[\begin{array}{c}\left(1-\frac{\zeta^2}{2}\right) J_0(k\rho\zeta) \\ -\frac{\zeta^2}{2} J_2(k\rho\zeta)\cos(2\theta)\end{array}\right] \frac{\zeta d\zeta}{\sqrt{1-\zeta^2}}\right\}\right). \tag{1.63}$$

From Equations (1.62) and (1.63), the intensity and the power flux are seen to be symmetric with respect to the x-axis, which means that they take the same values for any pair of points with the polar coordinates (ρ, θ) and $(\rho, -\theta)$. It can also be inferred from Equation (1.62) that for a specified ρ the intensity is maximal at points $(\rho, 0)$ and (ρ, π), being minimal at $(\rho, \pi/2)$ and $(\rho, 3\pi/2)$. This is the reason why an elliptic focal spot is generated with the major axis being on the x-axis (light is polarized in the xz-plane). It is noteworthy that the radial symmetry breakdown is defined by the second term in the intensity expression and by the second multiplier in the power flux expression. If the NA is small, the multiplier $\zeta^2/(1-\zeta^2)^{-1/2}$ entering the second integral in Equation (1.62) is near-zero, preventing the second term from making a significant contribution to the intensity. This explains why the focal spot has a circular shape. If, however, tight focusing occurs, an essential proportion of the angular spectrum belongs to the waves with the value of ζ close to unity. In this case, the increased contribution of the second term may become several times greater than that of the first term, thus causing asymmetry, with the focal spot acquiring an elliptic or even a bone or a dumbbell form.

For the power flux, the relation is different. The factor $\zeta/(1-\zeta^2)^{-1/2}$ affects the symmetric and asymmetric parts of the expression in the square brackets in Equation (1.63) in the same way. For a small NA ($\zeta \ll 1$), the power flux is determined by the radially symmetric term $(1-\zeta^2/2)J_0(k\rho\zeta)$. Because of this, when measuring the focal spot's intensity, rather than the power flux, it has a circular form. When the NA is high and ζ is close to unity, the contribution of both terms in the square brackets is approximately the same. However, the second term's contribution cannot exceed that of the first term by very much. Besides, in the near-focus region ($\rho=0$), the zero-order Bessel function exerts a greater impact when compared with the second-order Bessel function. Thus, the resulting focal spot is closer to the circle than that measured in terms of intensity.

By way of illustration, let us analyze a Bessel beam. Then, its angular spectrum takes a circular form:

$$A(\zeta) = \delta(\zeta - a), \tag{1.64}$$

where $\delta(x)$ is the Dirac delta function and α is the Bessel beam parameter. In this case, with the integrals in Equations (1.62) and (1.63) being present no more, the intensity and the power flux are easily derived from

$$I = \left[2\pi a J_0(ka\rho) \right]^2 + \left[\frac{2\pi a^2}{\sqrt{1-a^2}} J_1(ka\rho)\cos\theta \right]^2, \tag{1.65}$$

$$S_z = -\frac{2\pi^2 k}{\omega\mu_0\mu} J_0(ka\rho) \left[\begin{array}{c} \left(1-\dfrac{a^2}{2}\right) J_0(ka\rho) \\ -\dfrac{a^2}{2} J_2(ka\rho)\cos(2\theta) \end{array} \right] \frac{a^2}{\sqrt{1-a^2}}. \tag{1.66}$$

To simulate the small-NA case, we take $\alpha=0.3$ (Figure 1.40). To simulate the tight focusing, we take $\alpha=0.8$ (Figure 1.41) and $\alpha=0.9$ (Figure 1.42). The intensity and the power flux for different values of α are shown in Figures 1.40–1.42: the x-axis is horizontal and the y-axis is vertical. The remaining simulation parameters are the wavelength $\lambda=532$ nm, the distance $z=10\lambda$, and the simulation domains $-5\lambda\leq x\leq+5\lambda$, $-5\lambda\leq y\leq+5\lambda$.

Figures 1.40–1.42 confirm the above assumptions. With increasing numerical aperture, when measured in terms of intensity, the focal spot at first has a circular form and then becomes elliptical, finally acquiring a dumbbell form. It should be

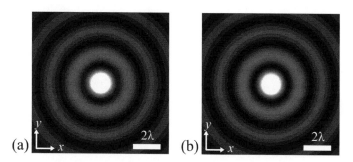

FIGURE 1.40 (a) The intensity and (b) the power flux for a Bessel beam at $\alpha=0.3$.

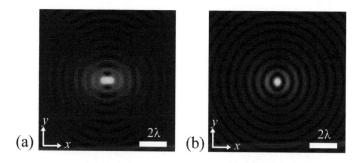

FIGURE 1.41 (a) The intensity and (b) the power flux for a Bessel beam at $\alpha=0.8$.

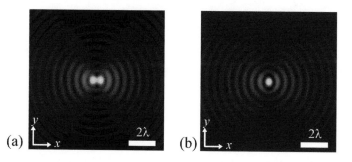

FIGURE 1.42 (a) The intensity and (b) the power flux for a Bessel beam at $\alpha=0.9$.

noted that in terms of power flux, the focal spot always remains nearly circular (in Figure 1.42(b), the focal spot is a weak ellipse elongated along the y-axis, as can easily be shown using Equation (1.66)). From Figures 1.41 and 1.42, it can be seen that the focal spot sizes in terms of intensity and power flux measured along the vertical axis are close to each other.

1.6.2 TIGHT FOCUSING OF LINEARLY POLARIZED LIGHT WITH A SUBWAVELENGTH BINARY AXICON

In the previous section, we simulated the propagation of a linearly polarized light beam with the initial amplitude in the form of a Bessel function using Equations (1.65) and (1.66). In this section, we employ the FDTD method to address the inverse problem of generating a linearly polarized Bessel beam by means of a binary micro-axicon. We demonstrate that the diffraction patterns have the same structure in both cases.

Let us simulate the propagation of a linearly polarized Gaussian beam of wavelength $\lambda = 532$ nm and waist radius $\omega = 7\lambda$ through a binary microaxicon of radius $R = 8$ μm and period $T = \lambda$ (see Figure 1.43), the refractive index $n = 1.52$, and the microrelief height $h = 532$ nm.

The simulation was conducted using the FDTD-method-based FullWAVE software, the grid quantization was $\lambda/50$ in space and $\lambda/100c$ in time, where c is the light speed in free space. It is interesting that when, instead of the intensity (light power density), one calculates the power flux (the absolute value of the Poynting vector's

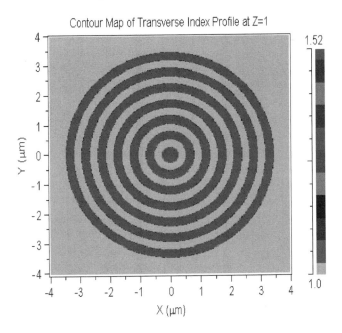

FIGURE 1.43 The contour map of the binary microaxicon of period $T = \lambda$ in the calculated field (the refractive index of the rings is $n = 1.52$ and the background has $n = 1$).

projection onto the optical axis), there appears a circular (rather than elliptical) focal spot, whose diameter is slightly larger than the smaller diameter of the elliptical focal spot for the intensity.

Figure 1.44(a) shows a two-dimensional (gray-level) intensity pattern in the immediate vicinity of the binary microaxicon of period $T = \lambda = 532$ nm (Figure 1.43), whereas Figure 1.44(b) shows the light flux in the same plane. From Figure 1.44, the power flux is seen to generate a circular focal spot with FWHM = 0.36λ, whereas the intensity is focused in an elliptic spot with the Cartesian axis diameters of FWHM$_x$ = 0.75λ and FWHM$_y$ = 0.30λ. The reason is that the E-field's longitudinal component E_z does not contribute to the Poynting vector's longitudinal component. The E_z component arises in the polarization xz-plane when the rays are converging in focus. In the perpendicular yz-plane, the E_z component does not arise when the rays are converging in focus. The focal spot patterns in Figures 1.44 and 1.42 are seen to be in a qualitative agreement.

1.6.3 TIGHT FOCUSING OF THE LINEARLY POLARIZED LIGHT USING A ZONE PLATE

1.6.3.1 Simulation

In this subsection, we analyze the focusing of a linearly polarized Gaussian beam of wavelength $\lambda = 532$ nm and radius $\omega = 7\lambda$ with a binary zone plate of focal length equal to the wavelength of the light under focusing, $f = \lambda$, radius 10.64 μm (20λ) and the refractive index $n = 1.52$. Figure 1.45 shows a template of such a ZP in the design field. The ZP has NA = 0.997. The focusing efficiency was calculated to be 42%.

The ZP radii (Figure 1.45(a)) were calculated by the well-known formula $r_m = (m\lambda f + m^2\lambda^2/4)^{1/2}$, where $f = \lambda = 532$ nm is the focal length and m is the zone radius number. The simulation was conducted using the BOR-FDTD method [80], with the comparison of the results derived for different values of calculation domain

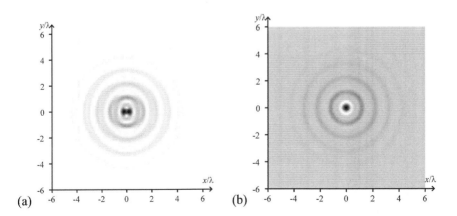

FIGURE 1.44 Patterns of (a) intensity (power density) and (b) power flux (the absolute value of the Poynting vector's projection onto the optical axis) in the surface vicinity (at distance $\lambda/20$), generated by the binary microaxicon of period $T = \lambda$.

spatial discretization. Figure 1.45(b) depicts the intensity pattern in the focus in pseudo-colors.

Figure 1.46 shows the focal spot's profiles in the Cartesian coordinates in the x-axis $(\varphi = 0)$ and the y-axis $(\varphi = \pi/2)$ for (a) the intensity and (b) the power flux (absolute value of the Poynting vector's projection onto the z-axis). From Figures 1.45(b) and 1.46(a) the focal spot is seen to be elliptical in terms of intensity. At the same time, from Figure 1.46(b) the focal spot is seen to be circular in terms of power flux. In Table 1.3, two upper rows show the focal spot size (FWHM) in terms of intensity (row 1) and in terms of power flux (row 2).

Shown in Figure 1.47 are the intensity profiles calculated by the R-FDTD method in MATLAB [80](curve 1) and by the FDTD method in the FullWAVE software (curve 2). The values of the depth of focus are shown in the last column of Table 1.3.

1.6.3.2   Experiment

The high-quality ZP was fabricated by a lithographic procedure in a ZEP resist (the refractive index $n = 1.52$). The AFM images of the ZP are shown in Figure 1.48:

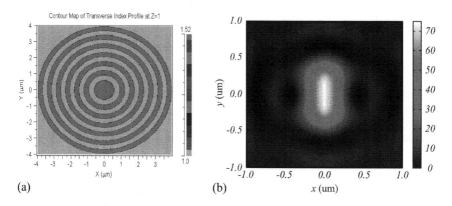

(a)   (b)

FIGURE 1.45   (a) Template of a ZP of focal length equal to the wavelength, $f = \lambda$; and (b) intensity pattern in the focal plane. The polarization is along the y-axis.

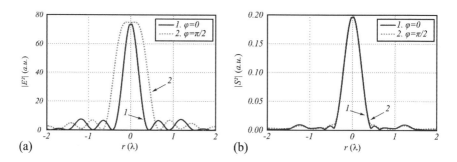

FIGURE 1.46   Patterns of (a) intensity and (b) power flux (the absolute value of the Poynting vector's projection onto the z-axis) in the focus at discretization step $\Delta r = \lambda/50$. The profiles are mapped along the x- $(\varphi = 0)$ and y-axes $(\varphi = \pi/2)$.

TABLE 1.3
Focal Spot Size

	FWHM$_x$ ($\varphi=0$), λ	FWHM$_y$ ($\varphi=\pi/2$), λ	DOF, λ
Intensity	0.42 ± 0.01	0.84 ± 0.01	0.86 ± 0.01
Modulus of the Poynting vector's projection onto the z-axis	0.45 ± 0.01	0.45 ± 0.01	–
Experiment on NSOM	0.44 ± 0.02	0.52 ± 0.02	0.75 ± 0.02

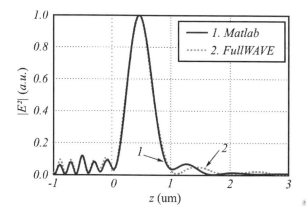

FIGURE 1.47 Comparison of the intensity profiles along the optical axis, generated by the ZP when modeling in R-FDTD MATLAB (curve 1) and FDTD FullWAVE (curve 2).

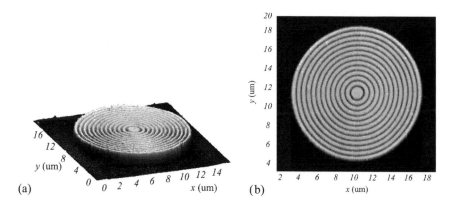

FIGURE 1.48 The AFM images of the ZP: (a) side view; and (b) top view.

(a) side views and (b) top view. The ZP parameters are relief depth – 510 nm, diameter –14 μm, and the outermost zone – $0.5\lambda = 266$ nm. The ZP has 12 rings and a central disk.

The propagation of a linearly polarized Gaussian beam of wavelength $\lambda = 532$ nm through the ZP of focus $f = \lambda$ was experimentally studied using a near-field scanning optical microscope NTEGRA Spectra (Figure 1.49).

Figure 1.50(a) depicts the experimental intensity profile on the optical axis generated by the ZP under study (curve, the axis on the left) and the values of the focal spot size (squares, the axis on the right). Figure 1.50(b) shows an example of the focal spot intensity pattern (as obtained directly from the microscope).

The averaged values of the ZP-generated elliptic focal spot size are given in row 3 of Table 1.3. From Figure 1.50(a), the intensity peak on the axis is seen to be shifted from the geometric focus plane, $f = \lambda = 532$ nm, toward the ZP ($z = 400$ nm). Note that

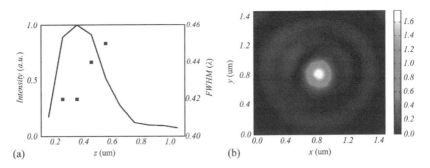

FIGURE 1.49 The NSOM used in the experiment.

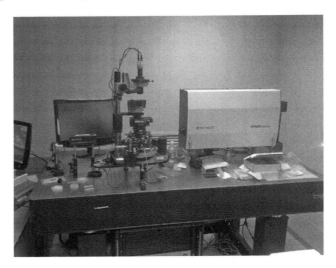

FIGURE 1.50 (a) The experimental intensity profile on the optical axis from the ZP in Figure 1.48 (curve, the axis on the left) and the focal spot smaller sizes (squares, the axis on the right); (b) the focal spot cross-section at the focal length $f = \lambda = 532$ nm (the vertical axis is in the polarization plane).

at the above-indicated distance ($z = 400$ nm) the focal spot's smaller size is the same for the experiment and the simulation: FWHM$=0.42\lambda$.

For comparison, Figure 1.51 depicts the profiles of simulated intensity (curve 1) and power flux (curve 3), and the experimentally measured distribution obtained by the NSOM (curve 2). The curves are seen to be nearly coincident (see column 2 of Table 1.3), with the difference between them being below the measurement error ($\pm 0.02\lambda$). With the difference being only noticeable at sidelobes, it still does not allow one to say unambiguously whether the device really measures the intensity or the power flux, because the sidelobes are somewhat lower than those of the experimental curve for the calculated power flux and somewhat higher for the calculated intensity.

Figure 1.52 shows the curve profiles in the focal spot on the y-axis (found in parallel with the polarization plane): the calculated distribution of the Poynting vector's

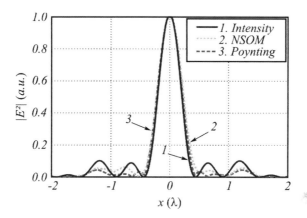

FIGURE 1.51 Comparison of the experimental and calculated distribution in the focal spot on the x-axis: calculated intensity profile (curve 1), experimental intensity profile (curve 2), and the calculated distribution of the Poynting vector's absolute value onto the z-axis (curve 3).

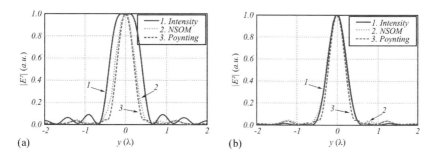

FIGURE 1.52 Comparison of the experimental and calculated distribution in the focal spot on the y-axis (found in parallel with the polarization plane): calculated distribution of the Poynting vector's absolute value onto the z-axis (curve 3), the experimental intensity distribution (curve 2), and the calculated intensity distribution (curve 1) taken as a superposition of (a) all components and (b) only transverse components.

absolute value onto the z-axis (curve 3), the experimental intensity distribution (curve 2), and the calculated intensity distribution (curve 2) taken as a superposition of (a) all components and (b) only transverse components.

Figure 1.52(a) suggests that the longitudinal intensity component is not measured in the course of the experiment (see column 3 of Table 1.3), because the total intensity curve (FWHM = 0.84λ) is wider than the experimental curve (FWHM = 0.52λ) by a value larger than the error of measurement ($\pm 0.02\lambda$). However, the experimental curve is, in its turn, also wider than the calculated curve for the power flux (FWHM = 0.45λ) by a value larger than the error of measurement, thus posing the question – what exactly is measured in the experiment? In Figure 1.52(b) a comparison is conducted of the experimental intensity (curve 2) and the transverse intensity $|E_x|^2 + |E_y|^2$ (curve 1), with their widths found to be the same (FWHM = 0.52λ). Because of this, it can be unambiguously inferred from Figure 1.52 that the near-field microscope NSOM, which has a metal-pyramid cantilever with a 100-nm hole (Figure 1.53), measures the transverse intensity $|E_x|^2 + |E_y|^2$ (energy density) rather than the power flux or the total intensity $|E_x|^2 + |E_y|^2 + |E_z|^2$. This is the reason why the focal spot ellipse in Figure 1.50(d) is less pronounced than the calculated ellipse in Figure 1.45(b).

The propagation of an electromagnetic field in a small hole in a metal screen is described within the Bethe–Bouwkamp theory [7, 112, 113]. According to the said theory, a linearly polarized plane wave incident at an angle to the metal screen that has a small hole of diameter $a \ll \lambda$ induces an electric dipole oriented perpendicularly to the hole and a magnetic dipole found in the hole plane. Therefore, for a small hole illuminated by a tilted plane wave $\mathbf{E}$, the far-field relationship

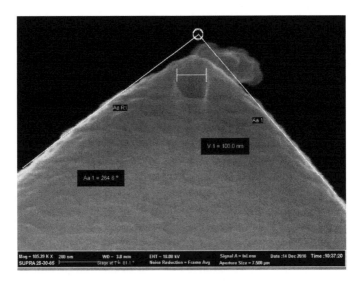

FIGURE 1.53 The electronic image of the pyramid-shaped metallic cantilever tip with a 100-nm hole used in the near-field scanning microscope (NSOM).

is described by the radiation of the electric **P** and magnetic **M** dipoles with the moments:

$$\mathbf{P} = -\frac{4}{3}\varepsilon_0 a^3 \left(\mathbf{E}\mathbf{n}_z\right)\mathbf{n}_z,$$

$$\mathbf{M} = -\frac{8}{3}a^3 \left[\mathbf{n}_z \times \left[\mathbf{E} \times \mathbf{n}_z\right]\right],$$

(1.67)

where $\mathbf{n}_z$ is the unit vector on the optical axis (perpendicular to the hole plane). From Equation (1.67), the electric dipole is seen to be formed only by the longitudinal component of the electric field **E**. Note, however, that a dipole oriented along the optical axis radiates in the transverse direction, not radiating along the optical axis itself. On the contrary, the magnetic dipole in Equation (1.67) is formed only by the transverse electric field components, because the internal vector product on the right-hand side of Equation (1.67) equals zero for the longitudinal electric field component. Thus, the longitudinal electric field component is not registered by a photo-receiver put on the optical axis at a distance from the small hole in the metal screen.

Summing up, the following results have been arrived at in this section. By decomposing the vector electromagnetic field in terms of plane waves, it has been generally shown that the linearly polarized incident light wave with high NA (with the divergence semi-angle being close to 90 degrees) forms a focal spot with transverse intensity (power flux) distribution in the form of an ellipse or a dumbbell elongated in a plane parallel with the incident light polarization plane. Meanwhile, the transverse distribution of the power flux (the Poynting vector's projection onto the optical axis) is in the form of a circle or an ellipse whose major semi-axis is in a plane perpendicular to the incident beam polarization plane. The FDTD method simulation has shown that subwavelength focusing of a linearly polarized wave, with a wavelength-period binary axicon, forms a focal spot with "dumbbell"-intensity distribution and circular power flux distribution (the absolute value of the Poynting vector's projection onto the optical axis) in the immediate vicinity of the axicon surface. The FDTD simulation has shown that when the linearly polarized light is focused with a glass binary zone plate of wavelength focus, the intensity focal spot is an ellipse with the smaller diameter FWHM $= (0.42 \pm 0.01)\lambda$, whereas the power flux focal spot is a circle with diameter FWHM $= (0.45 \pm 0.01)\lambda$. The experiment conducted on a near-field microscope with a 100-nm-hole metallic cantilever has shown that when illuminated by a linearly polarized Gaussian beam, the wavelength-focus binary zone plate forms a focal spot in the form of a weak ellipse of size FWHM$_x = (0.44 \pm 0.02)\lambda$ and FWHM$_y = (0.52 \pm 0.02)\lambda$ along the Cartesian axes and the depth of focus DOF $= (0.75 \pm 0.02)\lambda$. The comparison of the experimental and simulation results suggests that the near-field microscope measures the transverse intensity (energy density), rather than the power flux or the total intensity. The conclusion that the small-hole metallic cantilever measures the transverse intensity $|E_x|^2 + |E_y|^2$ follows from the Bethe–Bouwkamp theory.

2 Near-Field Subwavelength Focal Spot

2.1 HYPERBOLIC PHOTONIC NANOJET

It has long been observed that the diffraction of light by micro-objects of near-wavelength size produces a near-surface focal spot [114]. In addition, with two competing processes – diffraction and focusing – proceeding simultaneously, the near-surface focal region becomes elongated, appearing as a photonic nanojet. The term "photonic nanojet" (PNJ) usually refers to a focal region found near a homogeneous microsphere [115, 116]. It has been numerically demonstrated that the PNJ with the minimal diameter of full width at half-maximum (FWHM) = 130 nm = 0.325λ (λ is the wavelength) is generated as a result of diffraction by a polystyrene microsphere of diameter 1 μm ($\lambda = 400$ nm) [115]. The experimentally measured minimal diameter of the PNJ was FWHM = 270 nm = 0.52λ [117]. The polystyrene microsphere had a diameter of 3 μm ($\lambda = 400$ nm), whereas the depth of focus (DOF), i.e. the PNJ length, was DOF = 3λ. The PNJs have found uses in 3D nanolithography [118], laser microsurgery [119], plasmonics [120], and micromanipulation [121]. Note that the PNJ can form different structures, e.g. a hollow light pipe [122]. The resulting structure depends on the type of incident light. It should also be noted that the PNJ can be generated when light is diffracted not only by the microspheres but also by other micro-objects. In Reference [109] the PNJs were experimentally observed as a result of diffraction by a micropyramid: pyramid base 2×2 μm, height $H = 400$ nm, FWHM = 580 nm = 1.4λ, DOF = 8 μm = 20λ, where $\lambda = 400$ nm. The maximal intensity of the PNJ was found to be 4.12 times that of the incident light. In Reference [121] the PNJ were generated by diffracting the beam by polymeric conical microaxicons. The microaxicon was of base diameter 200 μm, height $H = 65$ μm, $n = 1.43$, $\lambda = 532$ nm, FWHM = 0.6 μm, and DOF = 240 μm. In Reference [123] the photonic nanojets were observed in a system of Si_3N_4 microdisks: the disks were of diameter 9 μm, height $H = 400$ nm, $n = 2.1$, $\lambda = 532$ nm, FWHM = 0.86λ, and DOF = 0.96λ. In Reference [124] it was shown by finite-difference time-domain (FDTD) simulation that a truncated cylindrical zone plate with focal distance $f = \lambda = 532$ nm forms a photonic nanojet with a diameter FWHM = 0.31λ, and DOF = 1.18λ.

It should also be noted that all the articles cited above studied the PNJs elongated along the optical axis as they had been generated with axially symmetric microoptics elements, like microspheres, microcones, micropyramids, microdisks, and zone plate.

In this section, based on the FDTD simulation and measurements conducted with a near-field scanning optical microscope (NSOM) that has a metal cantilever tip with a 100-nm aperture, the diffraction of a linearly polarized plane wave by a corner phase fused-silica microstep of height equal to two incident wavelengths is shown to generate an elongated region of enhanced intensity, termed as a hyperbolic PNJ. The hyperbolic PNJ has a length of about $DOF = 9.5\lambda$ and the smallest diameter $FWHM = (1.8 \pm 0.15)\lambda$ at distance $z = 5.5$ μm. In the course of propagation, not only does the focal spot get offset from the optical axis but it also changes the geometry, in a similar way to a light beam with astigmatism. The focal spot size close to the phase step ($z = 0.5$ μm) is found to be $FWHM = (1.3 \pm 0.15)\lambda$. The diffraction field patterns from the corner phase step and a corner slit in the opaque screen for the plane waves are also shown to be similar.

2.1.1 SIMULATION

When fabricating the surface microrelief by means of the e-beam or photolithography a binary microrelief in the form of a step, or a corner step, with one or two vertical sidewalls can be fabricated most easily and accurately. In the latter case, the aspect ratio defined as the ratio of the etch depth to the aperture diameter equals zero for any etch depth since the "aperture diameter" is infinite. Figure 2.1(a) shows a schematic view of the corner phase step under study that has a height of $H = 2\lambda$ ($\lambda = 633$ nm, refractive index $n = 1.5$). Figure 2.1(b) shows an atomic force microscope (AFM) image of the corner phase step fabricated by microlithography and ion etching techniques in a fused silica. The step's length on the right horizontal axis is 10 μm, on the left horizontal axis is 5 μm, and the height on the vertical axis is 1.2 μm.

When such a step is illuminated from the substrate side with a linearly polarized plane laser wave, the output field from the corner's apex will form an elongated enhanced-intensity region. This region is similar to the photonic nanojet generated as a result of laser light propagation through microspheres [115, 116]. The parameters of the said focusing region can be estimated by numerically solving Maxwell's equations using the FDTD simulation in the FullWAVE software. Figure 2.2(a) shows the time-averaged intensity of light $z = 3.5$ μm away from the step surface (or $z = 4.766$

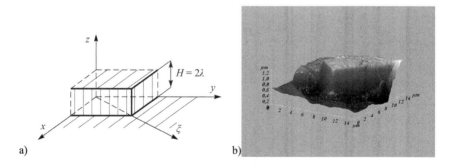

FIGURE 2.1 (a) Schematic view of the corner phase step; and (b) an AFM image of phase step fabricated by e-beam lithography in a fused silica of size $5 \times 10 \times 1.2$ μm.

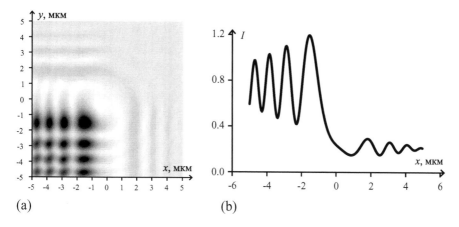

FIGURE 2.2   (a) The time-averaged intensity of linearly polarized light $z = 3.5$ μm above the step; and (b) intensity profile along a horizontal axis which is parallel to the x-axis and traverses the maximal intensity value ($y = -1.5$ μm). The corner step occupies the area ranging from –5 μm to 0 on the x-axis and from –5 μm to 0 on the y-axis.

μm from the substrate). The step size is 5 μm along the x- and y-axes, the step height is $H = 2\lambda$, where $\lambda = 633$ nm is the incident wavelength and $n = 1.5$ is the refractive index of the material. There is a plane wave incident on the step along the z-axis. The incident light polarization plane is coincident with the corner step's diagonal plane. Note, however, that the diffraction pattern in Figure 2.2 is hardly changed while the angle the polarization plane makes with the x-axis is varied from zero to 45°.

The calculation parameters were as follows: the space step is $\lambda/20$; and the time step is $\lambda/(20c)$, where c is the speed of light in free space. The simulated diffraction pattern in Figure 2.3 has the following parameters: the maximal intensity at point $(-1.52$ μm; -1.52 μm) is $I_{max} = 4.8I_0$, where I_0 is the intensity of the incident plane wave; the distance from the corner apex to the focal spot center along the x-axis is $T_0 = 2.4\lambda$; the distance from the major intensity maximum to the first sidelobe is $T_1 = 2.1\lambda$; and the major focal spot size is FWHM $= 1.9\lambda$. The intensity sidelobes

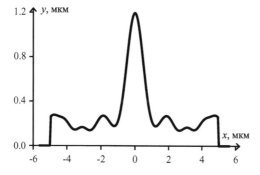

FIGURE 2.3   The calculated intensity cross-section in the diagonal plane drawn at angle 45° when the plane wave is diffracted by the corner phase step of Figure 2.2(a).

in Figure 2.2(b) amount to 90% of the major peak. Figure 2.3 depicts the diagonal cross-section of the intensity in Figure 2.2(a) taken at an angle of 45° along the bisector. The focal spot size for the said cross-section is FWHM = 1.9λ, which is the same as in Figure 2.2(b), whereas the sidelobes amount to 25% of the major peak intensity.

The diffraction pattern in Figure 2.2(a) can be approximately described in terms of the Fresnel scalar diffraction of a plane wave by an amplitude screen that transmits light only at $x < 0$ and $y < 0$. In this case, solving the problem of diffraction by an opaque semi-infinite screen with a rectilinear boundary, the complex amplitude of light at distance z behind the screen is given by

$$E(u,v,z) = \left\{ \frac{1}{2} - \sqrt{\frac{-i}{2}} \left[C(\omega) + iS(\omega) \right] \right\} \left\{ \frac{1}{2} - \sqrt{\frac{-i}{2}} \left[C(\psi) + iS(\psi) \right] \right\}, \qquad (2.1)$$

where $\omega = \sqrt{2/(\lambda z)}u$, $\psi = \sqrt{2/(\lambda z)}v$, and where u, v are the Cartesian coordinates at distance z and $C(\psi)$ and $S(\psi)$ are the Fresnel integrals:

$$C(\omega) = \int_0^\omega \cos\left(\frac{\pi t^2}{2} \right) dt, S(\omega) = \int_0^\omega \sin\left(\frac{\pi t^2}{2} \right) dt \qquad (2.2)$$

Figure 2.4 shows the intensity pattern calculated as the squared modulus of the amplitude in Equation (2.1), assuming the wavelength $\lambda = 633$ nm and the distance traveled by light $z = 4.766$ μm. At the same distance from $z = 0$, the intensity is calculated in Figure 2.2.

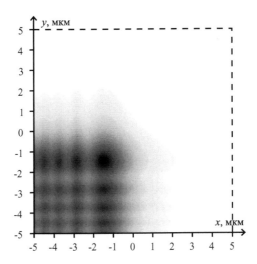

FIGURE 2.4 Intensity of the scalar light field in Equation (2.1) when light is diffracted by a semi-infinite aperture transparent to light in the region $x < 0$, $y < 0$, the incident wavelength is $\lambda = 633$ nm, distance traveled is $z = 4.766$ μm. Black shows the maximal intensity, white the zero intensity. The dashed line shows the computation region boundary.

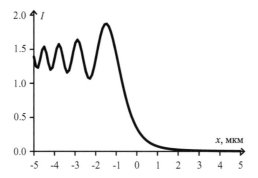

FIGURE 2.5 Intensity profile along the axis parallel to the horizontal axis in Figure 2.4 drawn through the maximal intensity point (−1.46 µm, −1.46 µm).

The major intensity maximum is at point (−1.46 µm; −1.46 µm). The horizontal and vertical intensity profiles are coincident and depicted in Figure 2.5.

The comparison of the diffraction patterns in Figure 2.2(a) and Figure 2.4 and the intensity profiles in Figure 2.2(b) and Figure 2.5 shows them to be similar, so that even the sidelobe coordinates are similar. For instance, the major intensity maximum calculated analytically has the coordinates (−1.52 µm; −1.52 µm), while the scalar approximation gives somewhat different coordinates (−1.46 µm; −1.46 µm). However, essential differences can also be noted: the intensity maximum is 4.8 times higher in Figure 2.2(a) and only 1.85 times higher in Figure 2.5 than the incident light intensity. Besides, the sidelobe contrast is seen to be nearly twice as large in Figure 2.2 as in Figure 2.5. The reason behind the differences is that different, though similar, tasks have been addressed, namely, diffraction of the plane wave by a corner phase step (Figure 2.2) and by a semi-infinite rectangular screen (Figure 2.4).

From Equation (2.1) it follows that the Cartesian coordinates of the major maximum can be derived from the relation

$$T_0 \approx 0.92\sqrt{\lambda z}. \tag{2.3}$$

The displacement of the PNJ along the corner step's hypotenuse will be $\sqrt{2}$ times larger than the value of T_0 in Equation (2.3). It is interesting to note that the scalar paraxial diffraction of a plane wave by a spiral phase plate (SPP) with topological charge $m = 1$ produces an optical vortex in the form of a bright ring, with its radius R defined by a very similar formula [124]:

$$R \approx 0.94\sqrt{\lambda z}. \tag{2.4}$$

Thus, it can be inferred that when light is diffracted by the SPP with transmission $\exp(-im\varphi)$, where φ is the polar angle in the SPP plane, the axial point can be interpreted as an amplitude edge or a phase step jump.

2.1.2 EXPERIMENT

The intensity distribution as a result of diffraction of a linearly polarized plane wave by the corner step of Figure 2.1 was experimentally measured using an NSOM

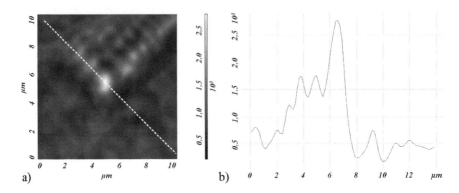

a)   b)

FIGURE 2.6 (a) NSOM image of the intensity distribution at distance $z=4.7$ μm from $z=0$; and (b) intensity profile along the step boundary (shown as white dotted line).

NTEGRA Spectra (NT-MDT) that had a metal pyramid cantilever tip with a 100-nm aperture. The distance from the step's top was also measured within 100-nm accuracy. Figure 2.6 shows the 2D intensity distribution at distance $z=3.5$ μm. Note that the Cartesian axes in Figure 2.6a are rotated by 45° clockwise relative to the axes in Figures 2.2(a) and 2.4.

The intensity profiles obtained by simulation (Figure 2.2(b)) and experimentally (Figure 2.6(b)) are seen to be in a qualitative agreement. The numerical values are also seen to compare well: the distance from the step edge to the major intensity maximum is $T_0 = 2.5\lambda$ (Figure 2.6(b)) and $T_0 = 2.4\lambda$ (Figure 2.2(b)). The intensity maximum is larger than the incident wave intensity by a factor of $I_{max}/I_0 = 4.8$ (Figure 2.2(b)) and $I_{max}/I_0 = 5.4$ (Figure 2.6(b)). The difference amounts to 11%. A larger intensity in the experiment may be due to not strictly vertical sidewalls of the phase step of height $H = 2\lambda$. Figure 2.7 depicts the intensity profile along the horizontal

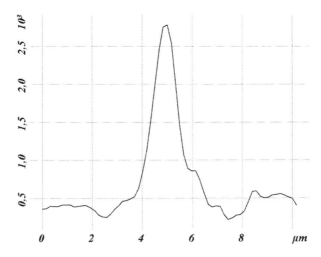

FIGURE 2.7 Profile of the intensity pattern in Figure 2.6(a) taken along the horizontal axis drawn through the focal point (4.6 μm, 4.6 μm).

axis drawn through the focus (which is not shown in Figure 2.6(a)). The focal spot size on the horizontal axis is FWHM = 1.76λ. This is close to the estimated value of FWHM = 1.9λ (Figure 2.2(b)), the difference being 7%.

The focal spot generated close to the corner phase step has different parameters at different distances on the z-axis. Table 2.1 gives the 2D intensity patterns (of size 2 μm × 2 μm) at different distances z above the step surface in the range of 0.5 μm to 3.5 μm. The focal spot size on the Cartesian x- and y-axes and the maximal focus intensity (in relative units) are also presented. The maximum intensity spot was observed 3 μm from the step surface, with the Cartesian axis size being $FWHM_x = 1.78\lambda$ and $FWHM_y = 2.18\lambda$. The step distance and beam diameter were measured with an error of about 100 nm. Table 2.1 shows that very close to the step ($z = 0.5$ μm) the focal spot is elongated along the x-axis. Then, while traveling further, the focal spot acquires a circular shape ($z = 2$ μm) before becoming an ellipse elongated along the vertical y-axis ($z = 3.5$ μm). We can infer that when a plane wave is diffracted from a corner phase step the resulting wavefront has the astigmatism that causes the focal spot to change its form.

The resulting PNJ of Table 2.1 has the full-width half-maximum DOF = 9.5λ. This is also evident from Figure 2.8, which depicts the intensity distribution in the $z\xi$-plane calculated using the FDTD technique for the diffraction of light from the corner phase step of Figure 2.1. Table 2.1 also suggests that with increasing distance z from the step, the focal spot center is displaced in the xy-plane. At $z = 0.5$ μm, the intensity maximum is at point (0.7 μm; 0.7 μm), whereas at $z = 3.5$ μm the maximum is at point (1.6 μm; 1.6 μm). Thus, over the length of 3 μm the NPJ experiences a shift from the incident beam optical axis by about $L = 1.27$ μm. From the diffraction pattern of Figure 2.8, the NPJ is also seen not only to experience a shift but also to travel along a curved path. The said curved path is described by the hyperbola of Equation (2.3). From Equation (2.3), the projection of the intensity maximum shift along the step's angle hypotenuse at distance $z = 3.5$ μm can be analytically estimated as ($z_1 = 0.5$ μm, $z_2 = 3.5$ μm):

$$L \approx 0.92\sqrt{2\lambda}\left(\sqrt{z_2} - \sqrt{z_1}\right) = 1.21 \text{ μm}. \tag{2.5}$$

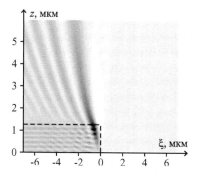

FIGURE 2.8 The FDTD-aided simulation of the intensity distribution in the $z\xi$-plane when a plane wave is diffracted from a corner phase step of Figure 2.1(a).

TABLE 2.1
The Focal Spot from a Corner Step (Figure 2.1)

Distance from the step surface, μm	Intensity pattern	Full-width half-maximum focal spot size		Maximum spot intensity I_{max}, a.u.
		On the x-axis FWHM$_x$, λ	On the y-axis FWHM$_y$, λ	
0.5		2.25	1.31	1660
1.0		2.00	2.13	2300
1.5		2.92	2.18	2310
2.0		2.19	2.17	2600
2.5		1.90	2.15	3010

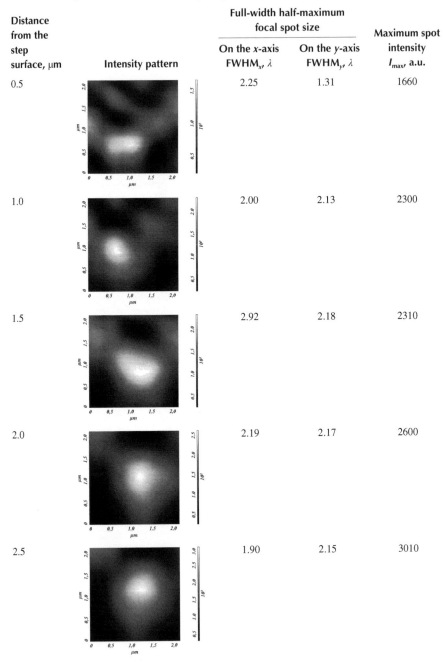

(*Continued*)

TABLE 2.1   (CONTINUED)

The Focal Spot from a Corner Step (Figure 2.1)

Distance from the step surface, μm	Intensity pattern	Full-width half-maximum focal spot size		Maximum spot intensity I_{max}, a.u.
		On the x-axis FWHM$_x$, λ	On the y-axis FWHM$_y$, λ	
3.0		1.78	2.18	3130
3.5		1.76	–	2520

From Figure 2.8, the projection of the NPJ off-axis shift can be numerically esti-mated as $L \approx 1$ μm. Thus, the values of the NPJ off-axis shift L derived experimen-tally ($L = 1.27$ μm), numerically ($L = 1$ μm), and analytically ($L = 1.21$ μm) compare well – showing a 20% difference.

Using the FDTD numerical simulation and an NSOM with a metal tip, it has been demonstrated that when a linearly polarized plane wave is diffracted from a glass corner phase step of height 2λ, where λ is the incident wavelength, the output beam is in the form of a low-divergence NPJ of full-width half-maximum length 9.5λ and diameter 1.8λ at distance 5.5λ, with the NPJ intensity maximum being 5 times larger than that of the incident wave. The difference between the numerically simulated and experimental results is 11%. We have shown that when illuminated with a plane wave, a corner phase step and an opaque screen with a rectilinear boundary produce identical diffraction patterns and similar NPJ.

2.2   ASYMMETRIC OPTICAL VORTEX GENERATED BY A SPIRAL REFRACTIVE PLATE

Laser vortex beams have been dealt with in quite a number of publications. The vor-tex beams are characterized by the annular symmetry, showing a well-known dough-nut form. As a rule, such beams are generated using a spiral phase plate (SPP), which is generally more energy effective in comparison with other optical elements (forked diffraction gratings, spatial light modulators, etc.). It is especially important in tasks involving uncontrollable sources of light, e.g. vortex coronagraph [125]. For the first time, the SPP was studied in 1992 [126]. The vortex beams have an orbital angular momentum similar to the Laguerre–Gaussian beams [127]. The paraxial diffraction of a Gaussian beam by an SPP was studied in Reference [128]. The investigation of the paraxial diffraction of an unbounded [129, 130] and bounded [131, 132] plane wave by the SPP was also reported. In the above-mentioned papers, the SPP was approximated by an infinitely thin transparency, with the analysis of the diffraction by the SPP conducted using the Fourier, Fresnel, or Kirchhoff integrals. Within the said theoretical approach, a radially symmetric light beam, with the amplitude in polar coordinates given by $E(r, \varphi, 0) = A(r)$, incident onto an mth-order SPP is transformed at distance z into the radially symmetric light beam with a spiral phase:

$$
E\left(\rho, \theta, z\right) = (-i)^{m+1} \frac{k}{z} \exp\left(\frac{ik\rho^2}{2z} + im\theta + ikz\right)
$$

$$
\times \int_0^\infty A(r) \exp\left(\frac{ikr^2}{2z}\right) J_m\left(\frac{k\rho r}{z}\right) r\,\mathrm{d}r,
$$

(2.6)

where (ρ, θ) are the polar coordinates in a plane perpendicular to the optical axis z, $k = 2\pi/\lambda$ is the wave number of light, λ is the wavelength of light in vacuum, and $J_m(x)$ is the Bessel function.

In this section, the SPP operation is studied without resorting to the thin transparency approximation. With the aid of three techniques (geometric optics, scalar wave theory, and the FDTD method), we show the refractive spiral phase plate with topological charge $m = 1$ and relief step to generate an asymmetric optical vortex (broken doughnut) for the near-field ($z < z_r$, z_r is the Rayleigh range) and middle-field ($z > z_r$) diffraction regions. For SPP with $m = 3$ we show it experimentally. Obtained results explain weakly asymmetrical experimental diffraction patterns in Reference [130] (Figures 2.10(c)–(e)), [133] (Figure 2.7(c)), [134] (Figure 2.5(a)), and [135] (Figure 2.3(a)).

2.2.1 Trajectory of Light Rays after Passing through the Refractive SPP

When studied in the geometric optics approximation, the SPP with a unit topological charge ($m = 1$) represents an element with the microrelief depth in polar coordinates given by

$$
h(r, \varphi) = \lambda\varphi / \left[2\pi(n - 1)\right], \quad \varphi \in \left[0, 2\pi\right),
$$

(2.7)

where n is the refractive index. If $n = 1.5$ the SPP's maximal depth is 2λ (Figure 2.9).

Assume that a light beam propagating in free space in parallel with the optical axis is incident onto the optical element at point $\mathbf{x} = (x, y)$. While propagating within the SPP, the beam's wave vector is $\mathbf{k}_1 = (0, 0, kn)$. Undergoing the refraction in the second (output) surface, the ray again propagates in free space in the direction of the

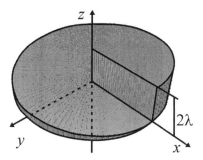

FIGURE 2.9 A first-order ($m=1$) SPP with a relief depth given by Equation (2.7).

unit vector **s**. At this stage, the wave vector will be given by $\mathbf{k}_2 = k\mathbf{s}$, with its coordinates derived from the refraction law:

$$\mathbf{k}_1 \times \mathbf{N} = \mathbf{k}_2 \times \mathbf{N}, \tag{2.8}$$

where **N** is the normal vector to the element's output surface, whose Cartesian components are given by $(-\partial h/\partial x, -\partial h/\partial y, 1)$. From Equation (2.8), we can derive the coordinates of the vector **s**:

$$s_{x,y} = \nabla_\perp h \left[1 + (\nabla_\perp h)^2 \right]^{-1} \left[n - \sqrt{1 - (n^2-1)(\nabla_\perp h)^2} \right], \tag{2.9}$$

where $\nabla_\perp h = (\partial h/\partial x, \partial h/\partial y)$. Using Equation (2.9), we can find the Cartesian coordinates of point $\mathbf{u} = (u, v)$ at which the ray is intersected with an arbitrary plane perpendicular to the optical axis and put at a distance of z away from the SPP's input plane:

$$\mathbf{u} = \mathbf{x} + \frac{z - h(\mathbf{x})}{s_z} \begin{pmatrix} s_x \\ s_y \end{pmatrix}. \tag{2.10}$$

For an SPP with the microrelief in Equation (2.7) ($n=1.5$), the normal vector is defined as $\mathbf{N} = \left(\lambda \sin\varphi/(\pi r), -\lambda \cos\varphi/(\pi r), 1 \right)$, where (r, φ) are the polar coordinates in the SPP plane. Then, Equation (2.9) takes the form:

$$s_{x,y} = \frac{\pi \lambda r}{(\pi r)^2 + \lambda^2} \left[n - \sqrt{1 - (n^2-1)\left(\frac{\lambda}{\pi r} \right)^2} \right] \begin{Bmatrix} -\sin\varphi \\ \cos\varphi \end{Bmatrix}. \tag{2.11}$$

From Equation (2.11) it is seen that as light propagates through the SPP, the closer the rays are to the optical axis, the stronger the refraction is, so that at $r < (\lambda/\pi)(n^2-1)^{1/2}$ the total internal reflection occurs from the output surface. Equation (2.11) also suggests that the rays that fall onto the left ($\varphi = \pi$) and right ($\varphi = 0$) SPP regions are, respectively, deflected downward ($s_x = 0$, $s_y < 0$) or upward ($s_x = 0$, $s_y > 0$); whereas the rays that fall onto the upper ($\varphi = \pi/2$) and lower ($\varphi = 3\pi/2$) SPP regions are, respectively, deflected to the left ($s_x < 0$,

$s_y=0$) and to the right ($s_x>0$, $s_y=0$). If it were not for the SPP microrelief step at $\varphi=0$, all the rays of the incident annular beam (r=constant, $\varphi \in [0, 2\pi)$) would have been deflected by the same angle. However, owing to the microrelief step, the beam loses its annular shape, acquiring a spiral shape. Shown in Figure 2.10(a) is such a spiral beam calculated by Equations (2.10) and (2.11) for the following parameters: wavelength is $\lambda=532$ nm, SPP topological charge is $m=1$, SPP material is glass with $n=1.5$, distance is $z=2.5\lambda$, and the distance of the incident rays to the optical axis is $r=0.36\lambda$. The correspondence between the rays of the incident annular beam and those of the resulting spiral beam is shown in Figure 2.10(b) and 2.10(c).

2.2.2 SCALAR THEORY OF DIFFRACTION BY THE REFRACTIVE SPP

The beam's asymmetry due to microrelief step can also be explained within the scalar wave theory of light. We shall apply the Huygens–Fresnel principle. The diffraction integral can be presented in two different ways. The first one calculates the optical path of rays that pass through the optical element in parallel with the optical axis, starting on a transverse plane prior to the element and ending on a transverse plane beyond the element, $z=z_{max}$ (Figure 2.11(a)). In this case, the diffraction integral in the Kirchhoff approximation takes the form:

$$E(\mathbf{u}) = \iint_{\Omega} \exp\left[ik(n-1)h(\mathbf{x})\right] \frac{\exp(ikR)}{R} \, d\mathbf{x}, \tag{2.12}$$

where $\mathbf{x}$ and $\mathbf{u}$ are the Cartesian coordinates in the SPP's front plane and in the observation plane, Ω is the SPP's region, and $R^2 = |\mathbf{u} - \mathbf{x}|^2 + (z - z_{max})^2$. The second one calculates the optical path of rays that pass through the element parallel with the optical axis, starting on the transverse plane prior to the element and ending on the element's output surface $h(\mathbf{x})$ (Figure 2.11(b)). In this case, the diffraction integral takes the form:

$$E(\mathbf{u}) = \iint_{\Omega} \exp\left[iknh(\mathbf{x})\right] \frac{\exp(ikR_1)}{R_1} \, d\mathbf{x}, \tag{2.13}$$

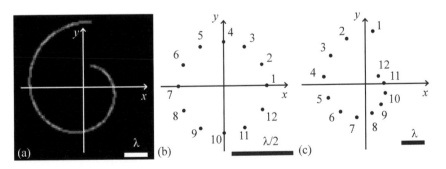

FIGURE 2.10 (a) Points of intersection of the rays having passed through the SPP with a transverse plane; (b) the ray distribution pattern in the (b) original and (c) resulting beams.

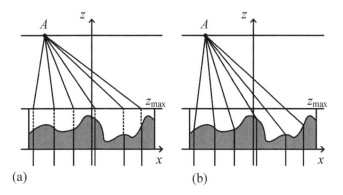

FIGURE 2.11 The calculation of the field (a) in the thin element approximation and (b) with regard for the refraction on the relief boundary.

where

$$R_1^2 = |\mathbf{u} - \mathbf{x}|^2 + [z - h(\mathbf{x})]^2.$$

Based on Equations (2.12) and (2.13), two fields generated when an annular beam of radius $r = 1.2\lambda$ passed through the first-order SPP (Figure 2.11) were calculated. The calculation parameters were identical to those used in Figure 2.10. The intensity and phase of the field derived from Equation (2.12) are shown in Figures 2.12(a) and (b), and from Equation (2.13) in Figures 2.12(c) and (d).

It can be inferred from Figure 2.12 that it is just when the SPP's microrelief step is accounted for that the asymmetric diffraction pattern occurs. The Rayleigh range for the case shown in Figure 2.12 is $z_r = (\pi r^2)/\lambda$. Therefore patterns in Figure 2.12 were obtained for the Fresnel number $F = r^2/(\lambda z) = z_r/(\pi z) \approx 0.6$. This means that despite the small radius of the illuminating beam, the patterns in Figure 2.12 are calculated for the Fresnel diffraction zone, i.e. almost for the far field.

In this subsection we used scalar theory but the polarization of the illuminating beam has not been taken into account. Therefore in the following subsection we numerically consider the vectorial case.

2.2.3 Numerical Simulation of Diffraction by SPP Using the Solution of Maxwell's Equations

The above-calculated asymmetric diffraction pattern can be verified by numerical simulation based on the rigorous finite-difference time-domain method. Shown in Figure 2.13(a) and (c) are the time-averaged intensities in the planes that are perpendicular to the optical axis and offset by $z = 2.5\lambda$ and $z = 15\lambda$ from the SPP's input surface. The remaining simulation parameters are the same: $\lambda = 532$ nm, $n = 1.5$. The incident linearly polarized Gaussian beam of waist radius $w = 2\lambda$ was focused onto an SPP of diameter 8λ. The intensity was calculated using the FDTD method with a spatial discretization of $\lambda/16$ for $z = 2.5\lambda$ and $\lambda/10$ for $z = 15\lambda$, temporal discretization values

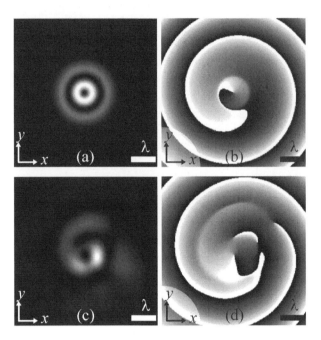

FIGURE 2.12 (a), (c) Amplitudes and (b), (d) phases of the light field formed by an annular beam having passed through the SPP at a distance of $z = 2.5\lambda$ ($\lambda = 532$ nm) away, which have been calculated using Equation (2.12) (a), (b) and Equation (2.13) (c), (d).

being, respectively, $\lambda/32$ and $\lambda/20$. The simulation region was $|x| \leq 8\lambda$, $0 \leq z \leq 20\lambda$. The simulation time was $20\lambda/c$, where c is the speed of light in free space. The amplitudes of the E_x component in the same planes are shown in Figures 2.13(b) and (d). From Figure 2.13, the phase is seen to become spiral at distance $z = 15\lambda$ in a similar way to conventional optical vortices, with the intensity distribution looking much more annular-like (although the asymmetry is still noticeable). The Rayleigh range for Figure 2.13(c) is $z_r = 4\pi\lambda$, and the Fresnel number is $F = 0.27$. Therefore the diffraction pattern of the optical vortex in Figure 2.13(c) is calculated almost for the far field (the far field can be considered as the case when $F \ll 1$).

The structure of the optical vortex depends on how the electric vector of the illuminating light wave is directed with respect to the line of break of the SPP relief. Shown in Figures 2.13(c) and (e) are two optical vortices, calculated for similar conditions but with different polarization. In Figure 2.13(c) the polarization plane coincided with the xz-plane, while in Figure 2.13(e) with the yz-plane. Comparison between Figures 2.13(c) and (e) shows that although the form of the optical vortex is polarization-dependent, the asymmetry is qualitatively the same in both cases.

In Figures 2.12(c) and 2.13(a), (c), and 2.13(e) there are areas with increased intensity on the light rings. Therefore diffraction patterns are similar to those obtained for spiral phase plates with fractional topological charge [136] (Figure 2.3(b)) or patterns obtained experimentally in Reference [137]. In Reference [136] SPP is considered as an infinitely thin optical element and fractional vortex beams appear because

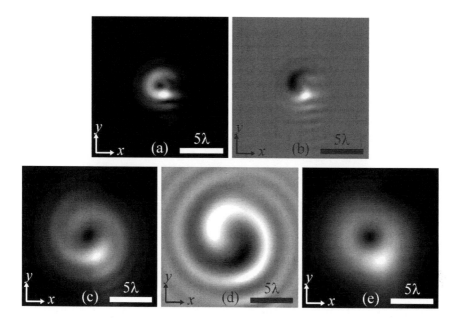

FIGURE 2.13 The diffraction pattern produced by a linearly polarized Gaussian beam transmitted through the SPP, as calculated by the FDTD method: (a), (c) time-averaged intensity and (b), (d) the instant E_x component taken at the instance $t = 20\lambda/c$ in the planes $z = 2.5\lambda$ (a), (b) and $z = 15\lambda$ (c), (d). For comparison an optical vortex with another polarization E_y is shown (e).

the periodicity of phase has been broken. In our case, the symmetry of the optical vortex is also broken, but the reason is the discontinuity of the relief of the SPP.

It is clear that if the SPP (Figure 2.9) is fabricated for another wavelength, e.g. for $\lambda = 633$ nm, which will be used in experiment, and if the height of the relief step is also 2λ, then nothing will be changed. Shown in Figure 2.14 are intensity distributions calculated by the FDTD method at various distances from the SPP with a topological charge of $m = 1$. The glass-made SPP ($n = 1.5$) was illuminated by a linearly polarized plane wave with a wavelength of $\lambda = 633$ nm and a diameter of $2r = 4$ μm. The polarization plane coincided with the xz-plane. The Rayleigh range was $z_r \approx 31\lambda$ and the Fresnel number for Figure 2.14(c) was $F \approx 0.5$. That is, the diffraction pattern in Figure 2.14(c) is calculated almost in the far field. It is seen in Figures 2.13 and 2.14 that the break of the light ring near the SPP (Figures 2.13(a) and 2.14(a)) disappears in the far field (Figures 2.13(c) and 2.14(c)) while asymmetry of the intensity distribution along the light ring remains. The character of asymmetry in Figures 2.13 and 2.14 allows detection of where the break line is located in an SPP relief and at which side of this line the SPP is thicker. For example, in case of Figure 2.14 SPP has been located as in Figure 2.13 and the line of the relief break was parallel to the x-axis. Therefore it can be concluded that in a far field the maximum intensity on the light ring of the optical vortex with a topological charge of $m = 1$ appears at the azimuthal angle of $\varphi \approx -\pi/2$, if the line of SPP relief break lies along the x-axis and the SPP thickness at $y < 0$ is larger than at $y > 0$. That is, the intensity maximum on

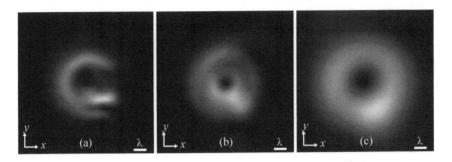

FIGURE 2.14 Intensity distributions of optical vortex calculated by the FDTD method at several distances z away from the SPP ($m=1$): (a) $z=2$ μm; (b); 5 μm; (c) 12 μm.

a light ring is localized at an azimuthal angle corresponding to the thickest part of the SPP relief. It allows us to distinguish the asymmetry of the optical vortex due to the displacement between the centers of SPP and the illuminating beam. In the latter case the local intensity maximum appears in an azimuthal direction coinciding with this displacement and has no relation with the azimuthal direction of the SPP relief step.

2.2.4 EXPERIMENTAL STUDY OF NEAR-FIELD DIFFRACTION BY SPP

Experimental researches have been made for an SPP with a topological charge $m=3$ (Figure 2.15(a)). Shown in Figure 2.15 are the optical microscope images of the central part of the SPP with a size of 90×90 μm (64×64 pixels) Figure 2.15(a) and intensity distributions of linearly polarized laser light with a wavelength of 633 nm, which was transmitted through the SPP and registered at various distances by a near-field microscope with metallic cantilever having a 100-nm hole in front of a charge-coupled device (CCD)-camera Figure 2.15(b)–(e). The SPP was fabricated by electronic lithography technology on a resist with a resolution of 5 μm. That is, on the SPP surface the minimal area with a constant thickness had the size of 5×5 μm. These "boxes" are seen in Figure 2.15(a). It is also seen in Figure 2.15(a) that thick parts of the SPP relief are located at those sides of three relief break lines that appear to be brighter. Due to the sampling one relief break line in Figure 2.15(a) is straight and the other two lines are polygonal. During the direct electronic beam writing of the relief height levels have been implemented. Therefore the maximal thickness of the SPP relief was equal to double the wavelength with a precision of 2–4%. It is also confirmed by measurements with the use of the white light interferometer Zygo NewView 5000. Shown in Figure 2.15(f) is a 3D view of the same SPP central part. Shown in Figure 2.15(g) is the dependence of the microrelief height on the azimuthal angle, measured at a distance of 80 μm from the SPP center. It is seen in Figure 2.15(g) that all three relief steps have a height of about 1.3 μm (root mean square (rms) error 4% from the desired height of 1.266 μm). The SPP has been illuminated by a focused Gaussian beam with a waist radius of $w=50$ μm, i.e. the Rayleigh range was equal to $z_r \approx 12$ mm while the Fresnel number for Figure 2.15(e) was $F \approx 10$. It is seen in Figure 2.15(e) that in contrast with Figures 2.15(b)–(d) there are no breaks of

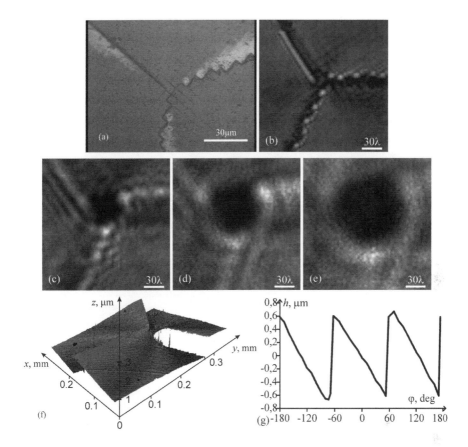

FIGURE 2.15 (a) Central part of the SPP with $m=3$ and intensity distributions near SPP at several distances z: (b) 100 μm, (c) 200 μm, (d) 300 μm, and (e) 400 μm. All patterns (b)--e) have the size of 90×90 μm. (f) 3D view of the same SPP central part and dependence of the relief height on azimuthal angle, measured by white light interferometer Zygo "NewView 5000" at a distance of 80 μm from the SPP center.

the optical vortex light ring, but there are three intensity maxima on this ring. These maxima are in connection with the three SPP ($m=3$) relief break lines.

Shown in Figure 2.16 are the results of comparison between the theory and experiment. In Figure 2.16(a) and (b) we show the intensity calculated by Equation (2.13) and by the beam propagation method (BPM). Figure 2.16(a) has been obtained for a Gaussian beam with waist radius $w=10\lambda$ and propagation distance $z=5\lambda$. Figure 2.16(b) has been calculated for $w=6\lambda$ and $z=10\lambda$. Shown in Figure 2.16(c) is the experimental diffraction pattern generated at a distance of 300 μm from the SPP ($\lambda=633$ nm) and coinciding with the pattern in Figure 2.15(d). It is seen in Figure 2.16 that all three patterns have a similar asymmetry: there are three local intensity maxima on the main light ring. All these three maxima are related with three breaks of SPP relief. The ring in Figure 2.16 appears like a triangle. Although physical sizes of the patterns in Figure 2.16 are different, the Fresnel numbers for them are of the

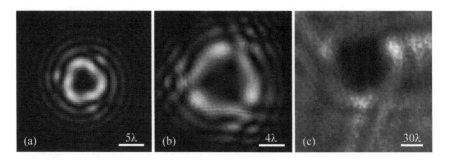

FIGURE 2.16 Intensity distributions of a Gaussian beam, diffracted by the SPP with topological charge $m=3$, (a) calculated by Equation (2.13), (b) calculated by BPM-method, and (c) experimental diffraction pattern.

same order of magnitude (all these numbers are greater than 1). For Figure 2.16(a) the Rayleigh range is $z_r=100\lambda$ and the Fresnel number is $F=20$, for Figure 2.16(b) the Rayleigh range is $z_r=36\lambda$ and the Fresnel number is $F=3.6$, and for Figure 2.16(c) the Rayleigh range is $z_r=12$ mm and the Fresnel number is $F=13$. Let us also note that random displacement between the centers of the illuminating beams and SPPs cannot lead to three maxima in Figure 2.15, but can only lead to different values of intensity at these maxima. When there is no displacement, these three intensity maxima must be equal (see Figures 2.16(a) and 2.16(b)).

2.2.5 ASYMMETRY OF OPTICAL VORTEX IN THE FAR FIELD

In this subsection we make a comparison between optical vortices in the far field that were generated by SPP with steps of relief. In Reference [130] the SPP with a topological charge of $m=1$ has been studied. This SPP has been fabricated on a resist with resolution of 5 μm and 32 levels of relief height. The height of the relief step had a difference of 2% from the desired height. The SPP was illuminated by a plane wave with radius of $r=1.25$ mm and wavelength of $\lambda=543$ nm, which had a 5% difference from the calculated wavelength ($\lambda=514$ nm). The intensity distributions of the optical vortex were measured in the Fresnel zone at a distance of z, equal to 120 mm, 300 mm, and 520 mm (see Figures 10(c), (d) and (e), respectively, in Reference [130]). The Rayleigh range for Figure 10(e) in Reference [130] is $z_r\approx 9$ m and the Fresnel number is $F\approx 6$. The pattern in Figure 10(e) in Reference [130] has the same type of asymmetry as in Figures 2.13(e) and 2.14(c).

In Reference [131] the SPP with a topological charge of $m=3$ has been fabricated on a resist also with high accuracy (an error of 4.3%) by the same technology as the SPP in Reference [130]. The SPP was illuminated by a Gaussian beam with a waist radius of $w=1.25$ mm and wavelength of $\lambda=633$ nm. The intensity of the optical vortex was registered by a CCD-camera in the focus of a spherical lens with a focal distance of $f=150$ mm (Figure 4 in Reference [131]). It is seen in Figure 4 in Reference [131] that inner part of the light ring has a weak asymmetry while two horizontal maxima in the intensity section (Figure 4(b) in Reference [131]) differ by 14% from each other.

In Reference [133] an SPP with a radius of 4.2 mm and with $m=1$ has been fabricated by polymer melding. The height of the relief step was 5.07 μm for a wavelength of $\lambda=813$ nm. Shown in Figure 7(c) in Reference [133] is the intensity distribution of the optical vortex registered in the far field by a CCD-camera. It is seen in Figure 7(c) in Reference [133] that the diffraction pattern has some asymmetry and an intensity maximum on the light ring is located at azimuthal angle of $\varphi=\pi$, while the SPP relief break line is at $\varphi=\pi/2$. Such asymmetry is in agreement with asymmetry in Figures 2.13(e) and 2.14(c).

In Reference [134] an SPP with a diameter of 8 cm and topological charge of $m=2$ has been fabricated by vacuum evaporation on fused silica for a wavelength of $\lambda=789$ nm. The number of relief levels was equal to 16. Shown in Figure 5 in Reference [134] is the intensity distribution of the optical vortex registered by a CCD-camera at the lens focus ($f=50$ mm). It is seen in Figure 5(a) in Reference [134] that the diffraction pattern is asymmetrical, and there are two local intensity maxima on the light ring, at $\varphi=\pi$ and at $\varphi=0$.

In Reference [135] a 24-level SPP with $m=1$ and a diameter of 500 μm has been fabricated for a wavelength of 633 nm on a resist SU-8. The fabrication accuracy was high, the measured height of the relief step was 1.04 μm and this value differed insignificantly from the desired 1.07 μm. Shown in Figure 3 in Reference [135] is the intensity distribution of the optical vortex in the far field. The intensity distribution in this figure is asymmetrical because the azimuthal inhomogeneity along the light ring is 10%.

In Reference [137] an SPP with the height of the relief step as 720 μm has been fabricated for a wavelength of 633 nm in polymethyl methacrylate (PMMA) ($n=1.49$). It has been placed into a cell with immersion liquid with a refraction index of $n=1.4922$ at a temperature of 25°C. Changing the temperature led to a change in the phase retardation, acquired by the laser beam after passing through the SPP. Shown in Figure 4 in Reference [137] are the experimental intensity distributions in a far field of optical vortices with various topological charges (both integer and fractional). It is seen in Figure 4 in Reference [137] that, in the top row at $m=-1$ or $m=1$, the intensity pattern of the optical vortex has a weak asymmetry.

So, in all the works listed in this subsection it has been experimentally detected (although not realized and not noticed) that despite the high accuracy of fabrication of an SPP with a relief step there is some asymmetry (about 10%) of the optical vortex in the far field. The light ring contains the number of local intensity maxima, equal to the SPP topological charge.

Thus, in this section the asymmetric character of the vortex light beam generated by a refractive spiral plate with a relief step has been demonstrated using three methods (ray-tracing, Kirchoff integral, and the FDTD-method) and experimentally. A new form of the Kirchhoff diffraction integral taking account of the specific features of the refractive optical element's microrelief has been derived. The SPP-aided diffraction pattern calculated using the said integral has been shown to have nearly the same form as that derived by the rigorous FDTD method. The character of the optical vortices' asymmetry in the Fresnel diffraction zone is almost the same for diffraction patterns obtained experimentally and by modeling. The review of experimental works [130, 131, 133–135] showed that in the far field the optical vortices,

generated by an SPP with a relief step, have weak azimuthal inhomogeneity along
the light ring.

2.3 PHOTOIC NANOJETS GENERATED USING SQUARE-PROFILE MICROSTEPS

The focusing of light into a subwavelength region has been a pressing problem in
nanophotonics. One of ways the problem has been addressed is by focusing light
with a dielectric microsphere (microball). The study on the microsphere-aided subwavelength focusing of light was first reported in Reference [138]. In Reference
[138], 0.5-μm quartz microspheres illuminated by an excimer KrF laser beam of
wavelength 248 nm focused the beam onto a silicon substrate, resulting in 100-nm-
wide melted hillocks. A similar mechanism for microrelief patterning was proposed
in Reference [139], where a 1-μm microsphere optically trapped in a 532-nm Bessel
beam was guided to a desired location and burned in a pit in the substrate by focused
pulsed laser light of wavelength 355 nm before being guided to the next location. The
focusing of light by means of microspheres was studied theoretically in Reference
[115]. In particular, a 1-μm microsphere (refractive index $n = 1.59$) illuminated by
a plane wave of wavelength $\lambda = 400$ nm was shown to form in a plane perpendicular to the polarization axis a focal spot of diameter FWHM $= 0.325\lambda$. In Reference
[115], the focal regions generated by the use of microspheres were called photonic
nanojets. The direct experimental observation of a photonic nanojet was reported in
Reference [117], where latex microspheres 1 μm, 3 μm, and 5 μm in diameter illuminated by a plane wave of wavelength 520 nm generated focal spots of diameters
0.62λ, 0.52λ, and 0.58λ, respectively. Thus, though having a subwavelength diameter,
the experimentally derived photonic nanojets did not demonstrate that the diffrac-
tion limit was overcome. One further characteristic of the photonic nanojet, namely,
its length, or depth of focus, was studied in References [140, 141]. In Reference [140]
it was numerically shown that the photonic nanojet could be lengthened by use of a
gradient-index microsphere, with the refractive index varying linearly from 1.43 to
1.59, with DOF $= 11.8\lambda$. The opposite task of attaining a shorter nanojet was posed
in Reference [141]. To this end, a microsphere of radius 2.5λ was illuminated by
a Gaussian beam focused with a wide-angle lens of numerical aperture NA ≈ 1.
The photonic nanojet was found to have a length of DOF $= 0.88\lambda$. Note that apart
from the microspheres, there have been articles on generating the photonic nanojets
using other dielectric micro-objects, e.g. microcylinders [142] or microdisks [123].
It should be also noted that sphere-aided focusing is challenging because the sphere
needs to be kept in place (e.g. using a light trap as in Reference [139]). Fabrication of
the focusing element on a substrate seems to be more suitable technologically. If the
rear-side of the substrate has an anti-reflection coating, the focusing efficiency will
be higher than with the same-diameter microsphere.

In this section, we study the focusing of a linearly polarized laser beam of wavelength $\lambda = 633$ nm using square-profile parallelepiped steps of height 695 nm fabricated of silica (refractive index $n = 1.46$) on a substrate. Such square steps of sides 0.4
μm, 0.5 μm, 0.6 μm, and 0.8 μm generate near-surface focal regions in the form of
photonic nanojets with their intensity nine times that of the incident light and focal

spot diameters equal to FWHM$=0.44\lambda$, 0.43λ, 0.39λ, and 0.47λ, respectively, which is below the diffraction limit of 0.51λ. For a square step of side 0.6λ, the focal spot diameters are found to be FWHM$_{min}=0.39\lambda$ and FWHM$_{max}=0.45\lambda$ in the experiment and FWHM$_{min}=0.40\lambda$ and FWHM$_{max}=0.49\lambda$ in the simulation. For a 0.6-μm step, the intensity maximum of the photonic nanojet is found to be 9.58 times that of the incident wave.

2.3.1 SIMULATION

The simulation was conducted using the FDTD method implemented in the FullWAVE software. The grid's spatial sampling was 0.012 μm, i.e. about $\lambda/53$. The simulation was conducted for the side of the square-profile silica step ($n=1.46$) varying from 0.4 μm to 0.8 μm with a 0.02-μm step (Figure 2.17) To approximate the experimental conditions, the step was assumed to be on a substrate. The rectangular columns (steps) were illuminated by a linearly polarized plane wave of wavelength $\lambda=633$ nm. The steps were taken to be of the same height of 695 nm (to be able to compare the simulation and experiment results).

The simulation results for the steps of width 0.4 μm, 0.6 μm, and 0.8 μm are presented in Figures 2.18 through 2.20. Shown in Figure 2.18 is the intensity pattern in the xz-plane perpendicular to the incident beam polarization, and the intensity pattern in the yz-plane parallel to the incident beam polarization is shown in Figure 2.19. The intensity pattern along the optical axis z is shown in Figure 2.10.

From the intensity patterns in Figures 2.18 and 2.19, the enhanced-intensity region generated directly behind the step surface is seen to be similar in shape to

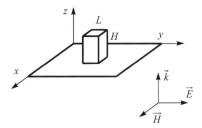

FIGURE 2.17 Schematic view of the step under study.

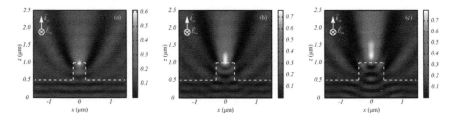

FIGURE 2.18 Intensity pattern in the plane perpendicular to the incident wave polarization (xz) for a step of width (a) 0.4 μm, (b) 0.6 μm, and (c) 0.8 μm. White dashed contour outlines the element's boundary.

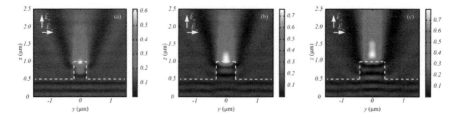

FIGURE 2.19 Intensity pattern in the plane yz parallel to the incident beam polarization for a step of width (a) 0.4 μm, (b) 0.6 μm, and (c) 0.8 μm.

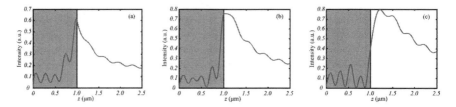

FIGURE 2.20 Intensity profile along the z-axis when using a step of width (a) 0.4 μm, (b) 0.6 μm, and (c) 0.8 μm. The black vertical line denotes the step's top surface.

the microsphere-aided photonic nanojets. From Figures 2.18 and 2.19, the resulting photonic nanojet is also seen to be elliptic, with its focus elongated along the input light polarization direction due to the longitudinal intensity component found in the plane of interest [143]. From the intensity profile along the z-axis in Figure 2.20 it is seen that if the step width is small (less than a wavelength) the near-surface intensity maximum is formed within the step. When the step's width (square side) reaches 0.68 μm, the intensity maximum shifts outside, with a further increase in the step's width resulting in a larger focal length (the distance along the z-axis from the step's edge to the intensity maximum). The near-surface focal spots (Figure 2.18) generated with the square steps with sides of 0.4 μm, 0.5 μm, 0.6 μm, and 0.8 μm are of size FWHM $= 0.44\lambda$, 0.42λ, 0.40λ, and 0.41λ, respectively.

From Figure 2.18 it is seen that inside the step there are several local maxima similar to the principal mode of a stepped-index planar waveguide. For instance, for transverse electric (TE)–polarization (the E-vector is perpendicular to the xz-plane) the principal mode is given by [144]

$$E_y = \begin{cases} \cos(ax), & |x| \le L/2, \\ \cos(aL/2)\exp\left[-\gamma\left(|x|-L/2\right)\right], & |x| \le L/2, \end{cases} \tag{2.14}$$

where L is the waveguide's width, and the parameters α and γ are related by the formula: $n^2 k^2 - a^2 = k^2 + \gamma^2$, where $k = 2\pi/\lambda$ is the wave number and n is the refractive index of the waveguide's material. The α parameter can be derived from the dispersion relationship, $\xi tg\,\xi = \sqrt{k_0^2 - \xi^2}$, where $\xi = aL/2$, $k_0 = (kL/2)\sqrt{n^2 - 1}$. From the dis-person relationship, we can approximately find that $\alpha \approx \pi/L$. Then, the full-width at half-maximum intensity of the principal mode can be derived from: $\cos^2(\alpha x) = \cos^2(\pi x/L) = 1/2$. We find that FWHM $\approx L/2$. For the square steps of side L equal to 0.4 µm, 0.5 µm, 0.6 µm, and 0.8 µm, we find that the mode widths (FWHM) are 0.32λ, 0.40λ, 0.47λ, and 0.63λ, $\lambda = 633$ nm. The focal spot size (0.42λ) and the mode width (0.40λ) are seen to have nearest values when the step size is $L = 0.5$ µm.

From Figure 2.18, the field is also seen to converge inside the step, forming a near-surface focal spot. The local maxima inside the step in Figure 2.18(b) are concave. This is because inside the step, light propagates with a larger phase speed near its edge than it does in its center. This can also occur because the wavefront of the plane incident wave is curved at the step edge. The wavefront gets curved in such a way that the radiation inside the step is directed from the edge toward the center, as was shown in Reference [145]. In particular, in Reference [145] the peripheral ray was shown to be shifting from the step edge toward its center along a square-root parabola given by

$$\Delta x = 0.92\sqrt{\frac{\lambda z}{n}}, \tag{2.15}$$

where Δx is the value of the local maximum shift inside the step. From Equation (2.15) we can find at which value of the square side both maximums (from the left and right step's edges) will converge at the center forming a focal spot, given a step height of H. To do so, put $\Delta x = L/2$ and $z = H$ in Equation (2.15), then,

$$L = 2 \cdot 0.92\sqrt{\frac{\lambda H}{n}} \cong 1.01 \, \text{µm}. \tag{2.16}$$

For the diffraction by a corner step, the intensity maximum at the output was shown to be 5 times the incident wave intensity in Reference [145]. Considering that in the present work the step under study is rectangular, there are two radiant fluxes converging in the focus, which implies that the focal intensity is expected to be 10 times the incident light intensity. This is what we have observed in the course of simulation. For a step of width $L = 0.6$ µm, the simulated intensity maximum of the photonic nanojet was found to be 9.58 times the intensity maximum of incident light. Figure 2.21 depicts the maximum intensity in the near-step focus as a function of the step's width L. The maximum value occurs for the step width of $L = 0.68$ µm. It is at this step width that the intensity maximum (focal spot) on the z-axis (Figure 2.20) occurs outside the step.

The fact that the focal intensity from the rectangular step is about twice that from the corner step [145], rather than being 4 times higher, can be clarified from Figure 2.19. Figure 2.19 suggests that when the E-vector of the transverse magnetic (TM)–wave lies in the step's side-surface plane, no convergence of the radiation occurs from the step's bottom toward its output surface, because the local inside-step maxima lines are straight.

It is seen from Figures 2.18 and 2.19 that nearly 1.5 periods of wavelength "fit" in the step. Thus, we can easily estimate the phase delay acquired by the radiation passing the step relative to the outside radiation:

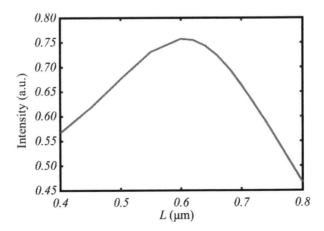

FIGURE 2.21 The near-step intensity maximum profile as a function of the step width (square side).

$$\Delta\phi = \frac{2\pi}{\lambda} H(n-1) = 1.01\pi. \tag{2.17}$$

From Equation (2.17), we can suggest that there occurs a near-π phase delay. It also accounts for the fact that light is focused at the step output, implying that the area of the focal spot FWHM is not larger than the area of the square-shaped output step. This is because at the square edges the radiation traveling outside it and outgoing from the step interfere in antiphase, Equation (2.17).

2.3.2 EXPERIMENT

An array of square-profile, equal-height micro-parallelepipeds of varying size was fabricated by photolithography (using an excimer 193-nm ArF laser) and the ion-beam etching of silica substrates. Figures 2.22 and 2.23 depict images of the fabricated elements' profiles obtained on an AFM Solver Pro.

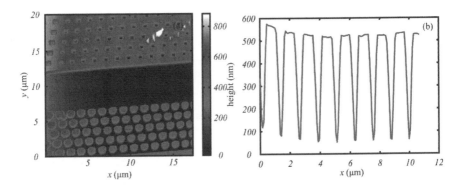

FIGURE 2.22 (a) AFM image of the microsteps under study with a 0.6-μm side, and (b) their selected profiles.

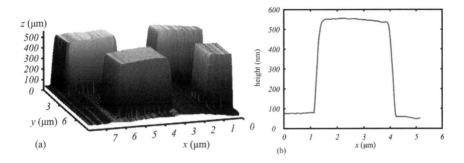

FIGURE 2.23   (a) AFM image of 2.5-μm side steps, and (b) an exemplified step's profile.

The experimental study was conducted using scanning near-field optical micros-copy (SNOM) on the NTEGRA Spectra (NT-MDT) microscope. The steps under analysis were illuminated from the substrate's side by a linearly polarized Gaussian beam of wavelength 633 nm with a waist radius of about 5 μm. The Gaussian beam was focused onto an individual step. Columns with sides of 0.4 μm, 0.5 μm, 0.6 μm, and 0.8 μm were characterized. Intensity patterns of the focal spots formed near the steps' output surface were obtained. The intensity was measured with a 15-nm step. Thus, the accuracy of the experimental measurement of the focal spot size (FWHM) was not lower than 0.01λ. An example of the resulting intensity distribution mea-sured with the SNOM is given in Figure 2.24.

Unfortunately, the SNOM arrangement is designed so that the illumination light polarization plane makes an angle of 45° with the Cartesian coordinates x and y (Figure 2.24(a)), with the slightly elliptical focal spot near the step's exit surface appearing to be elongated along the polarization direction. The focal spot intensity profiles in Figure 2.24(b) and (c) were measured along the minimal and maximal focal spot's diameters, so that at the nanojet's initial section they were found to be $FWHM_{min} = (0.39 \pm 0.01)\lambda$ and $FWHM_{max} = (0.45 \pm 0.01)\lambda$. In Figure 2.25, the exper-imentally measured values of the (a) minimal and (b) maximal focal spot diameter at the nanojet's cross-section near the exit surface versus the varying-width steps

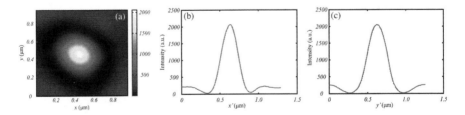

FIGURE 2.24   Intensity distribution near the exit surface of a 0.6-μm side step, which was experimentally obtained under illumination with a linearly polarized 633-nm laser light: (a) half-tone intensity pattern; (b) and (c) intensity profiles for the minimal and maximal focal spot's diameter.

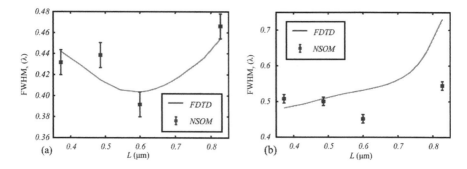

FIGURE 2.25 The (a) minimal and (b) maximal FWHM size of the photonic nanojet's cross-section near the exit surface vs. the side of the step's square cross-section. The curve shows the simulation results, the squares with vertical line show the experimental results.

are marked with squares with vertical lines. Figure 2.25(a) shows the focal spot size for the FWHM intensity in a plane perpendicular to the incident beam polarization. Figure 2.25(b) shows the measured (squares) and simulated (curve) size of the focal spot's transverse component in a plane parallel to the incident light polarization. The curves in Figure 2.25 were obtained using the FDTD simulation, whereas the squares correspond to the SNOM-aided experiment, with the vertical lines showing the experimental error.

From Figure 2.25 it is seen that in the plane perpendicular to the incident light polarization direction, the minimal photonic nanojet's (focal spot) diameter is achieved for a step width of 0.6 μm both in simulation, $FWHM_{min} = 0.40\lambda$, and in experiment, $FWHM_{min} = 0.39\lambda$. The discrepancy between the experimental and simulated curves in Figure 2.25 is not larger than 14% and can be assigned to imprecise positioning of the microscope's cantilever in the immediate step's surface vicinity. Even so, the simulated curve is found within the range of values defined by the experimental squares. Figures 2.26 and 2.27 illustrate the SNOM-aided intensity profiles for the square-step side of 0.5 μm (Figure 2.26) and 0.6 μm (Figure 2.27). For comparison, the simulated intensity profiles are also presented in the plots. The plots in Figures 2.26 and 2.27 suggest that the experimentally derived spot diameters agree with the transverse, rather than total, intensity distribution. The fact that a pyramid metal cantilever in the SNOM is primarily sensitive to the transverse intensity component, being insensitive to the longitudinal component was earlier noted by the present authors in Reference [146].

2.3.3 Comparison with Microsphere-Aided Focusing

To evaluate the difference between the characteristics of the photonic nanojets generated by means of square-shaped steps and microspheres [115, 117, 138–140], we simulated focusing a linearly polarized plane wave using microspheres with diameters equal to the square-step sides of 0.4 μm, 0.6 μm, and 0.8 μm respectively.

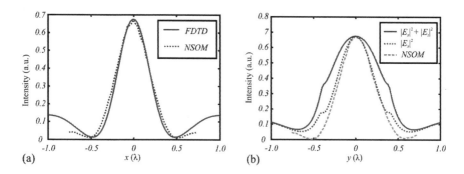

FIGURE 2.26 Intensity profiles directly behind the step surface with the square side of 0.5 µm in the plane (a) perpendicular and (b) parallel to the input beam polarization axis.

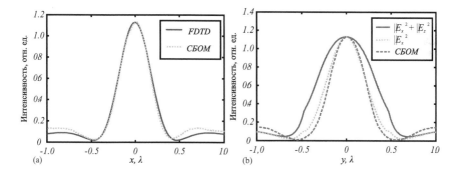

FIGURE 2.27 Intensity profiles directly behind the step surface with the square side of 0.6 µm in the planes (a) perpendicular and (b) parallel to the input beam polarization axis.

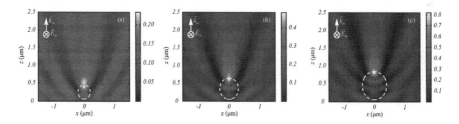

FIGURE 2.28 Intensity pattern in the plane perpendicular to the input radiation polarization (xz) for a microsphere of diameter (a) 0.4 µm, (b) 0.6 µm, and (c) 0.8 µm. The microsphere's boundary is shown as a dashed circle.

The simulation parameters were analogous to those for step-aided focusing. The simulation results are shown in Figures 2.28 through 2.30 (which are similar to the respective Figures 2.18 through 2.20 for the steps). Table 2.2 shows the comparison results for the parameters of the step-aided and microsphere-aided photonic nanojets. The photonic nanojet's depth of focus was calculated as the full-width

FIGURE 2.29 Intensity pattern in the plane parallel to the input radiation polarization (yz) for a microsphere of diameter (a) 0.4 µm, (b) 0.6 µm, and (c) 0.8 µm. The microsphere's boundary is shown as a dashed circle.

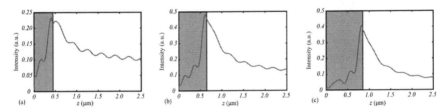

FIGURE 2.30 Intensity profile along the z-axis for a sphere of diameter (a) 0.4 µm, (b) 0.6 µm, and (c) 0.8 µm. The sphere's boundary is shown as a vertical line.

at half-maximum of intensity outside the step (i.e. if the maximum intensity was found inside the step, the half-maximum was counted from the surface). Besides, the step's surface was assumed to be the left half-maximum's boundary (Figure 2.30). The photonic nanojet's diameters were calculated near the element's surface (0.02λ) above the surface, which amounts to a single step of grid in the FDTD method.

Table 2.2 suggests that: (1) in the plane perpendicular to the polarization plane, the step- and microsphere-aided focal spots have similar diameters; but (2) the microsphere-aided focal spots show a wider range of variations (from 0.39λ to 0.53λ) than the step-aided spots (from 0.40λ to 0.44λ); except for one case, all the diameters were below the diffraction limit of 0.51λ; (3) the microsphere-aided focal spots are more elliptical than the step-aided focal spots: at $D=L=0.4$ µm, the diffraction by the step results in

TABLE 2.2
Comparison of the Optical Nanojet Parameters: Silica Microsheres vs. Square Steps ($\lambda = 633$ nm)

Photonic nanojet's parameter	A 695-nm high step with a square-section side L			A microsphere of diameter D		
	$L=0.4$ µm	$L=0.6$ µm	$L=0.8$ µm	$D=0.4$ µm	$D=0.6$ µm	$D=0.8$ µm
FWHM$_x$ (λ)	0.44	0.40	0.41	0.53	0.43	0.39
FWHM$_y$ (λ)	0.44	0.49	0.55	0.74	0.54	0.49
DOF (λ)	0.55	0.65	1.20	1.97	0.93	0.59

a circular focal spot of diameter FWHM=0.44λ, whereas the microsphere-aided focal spot is elliptical, measuring FWHM$_{min}$=0.53λ and FWHM$_{max}$=0.74λ; (4) with the microsphere's diameter and the square-step's side increasing, the DOF measured as intensity half-maximum behaves differently, increasing from DOF=0.55λ to DOF=1.20λ for the step and decreasing from DOF=1.97λ to DOF=0.59λ for the microsphere.

Based on the FDTD simulation and SNOM-aided experiment with a pyramid metal cantilever having a nanohole, we have investigated the sharp focus of a linearly polarized laser light of wavelength 633 nm by means of dielectric fused-silica square steps of varying size and the same height, fabricated on a silica substrate with $n=1.46$. The following results have been obtained.

It has been experimentally shown that microsteps on a silica substrate with a square section of side 0.4 µm, 0.5 µm, 0.6 µm, and 0.8 µm, all of the same height 695 nm, illuminated from the substrate side by a linearly polarized laser light of wavelength $\lambda=633$ nm, produce an enhanced-intensity region (photonic nanojet) near the substrate's surface. The nanojets are characterized by nine times the incident light intensity and FWHM diameters of 0.44λ, 0.43λ, 0.39λ, and 0.47λ, which are below the diffraction limit of 0.51λ. The least-width photonic nanojet has been experimentally observed for a 0.6-µm-side step, measuring FWHM$_{min}$=(0.39$\pm$0.01)λ and FWHM$_{max}$=(0.45$\pm$0.01)λ. The discrepancy between the simulated and experimental values was found not to exceed 9% for the said step, and 14% for all the steps measured. The error of the FWHM diameter was not larger than 0.01λ (or 3%) both in the simulation and in the experiment. A step of square side 0.6 µm has been shown to form a nanojet with a maximum intensity 9.58 times the input light maximum intensity. When the square side is smaller than the incident wavelength, the focus is inside the step. When the square side is larger than the incident wavelength, the focus is outside the step, making it look like an optical candle. The comparative simulation of microsphere-aided focusing, with the microsphere diameters varying from 0.4 µm to 0.8 µm, and the micro-parallelepiped-aided focusing, with a fixed height of 695 nm and square side varying from 0.4 µm to 0.8 µm, has shown that: (1) in the plane perpendicular to the polarization plane, the step- and microsphere-aided focal spots have similar diameters; but (2) the microsphere-aided focal spots show a wider range of variations (from 0.39λ to 0.53λ) than the step-aided spots (from 0.40λ to 0.44λ); except for one case, all the diameters were below the diffraction limit of 0.51λ; (3) the microsphere-aided focal spots are more elliptical than the step-aided focal spots: at $D=L=0.4$ µm, the diffraction by the step results in a circular focal spot of diameter FWHM=0.44λ, whereas the microsphere-aided focal spot is elliptical, measuring FWHM$_{min}$=0.53λ and FWHM$_{max}$=0.74λ; (4) with the microsphere's diameter and square-step's side increasing, the DOF (photonic nanojet's length) measured as intensity half-maximum behaves differently, increasing from DOF=0.55λ to DOF=1.20λ for the step, and decreasing from DOF=1.97λ to DOF=0.59λ for the microsphere.

2.4 MODELING THE RESONANCE FOCUSING OF A PICOSECOND LASER PULSE USING A DIELECTRIC MICROCYLINDER

The study of subwavelength focusing of laser light by means of microparticles – including microspheres and microcylinders of near-wavelength size – has recently been given much attention [147–151]. The subwavelength focusing with the aid of multilayer microspheres [147], spheroidal microspheres [148], and two-layer micro-spheres [149, 150] has been reported. The minimal focal spot size at full-width half-maximum of intensity reported was FWHM=0.4λ [148], whereas the depth of focus achieved was DOF=20λ [149] and DOF=22λ [150]. Subwavelength focusing with the aid of an elliptic [151] and multilayer [152] microcylinder has also been reported. The minimum size of the focal spot thus obtained was FWHM=0.46λ [152, 153]. Microsphere-aided resonance focusing of light was proposed in Reference [154]. Using a dielectric microsphere in combination with a nanometer metal bead [154], resonance focusing into a focal spot of size FWHM=0.25λ has been achieved, whereas the dielectric-microsphere-aided resonance focusing has enabled obtaining a focal spot with FWHM=0.40λ [153]. Resonance focusing of a laser TE-wave by means of a polyester microcylinder (refractive index $n=1.59$) has been analyzed based on a Bessel function series expansion [155]. For the whispering gallery mode (WGM) with mode number 18, an external focus of size FWHM=0.22λ was attained. Various methods for research of forming a WGM in a dielectric cylinder are considered in References [156, 157]. The FDTD method and the series solution are compared in Reference [156]. In Reference [157] the authors explore the formation of the WGM in a dielectric cylinder by using the Mie theory and spectral element method (SEM). Neither article has investigated the characteristics of the focal spot and temporal dynamics of the formation of the WGM.

In this section, using the FDTD method and FullWAVE software, we study temporal variations of parameters of a focal spot resulting from the resonance focusing of a plane TE-wave with a dielectric microcylinder. We show that as the picosecond pulse (ps pulse) propagates through the microcylinder, the subwavelength focus decreases in size and increases in intensity, also featuring the increase of the mode energy stored inside the cylinder. Once the ps pulse has passed through it, a WGM is excited in the microcylinder, being coincident with the mode in a circular wire with zero propagation constant. With time, the energy of the WGM leaks. In the stationary case (when the pulse duration tends to infinity), by solving the Helmholtz equation using a finite-element method implemented in the COMSOL software, the focal spot parameters have been shown to be nearly identical to the analytical results [155]: for the resonance radius $R=2.1749\lambda$ of the cylinder the focal spot size was FWHM=0.226λ, whereas the maximum intensity of the focus was 60 times the incident light intensity.

2.4.1 AN ANALYTIC ESTIMATION OF THE MICROCYLINDER RESONANCE RADIUS

The problem of the diffraction of a non-paraxial Gaussian beam by an infinite dielectric circular cylinder was analytically solved in Reference [158]. The solution to the Helmholtz equation for a TE-wave traveling in the microcylinder, derived by the separation of variables in polar coordinates, is given by an expansion in a Bessel series:

$$E_y(r,\phi) = \sum_j i^j b_j C_j J_j(knr) e^{ij\phi}, \qquad (2.18)$$

where k is the wave number of incident light, n is the refractive index of the cylinder, and the coefficients b_j and C_j are defined as

$$C_j = \frac{\omega_0 \sqrt{\pi}}{\lambda} \int_{\mathbb{R}} \exp\left(-\frac{k^2 q^2 \omega_0^2}{4} - ik\left(px_0 + qy_0\right) - ij \arcsin q\right) dq, \qquad (2.19)$$

$$b_j = \frac{J_j(z)H_j^{(1)'}(z) - J_j'(z)H_j^{(1)}(z)}{J_j(nz)H_j^{(1)'}(z) - nJ_j'(nz)H_j^{(1)}(z)}, \qquad (2.20)$$

where $z = kR$, R is the cylinder radius, $p = \sqrt{1-q^2}$, $J_m(x)$, $H_m^1 = J_m(x) + iY_m(x)$ are Bessel and Hankel functions, and $Y_m(x)$ is the Neumann function, $J_j'(z)$, $H_j^{(1)'}(z)$ are derivatives. The resonance radius can be estimated using a simple technique based on the assumption that an integer number m of wavelengths should fit in a circumference of radius $R_1 < R$ (R_1 is the radius on which the intensity maxima of the mode with number m are found):

$$2\pi R_1 = m\left(\frac{\lambda}{n}\right). \qquad (2.21)$$

From Equation (2.21), we find that $R_1 = 1.8018\lambda$. The ratio R/R_1 equals the ratio of the first roots of the Bessel function and its derivative, so that

$$R = \left(\frac{\gamma_m}{\gamma_m'}\right)R_1 \approx \left(\frac{m + m^{1/3}1.855757}{m + m^{1/3}0.808618}\right)R_1 \approx 2.0476\lambda \qquad (2.22)$$

The discrepancy between the resonance radius in Equation (2.22) and a more accurate value of $R = 2.1749\lambda$ [155] is as low as 5%. However, this is too big a discrepancy, as with just a 2% error in radius estimation, the resonance will occur no more [155].

2.4.2 RESONANCE FOCUSING IN THE STATIONARY CASE

In this subsection, we discuss the numerical simulation results obtained via solving the Helmholtz equation by a finite-element method using the COMSOL Multiphysics 4.3 software. A plane TE-wave of unitary amplitude is incident on a microcylinder from left to right (Figure 2.31). The microcylinder resonance radius is $R = 2.1749\lambda$. Figure 2.31 depicts a 2D intensity distribution. From Figure 2.32, the best focusing condition is at radius $R = 2.1749\lambda$. The Q-factor of the cavity [153] is defined as the ratio of the resonance frequency to the attenuation circuit bandwidth ($Q = \nu/\Delta\nu$). From Equation (2.21), the frequency is seen to be related with the resonance radius R, and thus, the Q-factor can be expressed as $Q = R/\Delta R$. From Figure 2.32, the FWHM resonance width is also seen to equal $\Delta R = 0.002\lambda$, with the resonance Q-factor being $Q = R/\Delta R = 1.087$.

A focal spot is generated near the microcylinder surface. Let the focal spot be described by the transverse intensity distribution $|E_y|^2$ found 2-nm away from the microcylinder.

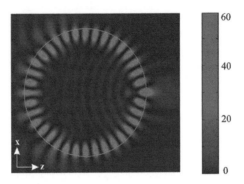

FIGURE 2.31 A stationary diffraction pattern in the *xz*-plane for diffraction of a plane TE-wave ($\lambda = 633$ nm) by a polyester microcylinder of resonance radius ($n = 1.59$ and $R = 2.1749\lambda$), simulated using the COMSOL software. The wave is incident from left to right.

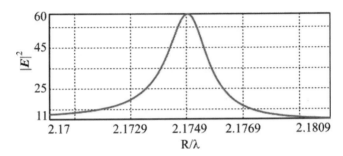

FIGURE 2.32 Maximum intensity $|E_y|^2$ at the focal spot vs. microcylinder radius. The resonance occurs at $R = 2.1749\lambda$.

By varying the microcylinder's cavity radius, the maximum focal spot intensity can be derived. Figure 2.32 depicts the maximum focal spot intensity versus the microcylinder radius.

Figure 2.33 shows the plot for $|E_y|^2$ across the focal spot at the resonance radius $R = 2.1749\lambda$.

In Figure 2.33, the focal spot size is FWHM = 143.16 nm = 0.2261λ. The longitudinal profile of intensity along the optical axis is shown in Figure 2.34. Note that in terms of full-width of half-maximum intensity, DOF = 189.43 nm = 0.299λ.

The diffraction efficiency in the focal plane is defined as the ratio of the focal spot power (Figure 2.32) to the power of the entire focal plane cross-section, amounting to 54%. The stationary resonance focusing has shown the focal spot size FWHM = 0.2261λ to be nearly identical to the theoretical estimate of FWHM = 0.22λ [155].

From Figure 2.33, the maximum intensity in the focal spot is seen to be 61 times the incident plane continuous wave intensity. Assuming the entire incident energy per period λ proportional to the microcylinder diameter $2R = 4.3498\lambda$ to equal $W = 4.3498\lambda^2$ and to be focused into a focal half maximum area $S = 0.05\lambda^2$ (of width FWHM = 0.2261λ (Figure 2.32) and length DOF = 0.299λ (Figure 2.34)), the

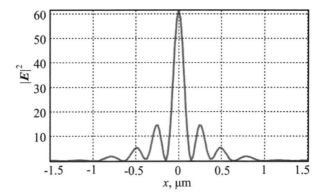

FIGURE 2.33 The cross-section of the intensity in Figure 2.31 drawn in parallel with the *x*-axis, perpendicularly to the optical *z*-axis and through the outside focus 2 nm away from the microcylinder surface.

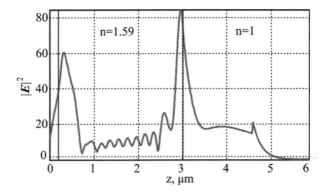

FIGURE 2.34 The profile of the intensity in Figure 2.31 along the optical *z*-axis.

maximum intensity in the focal spot is supposed to equal $I_{max} = W/S = 52.2$. This value is close to the maximum intensity in Figure 2.33. To get a deeper insight into the mechanisms behind the excitation of WGM in the microcylinder and the energy stored in it, the next section analyzes the diffraction of a ps pulse by a microcylinder with resonance radius.

2.4.3 TEMPORAL VARIATIONS OF THE RESONANCE FOCUS PARAMETERS

Below, we simulate the propagation of a plane TE-wave through a microcylinder within a 1.06-ps interval (TE-pulse) and 2.12-ps (continuous wave). The simulation is conducted in FullWAVE by solving Maxwell's equations with the aid of the FDTD method. The simulation parameters are the incident wavelength, $\lambda = 0.633$ μm, the microcylinder radius, $R = 2.1749\lambda$, the refractive index of environment equal to 1, and the refractive index of the microcylinder, $n = 1.59$. The discretization step in the spatial domain is 0.002 μm and in the temporal domain is 0.0001 μm. A 1.06-ps pulse corresponds to 500 wavelength periods. That is, having traveled 500 periods, the pulse passes through the entire microcylinder, going on to travel along the z-axis. Figure 2.35 depicts diffraction patterns at different instants as the pulse travels in the cylinder.

Figure 2.35 suggests that for the initial 25 periods, non-resonant focusing occurs (focal spot size is FWHM = 0.33λ). Once the pulse has traveled 500 periods, being on the brink of leaving the cylinder, a process of mode excitation is observed in the form of local intensity maxima along the microcylinder surface in Figure 2.35(b), with the focal spot size decreasing to FWHM = 0.26λ. Shown in Figure 2.35(c) is the diffraction pattern formed after the pulse has left the microcylinder. It is seen that the cylinder's circumference has fitted exactly 36 half-periods of the mode. The focus is formed on the optical axis near the surface of the cylinder (Figure 2.35(b)) as a result of constructive interference of two waves: the wave passing through the cylinder

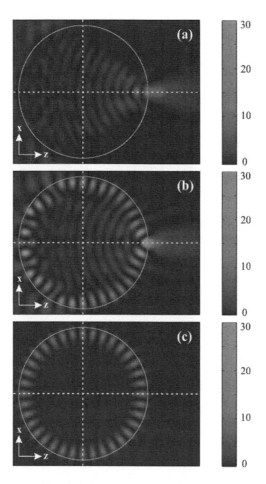

FIGURE 2.35 Diffraction pattern "snapshots" as a 1.06-ps TE-pulse (500 periods) travels in a microcylinder ($R = 2.1749\lambda$) at instants corresponding to (a) 25, (b) 500, and (c) 525 periods.

(Figure 2.35(a)) and the surface wave which is the WGM (Figure 2.35(c)). The size of the focus depends on the resonant cylinder radius (or number of the WGM) and the ratio of the intensities of these two waves, which are summed at the focus. This ratio depends on the number of the passed periods. The WGM can be described as a mode in the uncladded circular waveguide, which has the only non-zero electric vector projection on the optical axis:

$$E_y(r,\varphi) = \exp(im\varphi)\begin{cases} J_m(knr), & r < R, \\ J_m(knR)K_m^{-1}(ikR)K_m(ikr), & r > R, \end{cases} \qquad (2.23)$$

where $m = 18$ (Figure 2.35(c)), $K_m(x)$ is the Macdonald function, and (r, φ) are the polar coordinates in the xz-plane. From Equation (2.23) we can estimate the radius R_1 at which the mode in Figure 2.35(c) has intensity peaks. This can be done by equating the Bessel function's argument to its first root $\gamma'_m \approx m + m^{1/3}0.808618$. Then, we find ($m = 18$):

$$R_1 = \left(\frac{\gamma'_m}{2\pi n}\right)\lambda \approx 2.0138\lambda \qquad (2.24)$$

Figure 2.36 depicts the intensity profile of the mode in Figure 2.35(c) along the z-axis.

In Figure 2.36, the left intensity peak is found at radius $R_1 = 2.0114\lambda$, whereas the right intensity peak is at $R_2 = 1.9918\lambda$. These radii are close to the theoretical estimations in Equation (2.24).

It should be noted that Equation (2.23) defines the mode in an uncladded circular fiber with a zero-value propagation constant (the wave vector's projection on the fiber y-axis). Thus, we conclude that mode (2.23) in Figure 2.35(c) is leaking, meaning that with time the energy stored in the cylinder leaves it perpendicularly to its surface. The processes are illustrated in dynamics in Figure 2.37.

Figure 2.37 suggests that while the pulse was traveling through the microcylinder, feeding energy to the mode ($T < 500$), the focal spot size FWHM (Figure 2.37a)

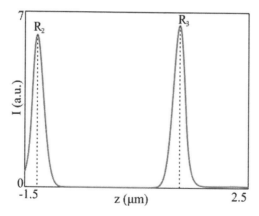

FIGURE 2.36 Intensity profile of the mode in Figure 2.35(c) along the z-axis.

and DOF (Figure 2.37(b)) were reducing, alongside the simultaneous growth of the maximal intensity I_{max} in the focus (Figure 2.37(c)). The growth of the maximal intensity in the focus means that a larger proportion of mode energy is being stored in the cylinder: I_{max} has shown a threefold increase from 10.34 to 30.9 over 1000 periods (dashed curve). The efficiency η has been deduced at a wavelength distance behind the microcylinder (to avoid the account of surface waves) as the ratio of focal spot power to the incident light power. From Figure 2.37(d), the efficiency is seen to hardly change, being approximately equal to $\eta = 9\%$ ($T < 500$). If the incident beam were to be continuous, the described pattern of the focal spot parameters would remain the same (dashed curve in Figure 2.37). After the pulse has passed through the microcylinder ($T > 500$), all plots in Figure 2.37 show a sharp jump (solid curve). The focal spot is seen to widen to FWHM$= 0.88\lambda$, with the depth of focus increasing to DOF$= 0.57\lambda$ (solid curve in Figure 2.37). In this case, it is only the WGM (surface standing wave) that accounts for the generation of the focal spot. Because the WGM is leaking, the maximal intensity of the focal spot in Figure 2.37(c) slowly decreases after 500 periods, with the mode (and focus) intensity decreasing ten times on the interval from 500 to 1000 periods (solid curve). Note that if a twice longer (2.12 ps) pulse were to be used, the focus size would decrease to FWHM$= 0.24\lambda$ during the 1000 periods it would take the pulse to pass through the microcylinder (dashed curve). The minimal focal spot size is achieved at the stationary case (at $T \to \infty$) and FWHM$= 0.226\lambda$ (Figure 2.32).

A relationship to estimate the resonator's Q-factor based on the radiation lifetime in the microcylinder proposed in Reference [159] is $Q = 2\pi\nu\tau$. Figure 2.37(c) suggests

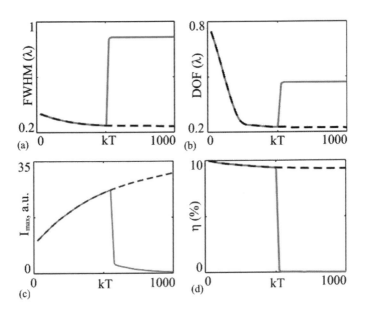

FIGURE 2.37 Focal spot parameters and efficiency η at radius $R = 2.1749\lambda$ vs. the number of periods T traveled by a 1.06-ps pulse: a) FWHM, b) DOF, c) I_{max}, d) η (dashed curve, continuous light; solid curve, pulsed light).

that after the pulse has passed through it, the radiation in the microcylinder shows an e-fold decrease after 250 periods, i.e. $\tau = 250/\nu$. Then, $Q = 2\pi\nu\tau = 1.570$.

It may seem that the focal spot can be made tighter by choosing a higher-order mode, e.g. by replacing the mode with number $m = 18$ by a mode with number $m = 19$. But this is not the case because as the mode number m increases, the resonance radius also increases. Thus, for the mode $m = 19$, the resonance radius is $R = 2.284\lambda$. Figure 2.38 shows intensity patterns generated by a 1.06-ps pulse traveling in a microcylinder of resonance radius $R = 2.284\lambda$.

Diffraction pattern snapshots in Figure 2.38(a), (b), and (c) show the focal spot sizes where (a) FWHM $= 0.55\lambda$, (b) FWHM $= 0.26\lambda$, and (c) FWHM $= 0.87\lambda$. Thus, after 500 periods the focal spot has tightened, becoming equal to that in Figure 2.35 (FWHM $= 0.26\lambda$). In Figure 2.38, exactly 38 half-periods of WGM (2.22) with $m = 19$ fit in the microcylinder.

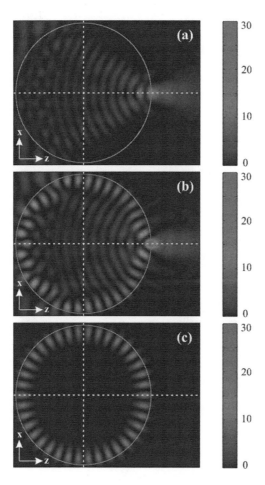

FIGURE 2.38 Intensity patterns generated by diffraction of the 1.06-ps pulse in the microcylinder ($R = 2.284\lambda$) at instants corresponding to a) 25, b) 500, and c) 525 periods.

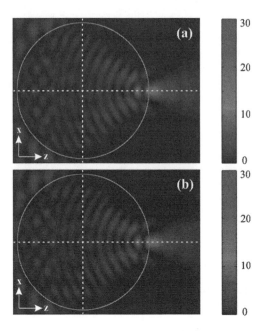

FIGURE 2.39 Snapshots of the intensity patterns generated by diffraction of a continuous beam in the non-resonance microcylinder ($R=2.2\lambda$) at instants corresponding to a) 25 and b) 325 periods.

In the case of a microcylinder with non-resonance radius, both the size and maximum intensity of the focal spot remain almost unchanged in the course of pulse propagation. Figure 2.39 depicts two diffraction patterns for a non-resonance microcylinder's radius ($R=2.22\lambda$) after 25 and 325 periods. The patterns are seen to be nearly identical with no WGM being excited inside. The size and maximum intensity of the focal spot are nearly the same in both cases: in Figures 2.39(a) and (b), FWHM$=0.345\lambda$ and FWHM$=0.351\lambda$, respectively, and $I_{max}=9.77$ and $I_{max}=9.34$ respectively.

2.4.4 Discussion of Results

The stationary case is established when the energy stored in the microcylinder per period becomes equal to the leaking energy. It becomes possible to accumulate energy in the microcylinder because a portion of incident light enters the microcylinder at larger-than-critical angle (of 39 degrees), reaching the opposite side at the same angle. Thus, owing to the total internal reflection (TIR), these rays get trapped in the microcylinder and start circulating along a circumference, experiencing multiple reflections at the surface. Since, due to symmetry, identical flows of light circulate clockwise and anticlockwise, the resulting pattern resembles a standing wave (Figure 2.35(c)). The said portion of light leaks from the cylinder owing to diffraction. At the initial stage of pulse propagation, an identical portion of light energy is stored inside for each period, being much higher than the leaking portion. As the inside energy is being built up during each period in equal portions, the proportion of leaking energy also increases. In the stationary case, the stored energy per period should be equal to the leaked energy. From Figure 2.37(c), the maximum intensity in the focus, per 1000 periods, is seen to grow but not reach saturation, and, thus, is far from being the stationary case. Knowing the per-period energy incident on the cylinder, $W = 4.35\lambda^2$, having estimated the energy stored inside per 500 periods, $W_1 = 2mI_0S_0 = 2 \times 18 \times 6 \times 0.394\lambda^2 = 85.1\lambda^2$ (assuming I_0 to be the maximum value of each intensity peak in Figure 2.36 and S_0 the FWHM-intensity area of each sidelobe in Figure 2.35(c)), and disregarding the energy leakage because of near-linear energy build up per 500 periods in Figure 2.37(c), we can estimate what proportion of incident energy W is stored in the microcylinder per period: $W_2 = W_1/525/W = 85.1/525/4.35 = 0.037$. Thus, per each period of wave, approximately 3.7% of incident light is stored in the microcylinder. From Figure 2.37(c) it is seen that after the pulse has passed through the microcylinder, the stored energy ceases to build up, just leaking: per 500 periods, the in-cylinder energy experiences a 10-times decrease. Thus, about 0.25% of the entire stored energy is leaking from the microcylinder per period.

So, using an FDTD approach, we have studied temporal variations of focal spot parameters for the resonance focusing of a plane TE-wave by a dielectric microcylinder with refractive index $n = 1.59$. It has been shown that as a picosecond pulse is traveling through the microcylinder, the focal spot gets tighter, whereas the focal spot intensity and WGM energy stored in the cylinder are increasing. After the pulse has passed through it, a leaking WGM remains in the microcylinder, which is identical with the mode in a wire circular fiber with zero propagation constant. With time, the WGM leaks from the cylinder.

Besides, using a finite-element method, we have modeled a stationary intensity pattern for the diffraction of a plane TE-wave by the dielectric microcylinder with refractive index $n = 1.59$. In the stationary case, the focal spot parameters have been found to be nearly identical with the analytical results [155]: for the resonance radius $R = 2.1749\lambda$, the focal spot size was FWHM $= 0.226\lambda$ and the maximal intensity in the focus (behind the cylinder surface) was 60 times the incident light intensity.

3 Subwavelength Focusing Light by Gradient Microoptics

3.1 HIGH RESOLUTION THROUGH GRADED-INDEX MICROOPTICS

Recent advances in microoptics and nanophotonics have made possible the focusing of coherent laser light into a subwavelength spot or the superresolution imaging of a point source of light. The subwavelength focusing beyond the diffraction limit of $0.5\lambda/n$, where λ is the wavelength in free space and n is the material refractive index at the focus, can be performed using a superlens [16]. In the 2D case instead of the conventional diffraction limit $0.5\lambda/n$ one must use $0.44\lambda/n$. This value can be obtained after replacing the Airy disk $2\,J_1(x)/x$ by sinc-function $\mathrm{sinc}(x) = \sin(x)/x$.

The superlens is a 2D planar plate made up of the metamaterial that comprises alternating metallic and dielectric layers. The electric permittivities of the layers are selected so that an effective refractive index of the composite material be equal to $n=-1$. Experiments on the superresolution through superlenses were reported in References [17, 18]. In the experiments, a superresolution of 0.4λ was achieved [17]. A similar experiment conducted in Reference [19] with a subwavelength silver layer operating as a superlens has shown that two lines separated by a 145 nm distance can be resolved when illuminated by ultraviolet (UV) light of wavelength 365 nm, thus producing a superresolution of 0.4λ.

A far-field hyperlens reported in Reference [29] was able to resolve two lines of width 35 nm spaced 150 nm apart for a 365 nm wavelength, again achieving a superresolution of 0.4λ. Note, however, that a hyperlens modeled in the form of a grating [160] was shown to achieve a superresolution of 0.05λ at the imaging plane found 1.5λ apart from the surface. Apparently, the absorption and scattering of light by metamaterial that occurs in real experiments was disregarded in modeling. This argument was indirectly verified by results reported in Reference [56], in which the laser light was focused with a zone plate made up of a gold film of thickness 100 nm. The focal spot size at half-maximum was estimated to be 0.35λ, whereas the experimentally measured size of the focal spot at half-maximum was found to equal λ.

Multilayer and anisotropic nanostructures allowing one to achieve a subwavelength resolution were analyzed in References [31, 161, 162]. For example, parameters of a 1D eight-layer Ag/SiO_2 structure of thickness 400 nm to focus light from a 0.4λ source into the same size focal spot were studied in Reference [161]. An anisotropic 2D nanostructure characterized by the dielectric permittivity tensor components $\varepsilon_x = 0.01 - i0.01$ and $\varepsilon_z = -100$, and a 400-nm thickness on the z-axis (for $\lambda = 700$ nm) was proposed in Reference [31]. The modeling has shown that such a structure is able to resolve two lines of width 3 nm placed 23 nm apart, providing a superresolution of 0.03λ. A nanostructure composed of two different anisotropic layers to resolve two narrow slits placed 50 nm apart when illuminated by a 1550-nm wavelength was studied in Reference [162] by the same authors.

Candidates for achieving the superresolution can be found among the photonic crystals (PhC). Modeling conducted in Reference [33] has shown that a 2D photonic-crystal slab with permittivity $\varepsilon = 12$ composed of a triangular array of circular holes of radius $r = 0.4a$ (a is the hole array period) can be used as an imaging lens for wavelength $\lambda = a/0.3$. In this case, a point source is imaged as a focal spot of size 0.3λ, whereas two point sources placed 0.5λ apart are resolved by the 20% criterion.

In recent experiments with a 2D photonic-crystal slab used as a superlens a point source of size 0.4λ was imaged [163]. The latest publications propose an improved variant of superlenses using a nanoshell [164] or a graded-index boundary of the negative-refraction material [165]. In Reference [166] an anisotropic layer was experimentally demonstrated to enhance and transform the evanescent surface waves into propagating light modes.

From the previous survey it follows that although theoretically superlenses allow arbitrary high resolution to be achieved, experimentally the values 0.3λ–0.4λ have been obtained [17, 19, 29, 33, 56, 163].

Such values of resolution can be obtained with the help of gradient-index (GRIN) optics as well. Gradient optics works like near-field optics: a gradient lens is placed near the object and the image emerges in the vicinity of the exit surface of the lens. Therefore the lens material affects the resolution. The limiting resolution that can be obtained with a gradient lens decreases n times in comparison with a conventional refraction lens and is $0.51\lambda/n$ in 3D and $0.44\lambda/n$ in 2D cases. If one uses a refraction lens with immersion then the index of immersion liquids does not exceed 1.5, while the gradient lenses can be made of silicon with index 3.47.

In this section we numerically demonstrate that graded-index Mikaelian lenses and Maxwell "fisheye" lenses, widely known in optics, may also be considered as candidates for subwavelength focusing. For a 2D Mikaelian lens, we show that a point source is imaged near the lens surface as a focal spot of size (full-width at half-maximum) FWHM$=0.12\lambda$. This value is close to the diffraction limit for silicon ($n = 3.47$) in 2D media FWHM$=0.44\lambda/n = 0.127\lambda$. This value is smaller than values earlier reported in References [33, 56, 161, 163]. We also show that a Mikaelian lens can resolve at half-maximum two point sources placed 0.3λ apart, which is smaller than reported in References [17, 19, 29, 33]. Analytical relationships for modes propagating in graded-index planar waveguides were derived in References [167, 168]. A general constraint of the above works has been the assumption on the existence of one [167] or two [168] turning points of the refractive index profile of the planar waveguide. We have derived extended analytical relationships for mode solutions in the graded-index planar waveguide that have no constraints on the number of turning points of the refractive index function.

3.1.1 SOLUTION OF THE HELMHOLTZ EQUATION FOR A 2D GRADED-INDEX WAVEGUIDE

Figure 3.1 gives a schematic representation of the problem. We consider the propagation of the transverse electric (TE) wave in a 2D graded-index medium with the refractive index $n = n(x)$, with the electric field vector directed along the y-axis.

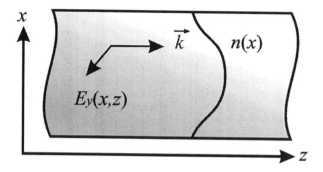

FIGURE 3.1 Schematic representation of the TE-wave propagating in a 2D graded-index waveguide.

The electric field amplitude $E_y(x,z)$ satisfies the Helmholtz equation [7]:

$$\left[\frac{\partial^2}{\partial x^2} + \frac{\partial^2}{\partial z^2} + k^2 n^2(x) \right] E_y(x,z) = 0, \tag{3.1}$$

where k is the wave number. The expansion of the electromagnetic wave amplitude in terms of the transverse modes of the graded-index medium is

$$E_y(x,z) = \sum_{n=0}^{\infty} C_n(x) \exp(i\beta_n z), \tag{3.2}$$

where $\beta_n = k_{zn}$ is the propagation constant of the nth mode. From Equation (3.2) it follows that the light field has an axial period T, so that $\beta_n = 2\pi n/T$. For example, in a graded-index medium with the quadratic-index profile, the modes are described by the Hermite–Gauss functions that form a countable basis [169]. Substituting Equation (3.2) in Equation (3.1) gives an equation of the amplitude of propagating modes in the graded-index medium:

$$\frac{d^2 C_n(x)}{dx^2} + p_n(x) C_n(x) = 0, \tag{3.3}$$

where

$$p_n(x) = k^2 n^2(x) - \beta_n^2. \tag{3.4}$$

Changing variables

$$C_n(x) = C_n(0) \exp\left[\int_0^x f_n(\zeta) d\zeta \right], \tag{3.5}$$

Equation (3.3) can be reduced to a nonlinear differential Whittaker equation [170] for the function $f_n(x)$:

$$\frac{df_n(x)}{dx} + f_n^2(x) + p_n(x) = 0. \tag{3.6}$$

Equation (3.6) can be solved by the expansion into the Taylor series of the functions $f_n(x)$ and $p_n(x)$:

$$f_n(x) = \sum_{m=0}^{\infty} C_m^{(n)} x^m, \quad p_n(x) = \sum_{m=0}^{\infty} p_m^{(n)} x^m, \tag{3.7}$$

where $C_m^{(n)}$ and $p_m^{(n)}$ are the unknown and known expansion coefficients of the corresponding functions. Substituting (3.7) into (3.6) yields recurrent relations $(m > 0)$ for the unknown series terms in Equation (3.7):

$$C_m^{(n)} = -m^{-1} \left(p_{m-1}^{(n)} + \sum_{s=0}^{m-1} C_s^{(n)} C_{m-1-s}^{(n)} \right), \tag{3.8}$$

where n is integer and $C_0^{(n)}$ are indefinite constants. Then, the amplitudes of the propagating modes in a graded-index waveguide are explicitly given by

$$C_n(x) = C_n(0) \exp\left(\sum_{m=0}^{\infty} C_m^{(n)} \frac{x^{m+1}}{m+1} \right), \tag{3.9}$$

where $C_m^{(n)}$ are derived from the recurrent relationships shown in Equation (3.8). The coefficients $p_m^{(n)}$ in Equations (3.7) are derived from

$$p_m^{(n)} = \frac{k^2}{m!} \frac{d^m n^2(x)}{dx^m} \bigg|_{x=0}, \quad m > 0,$$

$$p_0^{(n)} = k^2 n^2(0) - \beta_n^2, \quad m = 0. \tag{3.10}$$

Finally, the amplitude of the TE-wave propagating in the 2D graded-index waveguide is given by

$$E_y(x,z) = \sum_{n=0}^{\infty} C_n(0) \exp\left[i\beta_n z + \sum_{m=0}^{\infty} C_m^{(n)} \frac{x^{m+1}}{m+1} \right]. \tag{3.11}$$

For each mode of Equation (3.11), there are two indefinite constants, $C_n(0)$ and $C_0^{(n)}$. Since the modes of Equation (3.9) are neither orthogonal nor normalized, so in order for the field of Equation (3.11) to be expanded in terms of the said modes (with the aim of finding the coefficients $C_n(0)$) we need to truncate both series in Equation (3.11) to finite sums, and then solve the sets of linear algebraic equations.

The constants $C_0^{(n)}$ need to be selected in a special way for each mode. By way of illustration, consider one particular case. Assume the propagating modes in a quadratic-index medium: $n^2(x) = n_0^2 - a^2 x^2$. In this case, $p_n(x) = p_0^{(n)} + p_1^{(n)} x + p_2^{(n)} x^2$, where $p_0^{(n)} = k^2 n_0^2 - \beta_n^2$, $p_1^{(n)} = 0$, $p_2^{(n)} = -k^2 a^2$, and $p_m^{(n)} = 0$ at $m > 2$. Let $C_0^{(n)} = 0$, then, $C_2^{(n)} = -\left(C_0^{(n)} C_1^{(n)}\right) = 0$ and $C_1^{(n)} = -p_0^{(n)} = \beta_n^2 - k^2 n_0^2$. For the remaining coefficients to equal zero, $C_m^{(n)} = 0$, at $m > 2$, it will suffice to put the third coefficient to be equal zero: $C_3^{(n)} = -p_2^{(n)}/3 - (C_1^{(n)})^2/3 = 0$. Whence, we obtain the following condition on the medium parameter α: $k^2 a^2 = (k^2 n_0^2 - \beta_n^2)^2$. Assume that $\beta_n = k n_0/\sqrt{2}$, then we obtain $a = k n_0^2/2$. Thus, we can infer that the mode of the quadratic-index waveguide, $n^2(x) = n_0^2 \left(1 - k^2 n_0^2 x^2/4\right)$, is described by the Gaussian exponential function:

$$E_y(x,z) = C(0) \exp\left(\frac{ik n_0 z}{\sqrt{2}} - \frac{k^2 n_0^2 x^2}{4}\right). \tag{3.12}$$

3.1.2 GENERAL SOLUTION FOR A SECANT-INDEX WAVEGUIDE

If there is a waveguide with the refractive index defined by a secant function on the transverse coordinate

$$n(x) = n_0 \, \mathrm{ch}^{-1}\left(\frac{k n_0 x}{\sqrt{2}}\right) \tag{3.13}$$

a particular solution of Equation (1.1) is given by [171]:

$$E_{1y}(x,z) = \exp\left(\frac{ik n_0 z}{\sqrt{2}}\right) \mathrm{ch}^{-1}\left(\frac{k n_0 x}{\sqrt{2}}\right). \tag{3.14}$$

In this case, the Helmholtz equation takes the form:

$$\left[\frac{\partial^2}{\partial x^2} + \frac{\partial^2}{\partial z^2} + k^2 n_0^2 \, \mathrm{ch}^{-2}\left(\frac{k n_0 x}{\sqrt{2}}\right)\right] E_y(x,z) = 0. \tag{3.15}$$

We will seek the general solution of Equation (3.15) in the form:

$$E_{1y}(x,z) = A(x) \exp(i\gamma z). \tag{3.16}$$

Substituting (3.16) into (3.15), we obtain

$$\frac{d^2 A(x)}{dx^2} + g(x) A(x) = 0, \tag{3.17}$$

where

$$g(x) = k^2 n_0^2 \, \mathrm{ch}^{-2}\left(\frac{k n_0 x}{\sqrt{2}}\right) - \gamma^2. \tag{3.18}$$

It is known [170] that Equation (3.17) has a general solution

$$A(x) = A_1(x)\left[C_1 + C_2 \int_0^x A_1^{-2}(\xi)d\xi\right], \tag{3.19}$$

where $A_1(x)$ is a particular solution of Equation (3.17), and C_1, C_2 are indefinite constants. In our case, the solution in Equation (3.14) can be chosen as the particular solution, i.e.

$$A_1(x) = \text{ch}^{-1}\left(\frac{kn_0 x}{\sqrt{2}}\right), \quad \gamma = \frac{kn_0}{\sqrt{2}}. \tag{3.20}$$

Then, the general solution for the secant-index waveguide is given by

$$E_{1y}(x,z) = \exp\left(\frac{ikn_0 z}{\sqrt{2}}\right)\text{ch}^{-1}\left(\frac{kn_0 x}{\sqrt{2}}\right)$$

$$\times \left\{C_1 + \frac{C_2}{kn_0\sqrt{2}}\left[\frac{1}{2}\text{sh}\left(kn_0 x\sqrt{2}\right) + \frac{kn_0 x}{\sqrt{2}}\right]\right\}, \tag{3.21}$$

where

$$C_1 = \left|E_{1y}(x=0,z)\right|, \quad C_2 = \left|\frac{dE_{1y}(x=0,z)}{dx}\right|. \tag{3.22}$$

From Equation (3.14), the mode width at half-maximum in the secant-index waveguide is

$$\text{FWHM} = \frac{\ln\left(3+2\sqrt{2}\right)}{\pi n_0\sqrt{2}} \approx \frac{0,4\lambda}{n_0}, \tag{3.23}$$

where λ is the wavelength of light in free space and n_0 is the refractive index on the waveguide axis.

3.1.3 PARTIAL SOLUTION FOR A QUADRATIC-INDEX WAVEGUIDE

In the previous subsection, we showed that the propagating mode of a quadratic-index waveguide with definite parameters can be described by the Gaussian exponential function (3.12). In this section, we will demonstrate that this remains true of any quadratic-index medium with arbitrary parameters:

$$n^2(x) = n_0^2\left(1 - w^2 x^2\right), \tag{3.24}$$

where w is an arbitrary constant. So, Equation (3.1) takes the form:

$$\left[\frac{\partial^2}{\partial x^2} + \frac{\partial^2}{\partial z^2} + k^2 n_0^2(1 - w^2 x^2)\right]E_y(x,z) = 0, \tag{3.25}$$

and the solution of Equation (3.25) will be sought-for in the form:

$$E_{2y}(x,z) = E_0 \exp\left(ipz - q^2x^2\right). \tag{3.26}$$

where

$$q^2 = \frac{wkn_0}{2}, \quad p = kn_0\sqrt{1 - \frac{w}{kn_0}}.$$

Thus, a particular mode solution of Equation (3.25) is given by

$$E_{2y}(x,z) = E_0 \exp\left(ikn_0z\sqrt{1 - \frac{w}{kn_0}} - \frac{wkn_0}{2}x^2\right). \tag{3.27}$$

Note that at $w = kn_0/2$ the solution in Equation (3.27) coincides with that in Equation (3.12). At $w = kn_0/2$, it follows from Equation (3.27) that the Gaussian mode width (diameter) at half-maximum is

$$\text{FWHM} = \frac{\sqrt{\ln 4}\,\lambda}{\pi n_0} \approx \frac{0.38\lambda}{n_0}. \tag{3.28}$$

Comparison of Equation (3.28) and Equation (3.23) suggests that the both modes (the secant and the Gaussian mode) have nearly the same width. The effective width of the quadratic-index waveguide is derived from the condition $n(x_0) = 1$, being given by

$$2x_0 = \frac{2\sqrt{n_0^2 - 1}}{\pi n_0}\lambda. \tag{3.29}$$

At $n_0 = 1.5$, from Equation (3.29), we find that $2x_0 \approx 0.48\lambda$. Thus, the effective width of a glass quadratic-index planar waveguide that can only propagate the Gaussian mode of Equation (3.27) equals nearly half the free-space wavelength. Note that presently such half-wave waveguides are actively used in applications [172, 173].

3.1.4 MODELING THE PROPAGATION OF LIGHT THROUGH GRADED-INDEX MICROOPTICS

First experiments on superresolution imaging with superlenses in the optical regime were reported several years ago [18, 29]. Thus, a superresolution of 0.4λ has been achieved in the experiment reported in References [18, 29].

Theoretically (disregarding the absorption of the material) any degree of super-resolution can be achieved with a superlens. In Reference [31] it was shown by modeling that a hyperlens that would form a magnified subwavelength image in the near field can be implemented as a plane-parallel layer. A 400-nm-thick anisotropic slab with dielectric permittivities $\varepsilon_x = 0.01 - i0.01$ and $\varepsilon_z = -100$ was shown to resolve two 3-nm slits (directed along the y-axis) spaced 23 nm apart in a metallic screen with dielectric permittivity $\varepsilon = 1 - i10^4$, illuminated by a 700-nm transverse magnetic (TM)-wave. The resulting superresolution achieved is 0.05λ.

In the following subsections, using a well-known finite-difference time-domain (FDTD) method, implemented in FullWAVE, we numerically show that the high resolution is also achievable with the aid of 2D graded-index microoptics.

3.1.5 HIGH RESOLUTION IMAGING USING A MIKAELIAN MICROLENS

Because of the diffraction in uniform space, two nearby point sources of light cease to be resolved at a distance much smaller than the wavelength of incident light. By way of illustration, Figure 3.2(a) depicts profiles of five original coherent Gaussian light sources of width $\lambda/200$, spaced $\lambda/50$ apart in the plane $z = 5$ nm. Shown in Figure 3.2(b) is the intensity profile of these sources obtained in the image plane $z = 30$ nm for the wavelength $\lambda = 1550$ nm. Figure 3.2 suggests that two nearby light sources cease to be resolved at a distance z approximately equal to the spacing between them ($z = \lambda/50$).

Shown in Figure 3.3(a) is a numerically simulated image of two 35-nm sources spaced 150 nm apart obtained with a Mikaelian microlens (ML) [174]. The refractive index of this 2D secant-graded microlens is

$$n(x) = n_0 \, \mathrm{ch}^{-1}\left(\frac{\pi x}{2L}\right), \tag{3.30}$$

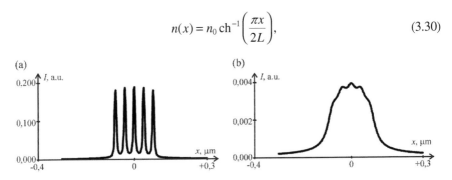

FIGURE 3.2 Light intensity profile near five Gaussian light sources of width $\lambda/200$, spaced $\lambda/50$ apart, at different distances: (a) $z = 5$ nm, (b) $z = 30$ nm, for $\lambda = 1550$ nm (intensity is plotted in relative units).

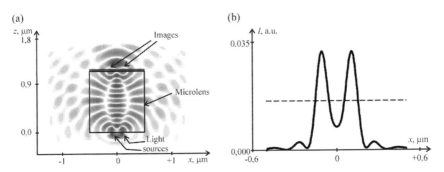

FIGURE 3.3 (a) The snapshot of the amplitude $E_y(x, z)$ in the ML; (b) averaged E-vector intensities of the TE-wave calculated directly on the lens "rear" side of the ML (the horizontal line indicates the half-maximum intensity plotted in relative units).

where $2L$ is the lens length. The lens width $2R$ is found from the condition $n(R) = 1$. The lens in Figure 3.3(a) has the axial refractive index $n_0 = 2.1$, width $2R = 1$ µm, and length $2L = 1144$ µm, operating at wavelength $\lambda = 365$ nm. These parameters are similar to those described in the experiment in Reference [29]. The distance between these sources is 0.41λ and they cannot be resolved by refraction lens. But the gradient-index lens Equation (3.30), which is placed close (20 nm) to the sources, works as immersion and allows an increased resolution by n_0 times, i.e. to resolve the two sources separated by distance $0.44\lambda/n_0$.

Figure 3.3(a) shows the instantaneous distribution of the electric field amplitude for the TE-wave (propagating from the bottom upward). The sources are seen to be imaged on the opposite side of the lens. Figure 3.3(b) shows a time-averaged intensity profile of the electric field, $I(x, z = z_0) = |E_y(x, z = z_0)|^2$, directly on the lens "rear" side, i.e. at a distance of $z_0 = 2L$ from the lens "front" side. In the experiment, the sources under imaging were put 20 nm before the lens "front" side. From Figure 3.3(b) it is seen that two sources spaced 150 nm apart (with center-to-center distance of 180 nm) can be resolved with confidence. The resolution achieved in Figure 3.3(b) is 0.41λ. From Figure 3.3(b) it can also be found that the point source imaged with the ML has a width at half-maximum of FWHM $= 100$ nm $= 0.27\lambda$. This value is comparable with the diffraction limit FWHM $= 0.44\lambda/n = 0.21\lambda$ in 2D media with index $n = 2.1$. This value of high resolution, 0.41λ, is very close to that reported in References [17, 18, 29] (0.4λ). Note that the source image width of 0.27λ agrees well with the minimal width of the propagating mode in a secant-graded waveguide: $0.4\lambda/n_0 = 0.27\lambda$ at $n_0 = 1.5$ (see Equation (3.23)).

The numerical aperture of the ML is $\mathrm{NA} = \left(n_0^2 - 1\right)^{1/2}\big/n_0 = 0.88$ for $n_0 = 2.1$. Therefore for such a lens the focal spot width at half-maximum intensity is FWHM $= 0.44\lambda/(n_0\mathrm{NA}) = 0.24\lambda$. This value is slightly less than obtained by simulation, FWHM $= 0.27\lambda$.

To improve the resolution of ML, as suggested by Equations (3.23) and (3.28), the axial refractive index was changed to $n_0 = 3.47$ (Si). The other simulation parameters were also changed (Figure 3.4(a)) to wavelength $\lambda = 1$ µm, lens width $2R = 6$ µm, and

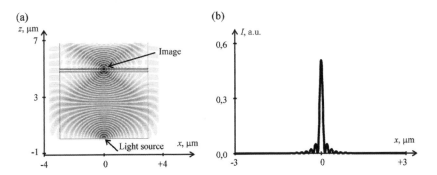

FIGURE 3.4   (a) Instantaneous pattern of the E-vector of the TE-wave in the ML from a point source found at the front plane; and (b) an averaged intensity pattern in the lens rear plane (with arbitrary units plotted on the y-axis).

lens length $2L = 4.92$ µm. The simulation step along the spatial axes in all examples considered was $\lambda/100$. The width of the Gaussian source at the original plane was $\lambda/20$. Shown in Figure 3.4(a) is an instantaneous pattern of the E-vector of the light wave in the ML calculated for the instance when the wave has travelled a 220 µm distance from the source. Figure 3.4(b) depicts an averaged intensity of the TE-wave at the ML output plane (Figure 3.4(a)). The computation has shown that the intensity of Figure 3.4(b) corresponds to the source image width at a half-maximum of FWHM $= 0.12\lambda$. However, the intensity (or light power density) provides no information as to the proportion of source power propagating along the z-axis. This information can be derived from the projection of the Poynting vector onto the optical axis, calculated at the ML output (Figure 3.4(a)), with the point source found at the lens input, as shown in Figure 3.5.

The central-maximum width of the flow of energy propagating along the z-axis in Figure 3.5 is equal to that in Figure 3.4(b), being equal to FWHM $= 0.12\lambda$. The value of the diffraction limit (in 2D case) that can be achieved when focusing light in the homogeneous medium is known to be equal to $0.44\lambda/n$, where n is the refractive index of the homogeneous medium. In our case, $n_0 = 3.47$, therefore the diffraction limit is FWHM $= 0.127\lambda$. The numerical aperture for the silicon ($n_0 = 3.47$) ML is $\mathrm{NA} = \left(n_0^2 - 1\right)^{1/2}\big/n_0 = 0.96$. Therefore the focal spot width at the half-maximum of intensity is FWHM $= 0.44\lambda/(n_0\mathrm{NA}) = 0.132\lambda$ for such lens. This value is greater than obtained by simulation FWHM $= 0.12\lambda$.

Presumably, this is because when focusing light at the two-media interface, there is also a contribution to the focus from inhomogeneous surface waves that form interference and diffraction patterns with deep subwavelength features [31, 162]. Figure 3.5 displays the fact that the surface wave plays a part when imaging the point source. In a certain region on the x-axis, the z-projection of the Poynting vector is negative, which means that near the output surface of ML there are both outgoing and incoming light waves, with a surface wave propagating along the microlens surface. As light propagates further in free space from the lens output surface, the light spot size quickly increases, becoming equal to the diffraction limit of 0.5λ at one wavelength distance from the surface.

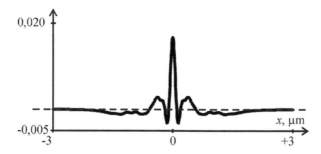

FIGURE 3.5 Profile of the z-projection of Poynting vector onto the x-axis calculated at the ML output (Figure 3.4(a)) with the point source at the input (in relative units).

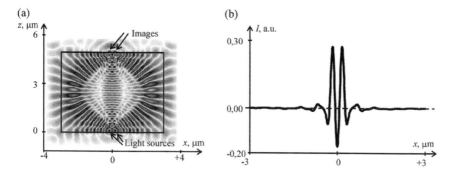

FIGURE 3.6 (a) Instantaneous distribution of the E-vector of the TE-wave in the ML of Figure 3.4(a) illuminated by two 50-nm sources with center-to-center distance 300 nm, placed 10 nm away from the input plane (bottom horizontal line); (b) relative time-averaged distribution of the z-projection of the Poynting vector calculated at 10-nm distance from the output plane (top horizontal line).

Note that the change of sign of the Poynting vector projection, similar to that shown in Figure 3.5, was earlier reported in Reference [161], being termed as the optical vortex and interpreted as resulting from the interference between the propagating wave and the enhanced surface plasmon. Note, however, that Reference [161] handled a 1D multilayered structure (a 1D photonic crystal).

Figure 3.6(a) shows the instantaneous distribution of the E-vector of the light wave in the Mikaelian lens (for the same parameters as in Figure 3.4(a)) illuminated by two 50-nm sources with center-to-center distance 300 nm, placed 10 nm away from the lens bottom plane. Shown in Figure 3.6(b) is the relative time-averaged distribution of the z-projection of the Poynting vector calculated at the 10 nm distance from the microlens output surface. It can be seen from Figure 3.6(b) that the resulting superresolution value is 0.3λ, which is smaller than that reported in References [17, 19, 29, 33]. Although the point-source image has width FWHM$=0.12\lambda$ (Figures 3.5 and 3.4(b)), only two point sources, separated by 0.3λ, are resolved with confidence (Figure 3.6(b)). This is because images of point sources interfere with each other. Let us note that such a gradient-index lens can be fabricated as a photonic-crystal lens [175].

3.1.6 High Resolution through a "Fisheye" Microlens

Alongside the ML discussed above, a high resolution can be achieved with other graded-index imaging optical elements for which the refractive index as a function of coordinates has been derived in an explicit analytical form. The possibility of obtaining an ideal image of the point source with help of Maxwell's "fisheye" lens is justified in Reference [176]. One such optical element is represented by Maxwell's "fisheye" [1], whose 2D refractive index is given in polar coordinates as

$$n(r) = n_0 \left[1 + \left(\frac{r}{R} \right)^2 \right]^{-1}, \qquad (3.31)$$

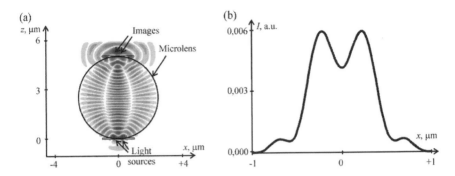

FIGURE 3.7 (a) An instantaneous distribution of the E-vector of the TE-wave in the 2D fisheye microlens simulated in FullWAVE, when illuminated from the bottom by two point sources; and (b) the relative time-averaged intensity distribution in the image plane.

where n_0 is the refractive index at the circle center and R is the element radius. According to Equation (3.31), the refractive index is halved at $r = R$. A disadvantage of this element is that the refractive index center-to-edge contrast cannot be larger than two, whereas, in the ML the refractive index contrast is a function of the material, e.g. for Si ranging from 3.47 to 1. However, with the fisheye having a circular symmetry, any source found on its surface will be perfectly imaged at the diametrically opposite point of its surface.

Figure 3.7(a) shows the profile of the E-vector of the TE-wave in the 2D fisheye microlens that has two nearby point sources on its surface. The simulation was performed for the refractive index at the lens center $n_0 = 3.47$, lens radius $R = 2.5$ μm, wavelength $\lambda = 1$ μm, two 0.05λ sources separated by a 440 nm distance, which is 0.44λ. Shown in Figure 3.7(b) is a time-averaged intensity pattern in the image plane (top horizontal line in Figure 3.7(a)). The two sources are seen to be resolved (the resolution being 0.44λ by the 20% Rayleigh criterion). Thus, the lens superresolution of 0.44λ is insignificant beyond the diffraction limit of 0.5λ.

When the microlens of Figure 3.7(a) is illuminated by a solitary source, the resulting intensity distribution is shown in Figure 3.8. The image size at half-maximum is

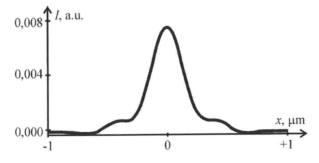

FIGURE 3.8 Relative time-averaged intensity pattern of the electric field in the image plane of the fisheye microlens of Figure 3.7(a) when illuminated by a single point source found on its surface.

FWHM = 0.3λ. From the comparison of Figures 3.8 and 3.5(b) we can infer that the fisheye forms a wider (by 2 times approximately) image of the point source when compared with the ML at similar parameters. Note that the magnitudes of high resolution obtained, 0.4λ (Figure 3.7(b)) and 0.3λ (Figure 3.8), are comparable with those reported in References [17, 19, 29, 33, 161, 163].

Although a point-source image has width FWHM = 0.3λ (Figure 3.8), two point sources must be separated by 0.4λ distance (Figure 3.7(b)) in order to resolve them by Rayleigh criteria. This is because images of point sources interfere with each other.

Thus, in this section we have derived mode solutions of the Helmholtz equation for an arbitrary graded-index planar waveguide, with the mode amplitude represented as the exponential of the Taylor series, whose coefficients are deduced from the recurrent relations. We have shown that the minimal mode width in quadratic- and secant-index planar waveguides amounts to 0.4λ of the wavelength in free space divided by the refractive index on the waveguide axis. We have shown by modeling in FullWAVE that graded-index ML and fisheye lenses are capable of high resolution imaging. We have shown that a point source is imaged through the 2D ML as a near-surface light spot of size FWHM = 0.12λ (λ is wavelength in free space), which is close to the diffraction limit for silicon ($n = 3.47$) in 2D media FWHM = $0.44\lambda/n = 0.127\lambda$ and smaller than values reported in References [33, 161, 163]. We have also shown that the ML is able to resolve at half-maximum two nearby sources placed 0.3λ apart, which is beyond the diffraction limit in free space of 0.44λ for the 2D case and smaller than values reported in References [17, 19, 29, 33].

3.2 SUBWAVELENGTH FOCUSING WITH A MIKAELIAN PLANAR LENS

In 1951 A.L. Mikaelian [174] found that the hyperbolic secant profile was optimal for focusing light. This cylindrical self-focusing waveguide, called SELFOC, is now widely applied in information optics. This gradient-index lens was studied in detail and is known as the Mikaelian lens [177–180]. The focusing properties of ML were first investigated experimentally in the microwave range [174]. In 1952 Mikaelian studied more complicated cylindrical media with the refractive index depending on two coordinates, and he showed that there exists among them an infinite number of self-focusing waveguides [181–183].

The task of light propagation in the Mikaelian waveguide (MW) and ML was solved in geometrical [174, 181–185], quasi-optical [186], and wave [187–190] approximations. Experimental results on focusing light with the ML were described in References [191, 192].

Recent years have seen an increased interest in gradient-index and photonic-crystal lenses, which are capable of providing subwavelength focusing of laser light [33, 37, 171, 193, 194] and are applied to single-molecule imaging [195] and ultracompact coupling of planar waveguides with different widths [175].

In this section, the TE-modes of the ML are expressed through the orthogonal Jacobi polynomials. We show that an arbitrary light field propagating in the MW will be periodically reproduced with the wavelength-independent Talbot-length period. We show in the general case and by an example that an arbitrary, even light field propagating in the MW has intensity extremums on the axis at intervals of quarter-Talbot lengths. For the ML, which is a truncated MW, it is shown that the fundamental mode is Fourier-invariant. This allows us to estimate the full-width half-maximum of the focal spot.

3.2.1   TE-Modes of the Mikaelian Lens

The planar ML is a 2D GRIN medium with the index $n(x)=n/\text{ch}(x/a)$, where x is the transverse coordinate (z is the longitudinal coordinate), a is the lens width across which the on-axis index n decreases 1.54 times. The Helmholtz equation for the electric vector projection of the TE-wave, $E_y(x, z)$, is

$$\left[\frac{\partial^2}{\partial x^2} + \frac{\partial^2}{\partial z^2} + \frac{k^2 n^2}{\text{ch}^2(x/a)}\right] E_y(x, z) = 0, \tag{3.32}$$

where k is the wave number. The mode solution of Equation (3.32) takes the form (TE-mode):

$$E_y(x, z) = \exp(i\beta z/a)\left(1 - y^2\right)^{\beta/2} P_m^{(\beta,\beta)}(y), \tag{3.33}$$

where $2\beta = [1+4(akn)^2]^{1/2} - (2m+1)$, m is a non-negative integer, β is the propagation constant of the mode, $y = \text{th}(x/a)$, and $P_m^{(\beta,\beta)}(y)$ are the Jacobi polynomials [84]. The modes in Equation (3.33) are mutually orthogonal and have a finite energy at $\beta > 0$. The two first modes of Equation (3.33) are given by $\Psi_0(x) = \text{ch}^{-\beta_0}(x/a)$, $\Psi_1(x) = \beta_0 \text{sh}[x/a]\text{ch}^{-\beta_0}[x/a]$, where $2\beta_0 = [1+4(akn)^2]^{1/2}-1$ or $\beta_0 = \beta + m$. An arbitrary solution of Equation (3.32) can be decomposed in terms of Equation (3.33):

$$E_y(x, z) = \exp(i\beta_0 z/a) \sum_{m=0}^{\infty} C_m \exp(-imz/a) \Psi_m(x), \tag{3.34}$$

where $\Psi_m(x) = (1-y^2)^{\beta/2} P_m^{(\beta,\beta)}(y)$, C_m are coefficients. From Equation (3.34) it follows that at $z_p = 2\pi ap$, $p = 1, 2, 3...$, the module of the complex amplitude in Equation (3.34) will be periodically repeated. The same is true of the even and odd functions $E_y(x, z)$, but in this case $z_p = \pi ap$. Therefore, the distance $L = 2\pi a$ is called the Talbot length. Note that the Talbot length is independent of the wavelength. This means that there is no chromatic aberration in the ML: the focal length is the same for all wavelengths, which was confirmed by the numerical simulation. Let us show that at distances $z_p = \pi ap/2$ there are extremums of the on-axis intensity ($x = 0$) for the even light fields (3.34). Indeed, the intensity of the light field (3.34) is

$$I\left(x=0,z\right)=\sum_{p=0}^{\infty}C_{2p}^{2}\varPsi_{2p}^{2}(0)$$

$$\pm 2\sum_{p<q}\left|C_{2p}C_{2q}\varPsi_{2p}(0)\varPsi_{2q}(0)\right|\cos\left[2(p-q)z/a\right].$$

(3.35)

The values C_{2p} and $\varPsi_{m}(0)$ are real, because the mode functions (3.33) are real at $z=0$ and we suppose that the light field $E_{y}(x,z)$ at $z=0$ is real and even. The requirement of the even light field is imposed to avoid the on-axis intensity being zero everywhere. The derivative of the intensity is

$$\frac{\mathrm{d}I(x=0,z)}{\mathrm{d}z}$$

$$=\pm\frac{4}{a}\sum_{p<q}\left|C_{2p}C_{2q}\varPsi_{2p}(0)\varPsi_{2q}(0)\right|(p-q)\sin\left[2(p-q)z/a\right].$$

(3.36)

From Equation (3.36) it follows that this derivative is zero at $z_{p}=\pi ap/2$, i.e. there will be intensity maxima and minima at these points of the optical axis. Because at $z_{p}=\pi ap$ the intensity is reproduced, at the intermediate points $z_{p}=\pi a/2+\pi ap$ the light field will be either collimated (intensity minimum) or focused (intensity maximum). This can also be illustrated by particular examples. For example, by using a reference series from Reference [61] the following expression can be obtained:

$$E_{y}(x,z)=\frac{\exp\left(i\beta_{0}z/a\right)}{\mathrm{ch}^{\beta_{0}}\left[x/a\right]}\sum_{m=0}^{\infty}\frac{\exp\left(-imz/a\right)P_{m}^{(\beta_{0}-m,\beta_{0}-m)}(y)}{\left(1-y^{2}\right)^{m/2}}$$

$$=\frac{\exp\left(i\beta_{0}z/a\right)}{\mathrm{ch}^{\beta_{0}}\left[x/a\right]}\left[1-\mathrm{sh}\left[x/a\right]e^{-iz/a}+\frac{\exp\left(-i2z/a\right)}{4}\right]^{\beta_{0}}.$$

(3.37)

The light field (3.37) can propagate in the MW because the function (3.37) satisfies Equation (3.32). The on-axis intensity of such a field takes the form:

$$I(x=0,z)=[17/16+\cos(2z/a)]^{\beta_{0}}.$$

(3.38)

It follows from Equation (3.38) that at $z=0$ there will be the on-axis maximum $I_{\max}=(5/4)^{2\beta_{0}}$, while at $z_{p}=\pi a/2$ minimum $I_{\min}=(3/4)^{2\beta_{0}}$. Based on the presence of the extremums of the on-axis intensity at points $z_{p}=\pi ap/2$, we can infer that the light-field amplitudes in the MW in planes separated by distance $z_{p}=\pi a/2$ are related by the Fourier transform. This is also confirmed by means of ray optics, because all the rays in the MW parallel to the optical axis (a normally incident plane wave) are gathered at distance $z=\pi a/2$ into one on-axis focal point [174]. Using this suggestion, we can estimate the focal spot size of the ML. Let at the ML input ($z=0$) there will be a fundamental mode $\Psi_0(x) = \text{ch}^{-1}(x/b)$ of width b different from the ML width a. There is a well-known integral [61]:

$$\int_{-\infty}^{\infty} \text{ch}^{-1}(x/b)\exp(idx\xi)dx = \pi b\,\text{ch}^{-1}(\pi db\xi/2),\qquad (3.39)$$

where ξ is the transverse coordinate in the focal plane. The integral (3.39) shows that the fundamental mode of the MW at $\beta_0=1$ is Fourier invariant. Let us find the parameter d in Equation (3.39). If a fundamental mode with width $b=a$ travels in the MS lens, its width remains unchanged upon propagation. Therefore, from Equation (3.39) we obtain that $d=2/(\xi a^2)$. The focal length of the ML is equal to one quarter of the Talbot length $f=\xi a/2$, then we have $d=1/(af)$. Another condition for the mode width a follows from $\beta_0=1$, i.e. $a=\sqrt{2}/(kn)$. Thus, we obtain the final expression for the parameter d: $d = kn/(f\sqrt{2})$. With the help of Equation (3.39), we find the focal spot width at half-maximum intensity: $|E_0(\xi,z=f)|^2 = 0.5(kb/2)^2$. Whence, the focal spot width at half-maximum intensity at distance $z=f=\xi a/2$ from the input plane $z=0$ is FWHM $= \lambda a \ln(3+2\sqrt{2})/\sqrt{2}\pi nb = 0.396\lambda a/nb$.

For $n=3.47$ (silicon), we obtain $FWHM=0.115\lambda a/b$. Note that the scalar diffraction limit in the 2D medium is $FWHM=0.44\lambda/n$. Therefore, for $b=a$, the focal spot width will be slightly less than the diffraction limit. The limiting width of the focal spot in the 2D medium $FWHM=0.44\lambda/n$ is obtained for the maximal numerical aperture (NA) NA$=n$, although the ML has a lower numerical aperture. The numerical aperture of the ML can be found from the ray equation in the GRIN medium, $n(x)\cos\theta(x) = \text{const}$, where θ is the angle between the tangent to the ray and the optical z-axis. Let the initial ray be incident onto the lens parallel to and at distance $x=R$ from the z-axis, where R is the radius of the ML that can be found from the condition $n(R)=1$: $R=\text{arcch}(n)$. Then, we obtain $n(x)\cos\theta(x) = n\cos\theta = 1$, where n, as before, is the refractive index on the optical axis ($x=0$) and θ is the angle between the optical axis and the tangent to the ray drawn to the ray-axis intersection point. From the above equations, we find $\cos\theta = 1/n$, thus, the numerical aperture of the ML is NA $= n\sin\theta = (n^2-1)^{1/2}$. It depends on the on-axis index only, e.g. for $n=3.47$ we have NA$=3.3$. Therefore, the diffraction limit for the 2D MS lens is $FWHM = 0.44\lambda/\text{NA} = 0.44\lambda/(n^2-1)^{1/2}$.

3.2.2 SIMULATION RESULTS

Figure 3.9 shows the results of FDTD simulation of focusing the TE-polarized fundamental mode of width $b\neq a$ with the ML: (a) the instantaneous pattern of the real part of the amplitude $E_y(x, z)$ and (b) the time-averaged intensity distribution $|E_y(x, z)|^2$ in the lens focal plane. The following simulation parameters were used:

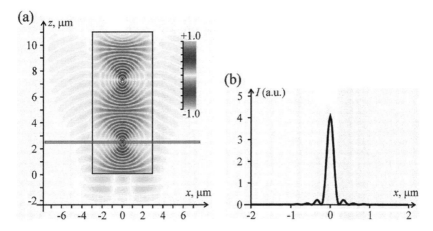

FIGURE 3.9 The results of FDTD simulation of focusing the TE-polarized fundamental mode of width $b \neq a$ with the ML: (a) the instantaneous pattern of the real part of the amplitude $E_y(x, z)$; and (b) the time-averaged intensity distribution $|E_y(x, z)|^2$ in the lens focal plane.

free-space wavelength $\lambda = 1.55$ μm, lens parameter $a = \lambda$; $b = 4\lambda$; the lens length along the z-axis from $z = 0$ to the first focus, $L_1 = \pi a / 2 = 2435$ nm; the lens width

$$W = 2a \ln \left[n_0 + \left(n_0^2 - 1 \right)^{1/2} \right] = 5.94 \text{ μm};$$ the on-axis refractive index is $n = 3.47$; and the

sampling step is 64 pixels per wavelength. The focal spot width (Figure 3.9(b)) at half-maximum intensity is FWHM $= 0.132\lambda$. The same focal spot width will be found in the immediate vicinity of the lens output plane if we cut away the lens upper part along the horizontal line in Figure 3.9(a). The diffraction limit of the ML in the medium given by FWHM $= 10.44\lambda / \left(n^2 - 1 \right)^{1/2}$ for silicon ($n = 3.47$) is FWHM $= 0.133\lambda$.

The ML can be used for superresolution imaging. In Figure 3.10(a) we show the imaging of two 50-nm sources with center-to-center distance 150 nm (using the light

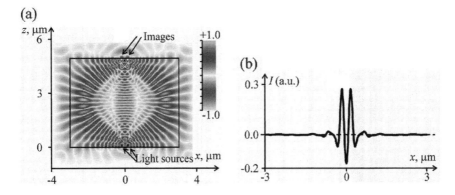

FIGURE 3.10 (a) The image of two 50-nm sources with center-to-center distance 150nm placed 10 nm before the ML made of silicon; and (b) the distribution of the z-projection of the Poynting vector in the plane placed 10 nm beyond the lens.

of wavelength 1 μm) placed 10 nm before the ML made of silicon ($n = 3.47$, lens width $W = 6$ μm, length $L = 4.92$ μm). The distribution of the z-projection of the Poynting vector in the plane placed 10 nm beyond the lens is shown in Figure 3.10(b). The ML can resolve two point sources at distance 0.15λ. Note that the Abbe's resolution limit is $\lambda/2$.

3.2.3 PHOTONIC-CRYSTAL MIKAELIAN LENS

The ML has been fabricated as a planar PhC lens, with the hole diameters increasing from the optical axis to the periphery. This PhC lens was applied to coupling two planar waveguides of different widths (Figure 3.11). Shown in Figure 3.11(a) is an electron microscope image of the ML (top view) fabricated as a PhC (with the hole diameters increasing from 180 nm at the center to 220 nm in the periphery) in a (200-nm-thick) silicon film combined with two planar waveguides of different widths (4.5 μm and 1 μm). In Figure 3.11(b) (top view) the light of wavelength 532 nm illuminates a silicon film with the lens at a small grazing incidence angle. It is seen from Figure 3.11(b) that the equally tilted fringes are formed because of the interference of waves reflected from the film and those passed through the film and reflected back from the wafer.

The phase shift between these waves is $2kd\left(n^2 - \sin^2 \theta\right)^{1/2}$, where d is the film width, θ is the grazing incidence angle at a given point of the film. If the film effective index n decreases (because of lens holes increases) then the incidence angle θ must decrease for the radicand to remain unchanged. The incidence angle decreases with increasing distance from the point source. Therefore, the interference fringes above the lens are bent as if they were visualizing a converging wavefront inside the lens. The resulting interference fringes are seen to be curved as arcs (Figure 3.11(b)), in a similar way as those shown in Figure 3.9(a).

Summing up, in this section we show that an arbitrary monochromatic TE-polarized light field propagating in a planar MW can be represented as a linear combination of orthogonal light modes with their amplitudes expressed by the Jacobi polynomials. We show that this light field will be periodically repeated with a period of Talbot length and focused with a period of half the Talbot length. This period is independent of the wavelength. The light field whose amplitude is proportional to and width is different from the corresponding characteristics of the fundamental mode of the MW is Fourier-invariant. In the focal plane of the ML, this field will preserve the shape of the fundamental mode, changing only in scale. The ML enables obtaining a focal spot of the width equal to the diffraction limit in the medium (for silicon, this limit is 0.133 of a wavelength). The numerical aperture of the ML depends only on the on-axis refractive index, being independent of the lens geometrical parameters. The ML can resolve two point sources at distance 0.15λ. The ML has been fabricated as a planar PhC lens in a silicon film for light of wavelength 1.55 μm. The converging wavefront in this lens was visualized using the light of wavelength 532 nm.

3.3 PHOTONIC-CRYSTAL LENS FOR COUPLING TWO WAVEGUIDES

In recent years, a lot of research has been focused on developing various micro- and nanophotonics devices designed to couple two waveguides of different types, for

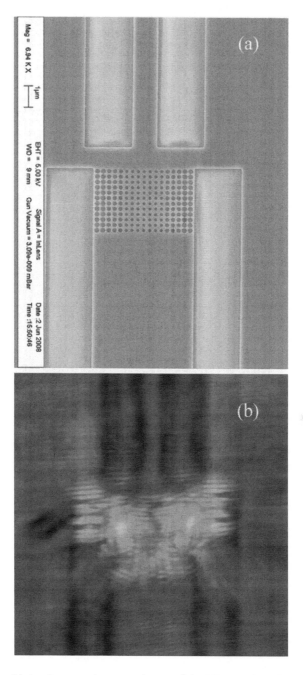

FIGURE 3.11 (a) An electron microscope image of the ML (top view) fabricated as a PhC in a silicon film combined with two planar waveguides of different widths; and (b) the light of wavelength 532 nm illuminates a silicon film with lens at a small grazing incidence angle.

example a conventional single-mode fiber with a wire or planar waveguide, or a planar waveguide with a PhC waveguide. Presently known nanophotonics devices for coupling two waveguide structures include: adiabatically tapered-ridge waveguides intended for coupling with PhC waveguides [196–202] (note that not only can the waveguide structures be joined output-to-input, but they can be overlapped in parallel to each other as well [203]); Bragg diffraction gratings in a waveguide [204–207] to output light (in this case, the Bragg-grating-aided fiber can be located on the surface of a planar fiber [208]); a parabolic micromirror put at an angle to couple light into a planar waveguide [209]; conventional refractive lenses or microlenses [210–213]; Veselago superlenses with negative refraction, which may be planar [35, 88, 214–219] or concave [34, 220, 221]; microwave coupling devices: superlenses [222, 223]; and PhC lenses [193]. Also, works dealing with the coupling of two different PhC waveguides were presented [224].

The tapered waveguides can have high coupling efficiency if the modes in the ridge waveguide and in the PhC waveguide are comparable in width. In this case, the coupling efficiency – defined as the ratio of the output energy to the input energy – can be as large as 80%–90% [196, 197, 199, 201]. If the width of a ridge waveguide (1.6 µm) is several times greater than that of a PhC waveguide (200 nm), the coupling efficiency is decreased to 60% [198]. For greater differences in width of the waveguides under coupling, the size of the adiabatically tapered waveguide portion becomes relatively large: when matching the mode of a single-mode fiber of core diameter 4.9 µm to the mode of a planar waveguide of width 120 nm, the taper's length is 40 µm [200], whereas the diameter of the 0.3×0.5 µm waveguide can be narrowed to the diameter of 75 nm at a distance of 150 µm [202]. It is worth keeping in mind that the optical mode size is much larger than the waveguide's physical dimensions.

Coupling devices to couple light from a single-mode fiber into a planar or PhC waveguide by means of on-waveguide diffraction gratings also have tapered portions. For instance, a Gaussian beam with a 14-µm waist can be narrowed to match a waveguide of width 1 µm by a taper as short as 14 µm [204, 205]. Note that the experimental coupling efficiency has been reported to be 35% [205] without a mirror layer and 57% [204] with a mirror on the waveguide's underside. A Gaussian beam of wavelength 1.3 µm has been coupled into a waveguide using an on-waveguide diffraction grating [205]. A similar coupling device comprising an on-Si-waveguide grating of period 630 nm with a 20–40 µm taper for the wavelength of 1.55 µm has been reported to have an experimental coupling efficiency of 33% (54% with the mirror) [206]. A higher-quality device comprising an in-Si diffraction grating of period 610 nm and width 10 µm designed to couple output light from a single-mode fiber into a wire waveguide of width 3 µm has been shown to have the experimental coupling efficiency of 69% [207]. A calculated coupling efficiency of over 90% has been reported for a J-coupler, designed to couple a wide waveguide (10 µm) with a PhC waveguide (420 nm) through a 15×20 µm parabolic mirror for the wavelength 1.3 µm [209].

Conventional refractive lenses and microlenses have also been successfully employed when solving coupling problems. For example, an Si waveguide of width 1–2 µm provided with an on-end lens can be coupled with an Si PhC waveguide with an estimated efficiency of 90% [210]. The modeling [212] has shown that a single-mode fiber of diameter 10.3 μm ($\lambda = 1.55$ μm wavelength) can be coupled with a PhC waveguide propagating a 0.19×0.27 μm mode using an Si focusing microlens of aperture radius 123 μm, achieving an 80% efficiency.

Of special note are coupling devices based on 2D superlenses, which utilize the negative-refraction phenomenon. A superlens with the effective refractive index of -1 can be obtained on the basis of photonic crystals. The superlens is used for imaging a point source. Notice that the original image is formed within the lens and the second image behind the lens at a distance of $2B-A$, where B is the thickness of a plane-parallel lens and A is the lens-to-source distance [214, 216]. It has been shown [35] that describing a 2D point source by the Hankel function $H_0(kr)$, where k is the wave number and r is the distance from the source to the observation point, yields a resulting image proportional to the Bessel function $J_0(kr)$. Thus, the superlens produces a spot of size FWHM $= 0.35\lambda$. The simulation of the 2D superlens performance [217] has shown that a lens comprising two layers of dielectric rods (permittivity $\varepsilon = 12.96$), operating at wavelength $\lambda = 1.55$ μm, has the refractive index of $n = -1$, imaging a point source put at distance $A = 0.26\lambda$ at about the same distance on the other side of the lens and producing the spot size of FWHM $= 0.36\lambda$. In some articles, a Veselago lens version with a concave surface instead of a plane-parallel PhC layer was considered [34, 220].

Another type of PhC lens was proposed in References [171, 225, 226]. The hole array of such a 2D PhC lens has a constant period but the hole sizes varies. The familiar gradient Mikaelian lens [227] operates by focusing all the rays parallel to the optical axis and incident perpendicularly to its plane surface at a point on the opposite plane surface. The refractive index of such an axisymmetric gradient lens is related with the radial coordinate as

$$n(r) = n_0 \left[\cosh\left(\frac{\pi r}{2L}\right) \right]^{-1}, \tag{3.40}$$

where n_0 is the refractive index on the optical axis and L is the lens thickness along the optical axis. In Reference [225], a 2D Mikaelian lens of aperture 12 μm was modeled. The reported efficiency was 55% [225]. In References [171, 226] a similar PhC ML, though with different parameters, was modeled. The focal spot size was FWHM $= 0.42\lambda$.

In this section, we have modeled, fabricated, and characterized a new, ultra-compact nanophotonic device that allows 2D waveguides of different width to be efficiently coupled by means of a PhC Mikaelian lens (PhC-ML). The device is fabricated on silicon-on-insulator (SOI), the input waveguide width is 4.5 μm, the output waveguide width is 1 μm, and the PhC-ML size is 3×4 μm. The lens comprises a 12×17 hole array, the array period being 250 nm and the hole diameter ranging from center to periphery from 160 to 200 nm. The device operates in a wavelength range of 1.5–1.6 μm. Depending on the input waveguide width, the calculated coupling efficiency is found to be in the range from 40% to 80%. The PhC-ML focuses light into a small focal spot in the air directly behind the lens, characterized by FWHM $= 0.32\lambda$, which is smaller than in References [171, 215, 217, 220].

3.3.1 SIMULATION

The PhC-ML under simulation comprises a 12×17 hole array in silicon (the effective refractive index of the silicon-on-insulator for a TE-wave is $n = 2.83$), the hole array period is 250 nm, the minimal hole diameter on the optical axis is 186 nm, and the maximal hole diameter on the lens edge is 250 nm. The thickness along the optical axis is 3 µm, the lens width (aperture) is 5 µm. The wavelength is $\lambda = 1.55$ µm.

The simulation was conducted using a finite-difference scheme FDTD for solving Maxwell's equations implemented in C++. Figure 3.12(a) shows the above-described 2D PhC-ML in SOI, and Figure 3.12(b) shows a half-tone time-averaged diffraction pattern of the plane TE-wave of intensity $I = |E_x(y,z)|^2$ (the x-axis is perpendicular to the plane of Figure 3.12). Shown in Figures 3.12(c) and (d) are the intensity distributions I on the optical axis-z and on the line parallel to the y-axis and passing through the focus. From Figures 3.12(c) and (d) it is seen that the focal spot size at half-intensity is FWHM $= 0.36\lambda$, whereas the axial focus size is FWHM $= 0.52\lambda$.

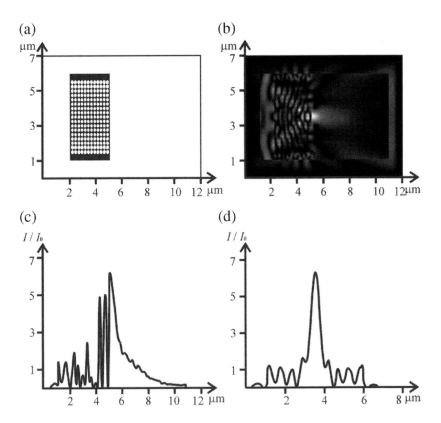

FIGURE 3.12   (a) A 2D PhC-ML comprising 12×17 holes in silicon of size 3×4 µm; (b) the diffraction field from a plane TE-wave or a 2D intensity distribution $|E_x|^2$, y, vertical axis, z, horizontal axis; (c) the axial intensity distribution; and (d) the intensity distribution in the focal plane.

Below, we model a PhC-ML with the same parameters as in Figure 3.12(a), which is located in the output of an Si waveguide of width 5 µm and length 5 µm (Figure 3.13(a)).

The time-averaged diffraction field of intensity $I = |E_x(y,z)|^2$ calculated by the FDTD method is shown in Figure 3.13(b) (wavelength 1.45 µm). Figure 3.13(c) shows the intensity distribution I on the optical axis. Shown in Figure 3.12(c) and (d) and Figure 3.13(c) and (d) is the ratio I/I_0 along the ordinate axis (where I_0 is the intensity of the beam lighting the in-waveguide PhC-ML). In comparison with Figure 3.12(c), the intensity in focus in Figure 3.13(c) is seen to have increased (6.5 and 7 respectively) and the intensity modulation amplitude within the lens is seen to have decreased. This is because the refractive index contrast between the lens and the waveguide (Figure 3.13(c)) is much less than that between the lens and the air (Figure 3.12(c)), resulting in a smaller amplitude of waves scattered from the medium interface. In Figure 3.13(d), the intensity distribution $I = |E_x(y,z)|^2$ in the lens focus in air on a line parallel to the y-axis shows that the focal spot size at half-intensity is FWHM $= 0.32\lambda$. From the comparison of Figure 3.13(d), it can be seen

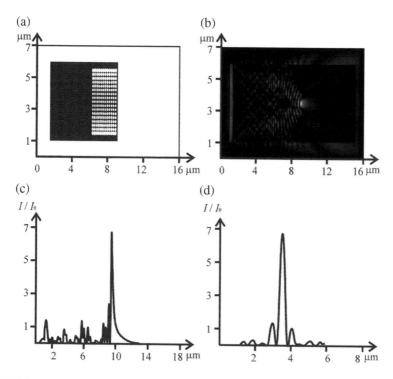

FIGURE 3.13 (a) A 2D PhC-ML in waveguide output; (b) the half-tone diffraction pattern of the plane TE-wave of amplitude Ex incident on the input of the 5 µm waveguide with a 3 µm lens at the output; and the intensity distribution (c) on the axis and (d) in the lens focus. The intensity units are arbitrary.

that alongside the focal spot decrease due to the in-waveguide PhC lens, the side lobes of the diffraction pattern in focus have also been decreased.

Note that in the 2D case, the scalar theory describes the diffraction-limited focus by the *sinc*-function: $E_x(y,z) = \text{sinc}\left[2\pi y/(\lambda \text{NA})\right]$, whence it follows that at a maximal NA = 1 the diffraction limit of the focal spot size at half-intensity is FWHM = 0.44λ. For a superlens [215], the limiting value of the focal spot is defined by the Bessel function $J_0(kr)$, producing the focal spot size at half-maximum of FWHM = 0.35λ. Hence, the lens in Figure 3.13(a) focuses light into a spot smaller than the diffraction limit. It can be explained because the focal spot is generated in the vicinity of the lens surface and the spot is partially created by surface (evanescent) waves with a wavelength smaller than λ.

In this and subsequent sections, the simulation by the FDTD method is implemented with sampling steps $\lambda/100$ in space and $\lambda/(50c)$ in time, where c is the speed of light in vacuum. Figure 3.14(a) depicts a device for coupling two 2D waveguides by means of a PhC-ML. The input waveguide has a width of 5 µm, the output waveguide is 0.5 µm wide, and both waveguides are 6 µm long. The PhC lens in Si (the effective index $n = 2.83$) comprises a 12×19 hole array of period 0.25 µm, the hole diameters being the same as in the previous sections. The wavelength is 1.55 µm.

The simulation was conducted by FDTD using FullWAVE 6.0. Note that the FullWAVE simulation engine calculates only instantaneous 2D distribution of light amplitude (Figure 3.14(b) and Figure 3.15(b)) while our C++ program calculates the 2D-field of time-averaged intensity (Figure 3.12(b) and Figure 3.13(b)). Figure 3.14(b) depicts the instantaneous amplitude distribution E_x of the TE-wave. The coupling efficiency was found to be 45%. The light was partially (about 20%) backscattered from the lens inside the input waveguide, with some light passing through the lens but missing the narrow waveguide. Shown in Figure 3.14(c) is a magnified fragment of the output amplitude distribution of Figure 3.14(b) of the narrow waveguide. For the narrow waveguide, the output intensity distribution on the transverse y-axis, $|E_x(y,z)|^2$, is shown in Figure 3.14(d), from which the modal spot size in air at half-intensity is equal to FWHM = 0.32λ. Note that, in the output waveguide of width 1 µm – the other conditions being the same – the focal spot size in the waveguide is FWHM = 0.21λ, where λ is the wavelength in vacuum. This is smaller than the earlier reported value of FWHM = 0.24λ [212].

The numerical simulation of the optical scheme stability (Figure 3.14(a)) has shown that with changing the medium index by 5% the coupling efficiency can change by 20%, and changing the hole diameters by 5% leads to a 17% changing in the coupling efficiency.

Figure 3.15 shows a 2D arrangement for coupling two coaxial waveguides separated by a space. The input waveguide comprising the PhC-ML is $W_1 = 4.6$ µm wide and the output waveguide is $W_2 = 1$ µm wide, the inter-waveguide spacing being $\Delta z = 1$ µm. The other parameters are $\lambda = 1.55$ µm, $n = 1.46$, and the PhC lens consists of a 12×17 hole array of period $a = 0.25$ µm with the hole diameters ranging from 186 to 250 nm. Shown in Figure 3.15(a) in white is the waveguide and in gray is the air. Figure 3.15(b) depicts the instantaneous amplitude distribution E_x of the TE-wave, calculated in FullWAVE 6.0 for the arrangement in Figure 3.15(a). Figure 3.15(c) depicts the coupling efficiency η (the ratio of the output power of the narrow

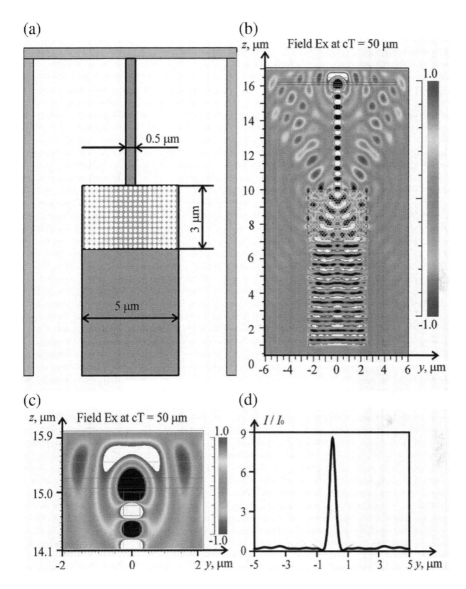

FIGURE 3.14 (a) A schematic diagram of coupling two planar waveguides with a PhC lens; (b) the instantaneous diffraction pattern of a TE-wave, calculated by the FDTD method using the FullWAVE 6.0; (c) a magnified fragment of the pattern at the output of the 0.5 μm waveguide; and (d) the output intensity distribution.

waveguide to the input power of the wide waveguide) as a function of the inter-waveguide spacing Δz. It can be seen from Figure 3.15(c) that the maximal coupling intensity of 73% is attained when the inter-waveguide spacing is 0.6 μm. Note that the inter-waveguide spacing is filled with the waveguide material rather than air.

Figure 3.15(d) depicts the coupling efficiency for the arrangement in Figure 3.15(a) as a function of the output waveguide width W_2 when the inter-waveguide spacing

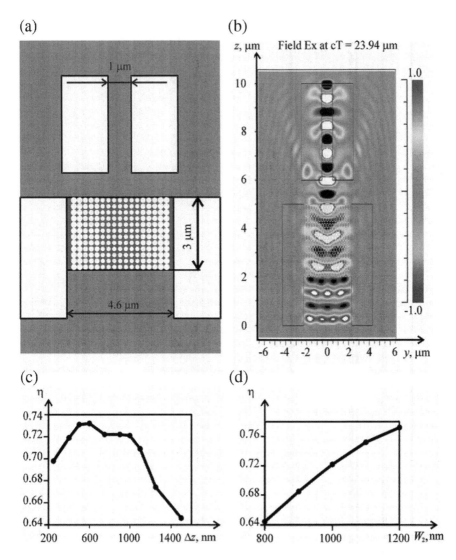

FIGURE 3.15 (a) The 2D arrangement for coupling two waveguides by means of a PhC lens for a $\Delta z = 1$ μm inter-waveguide spacing (white, material; black, air); (b) the instantaneous intensity distribution of the TE-wave, calculated in FullWAVE; and the coupling efficiency as a function of (c) the inter-waveguide spacing Δz and (d) the output waveguide width W_2.

is $\Delta z = 1$ μm. From Figure 3.15(d), the coupling efficiency is seen to increase nearly linearly with increasing width W_2 of the output waveguide.

3.3.2 FABRICATION OF TWO 2D WAVEGUIDES COUPLED WITH A PhC LENS

The devices were fabricated in SOI (220 nm thick silicon on 2 um of buried oxide). The pattern was exposed in ZEP520A resist (at a voltage of 30 kV) using a hybrid

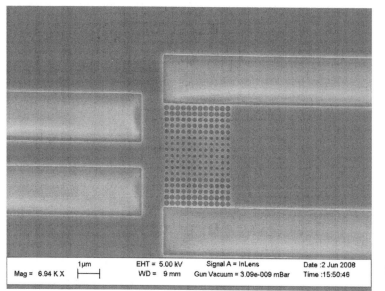

FIGURE 3.16   A SEM (×7000) photograph of two waveguides fabricated in the Si film and coupled with the PhC lens.

ZEISS GEMINI 1530/RAITH ELPHY electron beam lithography system and developed using xylene with ultrasonic agitation. The pattern was transferred into the silicon using reactive ion etching (RIE) in a mixture of gases CHF_3 and SF_6. The process is based on that of [228]. The holes in the PhC lens had diameters ranging from 160 nm to 200 nm and were completely etched through the silicon layer. The total length of the sample (the length of two waveguides) was 5 mm. Several similar structures with different value of the inter-waveguide spacing ($\Delta z = 0$ μm, 1 μm, and 3 μm) and different mutual location of the waveguide axes ($\Delta x = 0$ μm, ±0.5 μm, and ±1 μm) were fabricated simultaneously on the same substrate. Figure 3.16 shows a magnified (×7000) SEM (top view) image of two fabricated waveguides located $\Delta z = 1$ μm and combined with the PhC lens. The sample in Figure 3.16 has the following parameters: the designed width of the waveguides is $W_1 = 4.5$ μm and $W_2 = 1$ μm, and the PhC lens comprises a 12×17 hole array of period 250 nm. The sensitivity to fabrication errors of the device is similar to that reported in Reference [229] and is of the order of 0.3 GHz/nm.

3.3.3   Characterization of Two Waveguides Combined with a PhC Lens

Figure 3.17 depicts an optical setup for studying the transmission spectrum of two waveguides combined with a PhC lens. A broadband light-emitting diode (LED) light source (1450–1700 nm) is connected to an optical fiber. At the fiber output the light is collimated and TE-polarized. Then, the light is focused with a micro-objective onto the surface of the input waveguide.

A second microlens put at the narrow waveguide output focuses the light onto the end of a multi-mode optical fiber connected with an optical spectrum analyzer

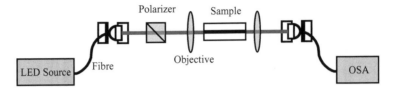

FIGURE 3.17 The optical setup used for studying the nanophotonic devices (comprising a pair of waveguides combined with a PhC lens).

(OSA). The total coupling efficiency is approximately 1%. Figure 3.18 shows the emission spectrum I_0 (a.u.) of a source that has a maximum at 1.55 μm. The intensity units are arbitrary. We give here the emission spectrum of the source in order to make it easier to understand the form of the experimental curves.

Figure 3.19 depicts the normalized spectra of the transmitted light I with respect to the intensity of the light source I_0 of the samples under study in the range 1.5–1.6 μm for the following values of inter-waveguide spacing Δz (a): 0 μm (curve 1), 1 μm (curve 2), and 3 μm (curve 3). Figure 3.19(b) depicts the off-axis displacements Δx with respect to the output waveguide optical axis (b): 0 μm (curve 1), −0.5 μm (curve 2), +0.5 μm (curve 3), −1 μm (curve 4), and +1 μm (curve 5). Figure 3.19(a) (curve 1) suggests that the transmission spectrum has 4 local maxima at approximately 1535 nm, 1550 nm, 1565 nm, and 1590 nm. Note that two of the maxima (at wavelengths of 1550 nm and 1565 nm) show the intensity 3 times larger than for the other two. This is most likely because at these wavelengths the intensity of the source (Figure 3.18) is several times smaller.

We find it difficult to account for the local maxima in the normalized transmission spectrum (Figure 3.19). It does not have any connection with the Fabry–Perot cavity inside the device in Figure 3.16, because a 15–20 nm period can arise at a cavity length of about 30 μm, while in our case the PhC lens thickness is just 3 μm. Though, it is possible that some fabrication errors lead to there appearing somewhere a 30 μm Fabry–Perot cavity. Likewise, there is no connection between the local maxima and the longitudinal shift of the focal spot because of changing incident wavelength (longitudinal chromatic aberration) [37], since in Reference [37] the

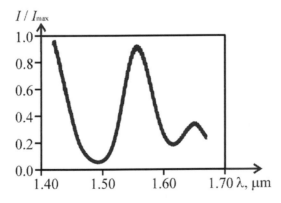

FIGURE 3.18 The emission spectrum of the light source.

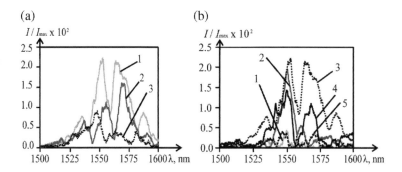

FIGURE 3.19 The normalized spectrum of the transmitted light with respect to the intensity of the light source measured using an optical setup of Figure 3.17 for the samples under study shown in Figure 3.16, (a) for the following values of the inter-waveguide spacing, $\Delta z = 0$ μm (curve 1), $\Delta z = 1$ μm (curve 2), and $\Delta z = 3$ μm (curve 3); and (b) for the following off-axis displacements with respect to the output waveguide optical axis, $\Delta x = 0$ μm (curve 1), $\Delta x = -0.5$ μm (curve 2), $\Delta x = +0.5$ μm (curve 3), $\Delta x = -1$ μm (curve 4), and $\Delta x = +1$ μm (curve 5).

maximal focal intensity was shown to be replaced by the minimal intensity when the light wavelength changed by 125 nm. This value is 10 times larger than in our case. It has been shown [229] that changing the hole diameters in the PhC waveguide by 20 nm leads to a 50-nm shift of the central wavelength (1550 nm). In our case, the PhC lens (Figure 3.16) consists of holes of three different diameters (160 nm, 180 nm, and 200 nm). Therefore, according to [229], the local intensity maxima (Figure 3.19) are expected to shift by 50–100 nm in our experiment.

When the inter-waveguide spacing is increased to $\Delta z = 1$ μm (Figure 3.19(a), curve 2), the transmission spectrum preserves its structure on average, but the local peaks are reduced and shifted toward the "red" spectrum region. For still larger inter-waveguide spacing of $\Delta z = 3$ μm (Figure 3.19(a), curve 3) not only do the local peaks become lower but acquire the "blue" shift as well. The "red" shift is about 10 nm, and the "blue" shift is also about −10 nm (for the peak near the central wavelength of 1.55 μm). As seen from Figure 3.19(b), when the output waveguide is displaced by 1 μm from the optical axis (curves 4 and 5) the output intensity is 8 times decreased (wavelength 1.55 μm). This means that the focal spot size produced by the PhC lens is less than 1 μm.

In this section, we have fabricated a silicon-on-insulator 2D photonic-crystal lens of size 3×4 μm at the output of a planar waveguide of width 4.5 μm for coupling with a planar waveguide of width 1 μm, with its input near the focus of the lens. A 1-μm off-axis displacement of the narrow waveguide resulted in an eightfold decrease in the output light intensity, implying that the focal spot size at the output of the in-Si lens has been less than 1 μm ($\lambda = 1.55$ μm). The simulation has shown that the photonics device proposed offers a maximal transmittance at a wavelength of 1.55 μm, providing the coupling efficiency between the waveguides of 73%. The measured transmission spectrum has been shown to have four local peaks in the range 1.50–1.60 μm: 1535 nm, 1550 nm, 1565 nm, and 1590 nm; these peaks become lower, being displaced toward the "red" or "blue" region with increasing inter-waveguide spacing. The focal spot size of the PhC lens in the air calculated at half-intensity level is FWHM $= 0.32\lambda$, where λ is the wavelength, which is less than in References [171, 215, 217, 220, 221, 229].

3.4   SUPERRESOLUTION MECHANISM IN A PLANAR HYPERBOLIC SECANT LENS

The combined use of a DOE and far-field focusing refractive optics makes it possible to achieve the superresolution, e.g. attaining the focal spot size at half-maximum intensity of FWHM $= 0.44\lambda$ [69], where λ is the wavelength of light in vacuum, thus beating the diffraction limit of FWHM $= 0.51\lambda$. Note, however, that the side lobes of the resulting diffraction pattern will amount to 20% of the maximum intensity of the focal spot. Although it is possible to obtain a still smaller focal spot size for the far-field diffraction pattern, the proportion of light energy directed to the sidelobes will accordingly increase and can eventually become comparable with or equal to the focal spot intensity [111]. Note that several diffraction limits of resolution are known in optics: the Rayleigh resolution limit of 0.61λ/NA [230], the Houston limit of 0.5λ/NA, and the Sparrow limit of 0.475λ/NA [231], where NA is the numerical aperture of the focusing system. For the purpose of this study, we shall define the resolution limit for the 3D fields as the half-maximum intensity of the squared Airy function (0.51λ/NA) and for the 2D fields as the half-maximum intensity of the *sinc*-function (0.44λ/NA).

To overcome the resolution limit without the increase of the sidelobes the optical element should be put closer to the light source. This domain of optics is dealt with by near-field microscopy [7]. If we eliminate from consideration metallic surfaces and surface plasmons [50], which enable attaining the superresolution of $\lambda/50$ due to a larger real part of the metal dielectric permittivity, and stay within the framework of refractive and gradient optics, the enhanced resolution can be achieved with the aid of near-field lenses, such as a solid-immersion lens (SIL) [232], a numerical aperture immersion lens [38, 233, 234], and a nano solid-immersion lens (nSIL) [40, 230].

Using a combination of the SIL (for wavelength $\lambda = 633$ nm) and a glass hemisphere LASFN9 of radius 5 mm and refractive index $n = 1.845$, the superresolution of FWHM $= 190$ nm $= 0.298\lambda$ was achieved experimentally [232]. Using an Si hemisphere and a NAIL of radius 1.6 mm, the superresolution of FWHM $= 250$ nm $= 0.23\lambda$ was experimentally achieved (for Si, the theoretical diffraction limit is FWHM $= 0.147\lambda$ at $n = 3.4$ and $\lambda = 1$ μm) [38]. In a later work [233], the achievement of the superresolution of FWHM $= 145$ nm $= 0.11\lambda$ ($\lambda = 1.3$ μm) by means of the annular aperture and an Si NAIL was reported. If the lens is illuminated by an annular beam, the focal spot is formed by the Bessel beam, for which the resolution limit is FWHM $= 0.36\lambda$/NA. It was experimentally demonstrated [230] that using the near-field optics (nSIL) and a glass hemisphere of radius 1–2 μm ($n = 1.6$) a focal spot of diameter FWHM $= 126$ nm $= 0.235\lambda$ ($\lambda = 532$ nm) could be obtained [40]. In Reference [104] with use of helical phase plate and SIL with numerical aperture NA $= 1.7$ a focal spot has been calculated with diameter of FWHM $= 0.31\lambda$. It should be noted that the near-field refractive optics is able to enhance the numerical aperture of the already converging beam, whereas to focus a beam propagating from the source requires the use of an additional optical component.

Near-field focusing can be done with a single optical microelement. For example, in Reference [123] a microdisk with 400-nm height and 10-μm diameter was used to experimentally obtain a focal spot of diameter FWHM = 0.86λ. In References [235] and [236] experiments with planar gradient Luneburg microlenses were described, although the focal spot diameters in those experiments were not record-breaking. Obtaining an image with a resolution of FWHM = 0.43λ with use of a planar photonic crystal was reported in Reference [237]. Also, the near-field image can be obtained with use of a superlens with FWHM = 0.28λ [238] and hyperlens with FWHM = 0.38λ [239]. In Reference [240] it was shown via numerical simulation that a spiral plasmonic lens with an 8-μm diameter allows focusing the plasmons into a spot with FWHM = 0.35λ. In Reference [241] a 3D plasmonic superlens with sinusoidal grating was designed. It allows resolving two lines (each 0.05λ wide) found 0.2λ apart. In References [109, 110] the focal spots were experimentally obtained on the nib of dielectric and metallic microcones.

Superoscillations-based imaging and focusing of light in the far field were considered in References [111] (FWHM = 0.17λ) and [242] (FWHM = 0.36λ). In Reference [101] a focal spot with FWHM = 0.58λ = 0.39λ/NA was experimentally generated at a 1-μm distance from a binary microaxicon. In Reference [106] the radially polarized light was focused into a spot with FWHM = 0.33λ with the help of an annular diaphragm and a micro-objective with numerical aperture NA = 1.4.

There are also works where the superresolution is achieved by the refinement of traditional optical elements and systems. In Reference [243] with use of cardioid annular condenser (NA = 1.4) a Richardson slide pattern image was obtained in coherent light with 90-nm resolution (0.2λ). In Reference [244] by using solid-immersion imaging interferometric nanoscopy a possibility was shown to achieve a resolution equal to the diffraction limit in medium FWHM = $\lambda/(2n)$, where n is the medium refractive index.

From the review above it is seen that the best resolution has been achieved with use of SIL microscopy (FWHM = 0.11λ) [233] and SOL (superoscillation lens) microscopy (FWHM = 0.17λ) [111].

In this section we design a dielectric planar binary near-field lens which allows one to obtain a focal spot with FWHM = 0.102λ. We also describe the mechanism by which the superresolution is achieved using the near-field optics, including a gradient-index planar lens. We show that inhomogeneous evanescent waves from a point source (first-type surface waves, $k < k_x < nk$, where k and k_x are the wave numbers in a vacuum and in the x-projection of the wavevector, respectively, and n is the refractive index of the medium) are partially tunneled into the (lens) medium and transformed into the medium modes, thus contributing to the source image alongside with the conventional propagating waves ($0 < k_x < k$). In the far-field optics, the first-type waves fail to achieve the image plane. The second-type surface waves ($k_x > nk$) transformed into the medium surface waves are propagating along the lens input plane, so that only their exponentially damped tails achieve the lens output (image) surface. We derive an integral representation of the TE-wave in a 2D medium propagating from an external light source (with the straight-line medium interface) and a similar representation of the light field behind a plane-parallel plate. The comparison of the numerical apertures of the near-field refractive lenses (SIL, NAIL) and a planar hyperbolic secant lens shows them to have similar values, with the difference being equal to 5% for Si-based elements. The simulation using the FullWAVE software shows that by combining the gradient-index hyperbolic secant lens with a subwavelength diffraction grating or replacing it by a binary analog the focal spot size can, respectively, be made 10% and 20% smaller than the diffraction limit in the given medium.

3.4.1 TUNNELING OF INHOMOGENEOUS WAVES FROM SOURCE INTO MEDIUM

For the 2D case, the electric field strength of a monochromatic TE-wave at distance z from the initial plane takes the form:

$$E_1(x,z) = \frac{k}{2\pi} \int_{-\infty}^{+\infty} \int_{-\infty}^{+\infty} E_0(x', z = 0)$$

$$\times \exp\left[-ik\xi(x'-x) + ikz\sqrt{1-\xi^2}\right] dx' d\xi. \tag{3.41}$$

A point source in the initial plane:

$$E_0(x, z = 0) = E_0 \delta\left(\frac{kx}{2\pi}\right), \tag{3.42}$$

where $\delta(x)$ is the Dirac delta-function, produces at distance z the field amplitude equal to the sum of plane waves and inhomogeneous evanescent waves:

$$E_1(x,z) = E_0 \int_{-\infty}^{\infty} \exp\left[-ik\xi x + ikz\sqrt{1-\xi^2}\right] d\xi. \tag{3.43}$$

Note that since the Hankel function of zero-order and first kind is given by [245]

$$H_0^1\left(k\sqrt{x^2+z^2}\right) = \frac{1}{\pi} \int_{-\infty}^{\infty} \frac{\exp\left[-ik\xi x + ikz\sqrt{1-\xi^2}\right]}{\sqrt{1-\xi^2}} d\xi, \tag{3.44}$$

Equation (3.43) can be expressed through the derivative of the Hankel function:

$$E_1(x,z) = -\frac{i\pi}{k} \frac{\partial}{\partial z} H_0^1\left(k\sqrt{x^2+z^2}\right). \tag{3.45}$$

If we put a medium interface at distance z from the source, the light will pass to a medium with refractive index n. Then, the E-vector amplitude at distance z from the source will be equal to

$$E_2(x,z) = E_0 \int_{-\infty}^{\infty} T_1(\xi) \exp\left[-ik\xi x + ikz\sqrt{n^2 - \xi^2}\right] d\xi, \tag{3.46}$$

where

$$
T_1(\xi) = \begin{cases}
\dfrac{2\sqrt{1-\xi^2}}{\sqrt{1-\xi^2}+\sqrt{n^2-\xi^2}}, & 0<|\xi|<1, \\[3ex]
\dfrac{2i\sqrt{\xi^2-1}}{i\sqrt{\xi^2-1}+\sqrt{n^2-\xi^2}}, & 1<|\xi|<n, \\[3ex]
\dfrac{2\sqrt{\xi^2-1}}{\sqrt{\xi^2-1}+\sqrt{\xi^2-n^2}}, & |\xi|>n.
\end{cases}
\tag{3.47}
$$

The values of $T_1(\xi)$ in Equation (3.47) describe the coefficients derived from Fresnel formulae for three different cases: conversion of the propagating plane wave into a propagating plane wave; conversion of the evanescent plane wave into a propagating plane wave; and conversion of the evanescent plane wave into an evanescent plane wave in the medium.

Actually, from Equation (3.47) it is seen that the propagating waves from a point source in the medium with $n=1$ and $0<|\xi|<1$, where $\xi=k_x/k$, entering the second medium at angles $0<\theta<\theta_1$, where $\theta_1=\arcsin(1/n)$, will then propagate in the medium with $n>1$ (Figure 3.20). The first-type surface waves from the source that have the wave number projection in the interval $1<|\xi|<n$ will enter the second medium at angles $\theta_2(\xi)=\arcsin(\xi/n)$ found in the range $\theta_1<\theta_2(\xi)<\pi/2$, because the maximal angle $\theta_2(\xi)$ equals $\pi/2$ at $\xi=n$. Being converted from surface evanescent into propagating, these waves will further propagate in the medium with $n>1$. The second-type surface waves from the source with $|\xi|>n$ will represent the surface waves of the medium, propagating along the medium interface.

3.4.2 NUMERICAL APERTURE OF A HYPERBOLIC SECANT LENS

Let us consider a 2D hyperbolic secant (HS) lens (or ML) whose refractive index is given by [171, 246]

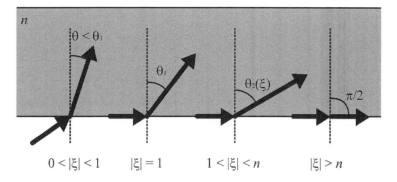

FIGURE 3.20 Different types of waves tunneling into media with index n.

$$n(x) = \frac{n}{\text{ch}\left(\dfrac{\pi x}{2L}\right)}, \tag{3.48}$$

where n is the refractive index on the lens axis, L is the lens length, and x is the coordinate in the transverse plane. If the lens (3.48) is of length L, all the rays parallel to optical axis intersect in the focal point at a distance of L. If the lens is of length $2L$, the on-axis point source is imaged at a distance of $2L$.

The numerical aperture of the HS lens can be derived from the ray equation in a gradient-index medium: $n(x)\cos\theta(x) = \text{const}$, where θ is the angle between the tangent to the ray and the optical z-axis. Assume that the ray is incident on the lens in parallel with and at distance $x = R$ from the optical axis, where R is the HS lens radius, which can be derived from the condition $n(R) = 1$: $R = \text{arcch}(n)$. Then, the ray equation takes the form: $n\cos\theta_0 = n(R)\cos\theta(R) = 1$, where θ_0 is the angle between the tangent to the ray and the optical z-axis at the point of intersection of the ray and the optical axis. Whence it follows that $\cos\theta_0 = 1/n$, i.e. the HS lens numerical aperture is $\text{NA} = n\sin\theta_0 = (n^2 - 1)^{1/2}$, where $\theta_0 = \arcsin[(n^2-1)^{1/2}/n]$. Then, the plane waves propagating at angles θ_1 and θ_2 smaller than θ_0 will contribute to the focal spot at the HS lens output. Let us find the maximal ξ_{max} for the light waves contributing to the HS-lens-aided focal spot. From the equality $\theta_2(\xi_{max}) = \theta_0$, we obtain $\arcsin(\xi/n) = \arcsin\left[\left(n^2 - 1\right)^{1/2}/n\right]$. From this equality it follows that $\xi_{max} = (n^2 - 1)^{1/2}$. For silicon and wavelength $\lambda = 1550$ nm, we obtain $\xi_{max} = 3.32$, because $n = 3.47$. Then, the minimal focal spot diameter at half-maximum intensity at the output of the planar HS lens is ($n = 3.47$):

$$\text{FWHM} = 0.44\frac{\lambda}{\text{NA}} = 0.44\frac{\lambda}{\sqrt{n^2 - 1}} = 0.44\frac{\lambda}{\xi_{max}} = 0.133\lambda. \tag{3.49}$$

The theoretical resolution limit that can be derived using near-field planar optics, such as SILs [232] and NAILs [38], is $n = 3.47$:

$$\text{FWHM} = 0.44\frac{\lambda}{n} = 0.127\lambda. \tag{3.50}$$

Equation (3.50) stems from the fact that the numerical aperture of the SIL and NAIL, $\text{NA}_{\text{SIL}} = n\sin\theta \leq \text{NA}_{\text{NAIL}} = (n^2 - \cos^2\theta)^{1/2}$, tends to $\text{NA}_{max} = n$ in the limit $(\theta \to \pi/2)$. The diffraction-limited focal spot of Equation (3.49) is only 5% smaller than the focal spot produced by the HS lens, Equation (3.50). Let us estimate the maximal angle at which the rays are propagating in the HS lens. The propagating plane waves from the source with the relative wave number projection found in the interval $0 < |\xi| < 1$ are transformed in the isotropic medium ($n = 3.47$) into the propagating waves travelling at angles in the interval $0 < \theta < \theta_1 = \arcsin(1/n) \cong 17°$, with the max-

imal angle (to the optical axis) at which the rays can propagate in the HS lens being
$\theta_0 = \arcsin[(n^2 - 1)^{1/2}/n] \approx 74°$.

Note that if a SIL or a NAIL is illuminated by an annular light beam the diffraction pattern in the focal plane will be described not by the Airy function (or the *sinc*-function for the 2D case) but by the zero-order Bessel function. Therefore, the theoretical resolution limit in the medium will be given by

$$\text{FWHM} = 0.36\frac{\lambda}{n} = 0.104\lambda. \tag{3.51}$$

3.4.3 Three Types of Waves Propagating in a Hyperbolic Secant Lens

Note, however, that not all waves entering the HS lens will go outside. Assume that waves from the source enter a bounded medium (with respect to the optical z-axis) in the form of a plane-parallel plate of thickness d. After passing the plate, the output waves will be described by the expression:

$$E_3(x,z) = E_0 \int_{-\infty}^{\infty} T_2(\zeta)\exp\left[-ik\zeta x + ikz\sqrt{1 - \zeta^2}\right]d\zeta, \tag{3.52}$$

where

$$T_2(\zeta) = \begin{cases} \dfrac{2\sqrt{(1 - \zeta^2)(n^2 - \zeta^2)}}{2\sqrt{(1 - \zeta^2)(n^2 - \zeta^2)}\cos A - i(n^2 + 1 - 2\zeta^2)\sin A}, & 0 < |\zeta| < 1, \\[4mm] \dfrac{2\sqrt{(\zeta^2 - 1)(n^2 - \zeta^2)}}{2\sqrt{(\zeta^2 - 1)(n^2 - \zeta^2)}\cos A - (n^2 + 1 - 2\zeta^2)\sin A}, & 1 < |\zeta| < n, \\[4mm] \dfrac{2\sqrt{(\zeta^2 - 1)(\zeta^2 - n^2)}}{2\sqrt{(\zeta^2 - 1)(\zeta^2 - n^2)}\,\text{ch}\,B - (n^2 + 1 - 2\zeta^2)\,\text{sh}\,B}, & |\zeta| > n, \end{cases} \tag{3.53}$$

$A = iB = kd(n^2 - \zeta^2)^{1/2}$. Similarly to T_1 in Equation (3.47), the magnitudes $T_2(\zeta)$ represent the coefficients derived from the Fresnel functions for three different cases. Equation (3.53) suggests that after passing through the plate, the plane waves propagating from the source ($0 < |\zeta| < 1$) will propagate behind the plate at the same angles (since the medium before and after the plate is air, $n = 1$). The first-type surface waves ($1 < |\zeta| < n$) are converted into the modes within the plane-parallel plate and again into the surface waves on the opposite (relative to the source) plate surface. Thus, these waves will not propagate in the space behind the plate. With the second-type surface waves ($|\zeta| > n$) being converted into the surface waves on the source-facing side of the plate, only their exponentially dumped "tails" reach the opposite side of the plate. Thus, the central rays from the source that propagate in the HS lens at angles to the optical axis smaller than $\theta_1 = \arcsin(1/n) \cong 17°$ for $n = 3.47$ can pass through and further propagate behind the lens with the length of $2L$ in Equation (3.48). The first-type surface waves from the source propagate in the HS lens as in a ring cavity, not leaving the lens. The second-type surface waves from the source remain as the HS-lens surface waves, experiencing partial scattering from the acute corners of the lens, because the lens is different from the plane-parallel plate, being limited by the transverse coordinates.

Thus, if we assume that the HS-lens-aided focal spot is generated only by the propagating waves with a maximal tilt of θ_1, the expected focal spot size is ($n = 3.47$; $\theta_1 = 17^0$) FWHM $= 0.44\lambda/(n\sin\theta_1) = 0.43\lambda$. However, taking into consideration the essential contribution of the first-type surface waves propagating in the HS lens with a maximal tilt to the optical axis of $\theta_0 = 74^0$ results in the focal spot diameter of FWHM $= 0.44\lambda/(n\sin\theta_0) = 0.132\lambda$. This figure is in good agreement with the simulation results [246] and with Equation (3.49), which, in fact, leads to the same result, although in a different way.

Figure 3.21 presents a characteristic plot of the focal spot size against the plane-wave spectrum width contributing to the focus according to Equation (3.48).

Thus, if the simulation shows that the HS-lens-aided focal spot size is smaller than the diffraction-limited size in the 2D medium (Equation (3.50)), we can infer that the second-type surface waves also contribute to the focal spot.

3.4.4 Numerical Apertures of Near-Field Optics

Below, we conduct the comparison of the NA of the near-field SIL and NAIL lenses with the NA of the HS lens. The comparison results were briefly discussed in the previous subsection. Assume the propagation in free space of a converging light beam with NA $= \sin\theta$, where θ is the maximal angle that the beam rays make with

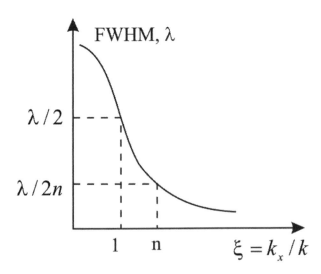

FIGURE 3.21 The point-source image diameter, Equation (3.48), vs. the plane-wave spectrum width, including the evanescent plane waves reaching the image plane.

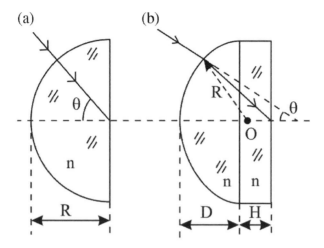

FIGURE 3.22   Diagram of incident and refracted rays for (a) SIL and (b) NAIL.

the optical axis. If such a beam enters a medium with refractive index n and planar interface, its numerical aperture will remain unchanged, $NA = n\sin\beta = \sin\theta$, where β is the maximal angle of the beam rays in the medium. The NA of the initial beam can be enhanced using a SIL, which comprises a hemisphere with refractive index n, with its spherical surface placed in such a manner that the rays are incident normally (Figure 3.22(a)). In this case, with the converging beam focused at the center of the hemisphere, its NA within the sphere will be defined by $NA_{SIL} = n\sin\theta$. Thus, using near-field optics, with the focus found on the hemisphere flat surface, the NA of the initial beam can be n times enhanced or the focal spot diameter n times reduced.

With another near-field lens, the NAIL, it becomes possible to achieve still higher values of the beam NA. In this case, a sphere segment (smaller than a hemisphere) is put in the path of the converging beam. The rays should fall on the spherical surface in such a manner that their angle with the optical axis is smaller than the angle between the axis and the normal drawn to the spherical surface at the point of ray intersection. The optical arrangement should also contain a cylinder of the same radius as that of the sphere and height H, made of the same material. The cylinder should be fitted in such a manner that the rays are focused just on the intersection of its rear plane with the optical axis. Assume that the sphere's radius is R and the segment height is $D < R$, then deriving the cylinder's height from the equation $D + H = R(1 + 1/n)$, we obtain the focal spot at the cylinder output plane (Figure 3.22(b)). It is seen from Figure 3.22(b) that the ray is refracted in the lens and falls on the optical axis at an angle larger than the angle θ of the ray incident on the lens. Considering that the NA of the incident beam on the NAIL is $NA = \sin\theta$, the NA of the beam travelling in the cylinder and converging on the cylinder's output surface is $NA_{NAIL} = (n^2 - \cos^2\theta)^{1/2}$. It can be shown that $NA_{NAIL} \geq NA_{SIL}$. The comparison of the above NA values with the HS-lens $NA = (n^2 - 1)^{1/2} = n\sin\theta_0$ shows that the difference is not large: the difference between the maximal NA of the near-field refractive optics $NA_{NAIL} = NA_{SIL} = n = 3.47$ and the NA of the HS lens $NA = (n^2 - 1)^{1/2} = 3.32$ is as small as 5%. It should be noted, however, that the SIL and NAIL are only able to gather light from the source, converging the first-type surface waves into the propagating waves without focusing. Additional refractive optics should be employed to perform the subsequent focusing. On the contrary, the HS lens is capable of both gathering and focusing light on its output surface.

3.4.5 FOCAL SPOT REDUCTION BY MODULATION OF REFRACTIVE INDEX

In recent papers [247, 248], it was numerically demonstrated that using the subwavelength gratings the surface waves from the source could be converted into the propagating waves, thus achieving the superresolution of $\lambda/20$. This was implemented by means of several stacked diffraction gratings of different subwavelength periods and arbitrary apertures [247] and a metallic subwavelength grating with high permittivity ($\varepsilon = -100$) [248]. However, the use of focusing or imaging optical elements was not considered in the above works.

The simulation using the FullWAVE software (RSoft) implementing the FDTD approach has shown that by combining the HS lens with a subwavelength grating or using a binary HS lens with subwavelength features the focal spot size can be reduced by 10% and 20%.

Figure 3.23(a) shows the (gray level) index profile of the HS lens (3.48) and (b) the intensity distribution at the lens output. The lens was illuminated by a TE-polarized plane wave, i.e. the electric field vector had only one component along the y-axis. The focal spot size in Figure 3.23(b) is FWHM = 191 nm = 0.123λ, $\lambda = 1.55$ µm. This value is somewhat smaller than the diffraction limit in silicon, Equation (3.50), serving to prove that second-type surface wave contribute to the HS-lens-aided focal spot.

Figure 3.24(a) depicts the same HS lens as in Figure 3.23, but in combination with a subwavelength diffraction grating put on the lens top (output) surface, and Figure 3.24(b) shows the intensity distribution at the lens output. The focal spot size

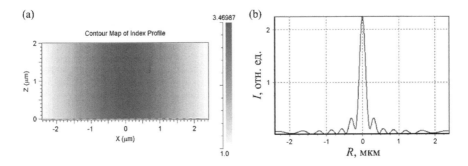

FIGURE 3.23 (a) Refractive index profile in the gradient-index HS lens (on the axis $n = 3.47$), horizontal dimension $2R = 4.8$ µm, vertical $L = 2$ µm. Light propagates vertically. (b) Transverse intensity profile $|E_y|^2$ at the lens output (10 nm apart).

(a)

(b)

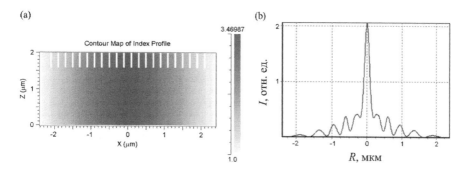

FIGURE 3.24 (a) Refractive index profile in the gradient-index HS lens ($n=3.47$), horizontal dimension $2R=4.8$ μm, vertical $L=2$ μm. On top of the HS lens, there is a diffraction grating of depth 0.4 μm, period 0.2 μm, and groove width 0.05 μm (the grooves filled with air). Light propagates from the bottom upward. (b) Transverse intensity distribution $|E_y|^2$ at the lens output (10 nm apart).

in Figure 3.24(b) is FWHM = 177 nm = 0.114λ. This is 8% smaller than the focal spot size in Figure 3.23(b) and 10% smaller than the diffraction limit of Equation (3.50).

3.4.6 Amount of Energy in the Focal Spot

Let us place in front of the lens (at a distance of $\lambda/16 \approx 97$ nm) a source of the bounded 2.4-μm wide plane wave (i.e. the wave width is half the lens width). Some portion of light will reflect from the front surface of the lens, not contributing to the focal spot. In order to determine how much of the light energy achieves the rear plane it is insufficient to measure the energy flux through it, because some portion of light is reflected from the rear plane and the total energy flux of this portion equals zero, though contributing to the focal spot. Therefore, let us replace the lens with a secant waveguide where all the light will propagate further instead of being reflected from the rear plane of the lens. We find that in this case 74.9% of the source energy fluxes through the surface of lens rear plane.

Now, let us consider an infinite plane wave (i.e. its width exceeds the simulation area) incident on the lens. Shown in Figure 3.25 are distributions of energy density (i.e. intensity $|E_y|^2$, Figure 3.25(a)) and density of energy flux (i.e. Poynting vector, Figure 3.25(b)) calculated in a plane located at a distance of $\lambda/50 \approx 31$ nm beyond the rear plane of the lens. The width of the central peak is FWHM = 192 nm in Figure 3.25(a), or 0.124λ, i.e. slightly below the diffraction limit. In Figure 3.25(b) the width of the central peak is FWHM = 213 nm (0.137λ).

The focal spot size that is smaller than the diffraction limit (0.44λ in the 2D case) can be explained by the existence of surface waves. In order to prove it let us consider the amplitude spectrum in a plane located at a distance of $\lambda/50$ beyond the lens. Shown in Figure 3.26 are normalized spectra of the amplitudes E_y in planes located at $2\lambda/50$ (curve a) and $\lambda/50$ (curve b) beyond the lens rear surface. It is seen from Figure 3.26 that the spectrum has non-zero values in the region $|\xi| \geq 1$, thus suggesting the existence of evanescent waves. Curve b in Figure 3.26 is slightly above

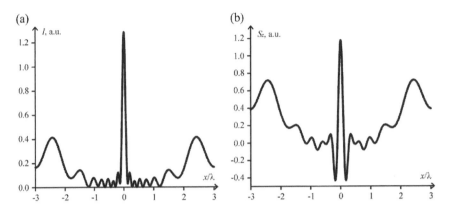

FIGURE 3.25 Time-averaged (500 values of the last 10 periods) density of energy (intensity $|E_y|^2$): (a) and the z-axis projection of energy flux (Poynting vector); and (b) in a plane located at a distance of $\lambda/50$ beyond the lens.

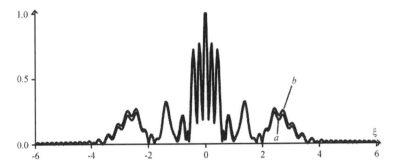

FIGURE 3.26 Spectra of the amplitude E_y in a plane at a distance of (a) $2\lambda/50$ and (b) $\lambda/50$ beyond the lens.

curve a because closer to the lens the evanescent waves have a stronger impact onto the focal spot.

In order to prove that the grating allows increasing the contribution of the evanescent waves into the focal spot, the angular spectrum of plane waves has also been calculated at a distance of $\lambda/50$ from the grating. The other simulation parameters were the same: the grating period was 300 nm, the grating depth was 300 nm and the grating fill factor was 50%. In this case, the focal spot at half-maximal intensity also equals FWMH$=0.114\lambda$. Figure 3.27 shows the resulting angular spectrum.

It is seen from Figure 3.27 that the main contribution to the focal spot comes not from a plane wave that is normally incident onto the lens, but from two tilted plane waves propagating at an angle of arcsin(0.4) $\approx 23°$ with the optical axis. Besides, the contribution of the surface waves increases, which follows from scarcely noticeable maxima in the region $|\xi| \geq 1$.

Measurements at a distance of λ from the grating show that 10% of the total intensity incident onto the lens front surface goes into the zero-th diffraction order.

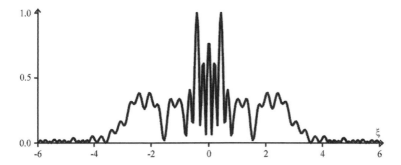

FIGURE 3.27 Angular spectrum of the plane waves of amplitude E_y at a distance of $\lambda/50$ from the grating.

3.4.7 Focusing Light with a Binary HS-Lens

Let us consider a 2D binary lens with the zone sizes such as to produce the effective refractive index described by Equation (3.48). Such a binary lens will be referred to as a binary HS lens.

Index profiles of the gradient-index HS lens and the corresponding binary HS lens are shown in Figure 3.28. The radius of the binary lens is broken down into intervals $[x_m, x_{m+1}]$, $m = 0, 1, 2, \ldots, M$, with a point $x_m < x_m^b < x_{m+1}$ found in each interval, such that on the interval $[x_m, x_m^b]$ the lens material is silicon, and on the interval $[x_m^b, x_{m+1}]$, air. The point x_m^b is chosen such as to approximate the refractive index of Equation (3.27):

$$\int_{x_m}^{x_{m+1}} n(x)\mathrm{d}x = n_0 \left(x_m^b - x_m \right) + 1 \cdot \left(x_{m+1} - x_m^b \right). \tag{3.54}$$

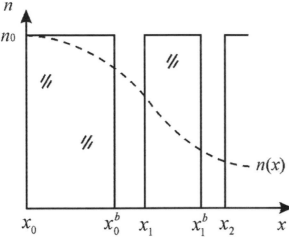

FIGURE 3.28 Refractive index profiles of the gradient-index HS lens (dashed line) and the corresponding binary HS lens (solid line). Shaded area denotes silicon.

Let us note that for the index distribution (3.48) the left part can be computed analytically, so from Equation (3.54) we can derive explicit expression for x_m^b:

$$x_m^b = \frac{n_0 x_m - x_{m+1}}{n_0 - 1} + \frac{1}{n_0 - 1} \frac{4 n_0 L}{\pi}$$

$$\times \left\{ \arctan\left[\exp\left(\frac{\pi x_{m+1}}{2L} \right) \right] - \arctan\left[\exp\left(\frac{\pi x_m}{2L} \right) \right] \right\}. \tag{3.55}$$

Shown in Figure 3.29(a) is a binary HS lens designed using Equation (3.54) on the basis of the lens in Figure 3.23(a), and in Figure 3.29(b) is the intensity distribution in the focal plane at the lens output. The focal spot width in Figure 3.29(b) is FWHM = 159 nm = 0.102λ. This value is smaller, by 17%, than the focal spot size in Figure 3.23(b) and 20% smaller than the diffraction limit of Equation (3.50). Thus, we can infer that in compliance with Figure 3.21, the focal spot at the binary HS lens output (Figure 3.29(b)) is formed with the 20% contribution of second-type surface waves ($k_x > nk$).

3.4.8 IMAGING WITH DOUBLE HS-LENS

Figure 3.30(a) shows the instantaneous pattern of the E-vector in the HS-lens (made of silicon, $n = 3.47$, width $2R = 6$ μm, and length $2L = 4.92$ μm) produced by two point sources (each is a Gaussian beam with 50-nm waist radius and free-space wavelength of $\lambda = 1$ μm) located 300-nm apart and 10-nm away from the lens input surface. The simulation was conducted by the FDTD method with a space sampling of $\lambda/100$ and simulation time of $200\lambda/c$, with c being the speed of light in vacuum. Figure 3.30(b) shows the time-averaged distribution of Poynting vector projection onto the optical z-axis, calculated 10-nm away from the HS-lens output surface. From Figure 3.30(b) the two sources are seen to be resolved and the superresolution value is 0.3λ. Although the point-source image has the width FWHM = 0.12λ, the two

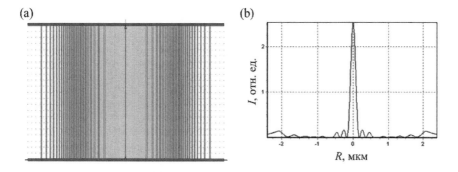

FIGURE 3.29   (a) Binary HS lens ($n = 3.47$), horizontal dimension, $2R = 4.8$ μm, vertical $L = 2$ μm, minimal groove 20 nm (filled with air). Light propagates vertically. (b) Transverse intensity distribution $|E|^2$ at the lens output (10 nm apart).

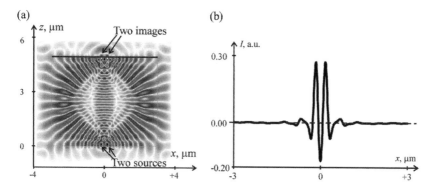

FIGURE 3.30 (a) Instantaneous pattern of the E-vector amplitude of a TE-wave in the HS-lens when two point sources (each is 50 nm wide), located 300 nm apart and 10 nm away from the input plane; (b) time-averaged distribution of Poynting vector projection onto the optical axis, calculated 10 nm away from the waveguide output plane (top horizontal line, the intensity is shown in the arbitrary units).

point sources are reliably resolved only when separated by a distance of 0.3λ. This is because the images of the two sources interfere with each other.

Note that such a gradient waveguide can be implemented as a photonic-crystal device [175].

Now we consider the energy flux (z-projection of the Poynting vector) instead of the intensity. In this case, two light sources can be resolved even if they are closer to each other. Figure 3.31 shows that two sources, separated by a distance of 0.25λ, are reliably resolved via the energy flux and scarcely resolved via the intensity. Sources separated by a distance of 0.15λ can be resolved only via the energy flux, as it can be seen from Figure 3.32. Light sources separated by a distance of 0.12λ can be resolved neither via the energy flux nor via the intensity (Figure 3.33).

In the case of the TE-wave, the energy flux can be measured in the following way. It is related with the intensity as follows:

$$S_z = \mathrm{Re}\left(E_y H_x *\right) = \frac{I(x,z)}{\omega \mu_0 \mu} \frac{\partial \psi}{\partial z},\qquad(3.56)$$

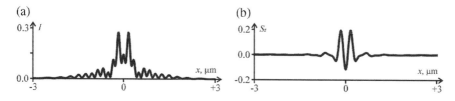

FIGURE 3.31 (a) Time-averaged intensity and (b) energy flux calculated 10 nm away from the secant waveguide output plane for two light sources with center-to-center distance of 250 nm.

(a)   (b)

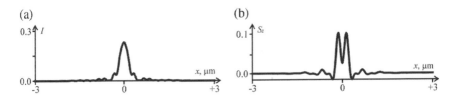

FIGURE 3.32   (a) Time-averaged intensity and (b) energy flux calculated at a distance of 10 nm from secant waveguide output plane for two light sources with center-to-center distance of 150 nm.

where ω is the frequency, μ is the magnetic permittivity, $I(x,z) = |E_y|^2$ is the intensity and $\psi(x,z)$ is the phase.

We have obtained the following results. The integral representation of the amplitude of the TE-wave propagating from a point source and behind a plane-parallel plate is given as a sum of three terms that describe three types of light waves: propagating waves and (inhomogeneous evanescent) surface waves of the first and second types. The comparison of the NAs of the near-field refractive lenses (SIL, NAIL) and a planar HS lens has shown them to be similar, with the difference for silicon being as small as 5%. Simulation using the FullWAVE software has shown that by combining the gradient-index HS lens with a subwavelength diffraction grating or replacing the lens with it binary analog the focal spot size can be reduced by 10% and 20% when compared with the diffraction-limited size in the medium. The HS-lens of length 2L has been shown to resolve two point sources with a center-to-center distance of 0.15λ via measuring the energy flux. A planar binary silicon microlens has been calculated that generates a near-surface focal spot of diameter FWHM = 0.102λ.

3.5   HYPERBOLIC SECANT SLIT LENS FOR SUBWAVELENGTH FOCUSING OF LIGHT

The tight focusing of light near the interface surface has been performed with the aid of nanoslits [249, 250]. In Reference [249] a focal spot of size ~λ/2 was generated using a "subwavelength generator" consisting of two 80-nm slits. Lenses with nanoscale slits made in gold were reported in Reference [250], in which the incident light having passed through 20-nm slits was focused into a focal spot of about a

(a)   (b)

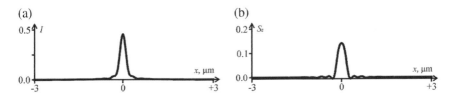

FIGURE 3.33   (a) Time-averaged intensity and (b) energy flux calculated at a distance of 10 nm from secant waveguide output plane for two light sources with center-to-center distance of 120 nm.

half-wavelength in width. In Reference [251], light confinement in metamaterial was discussed, with the waves passing through two or more slits. Having passed through 20-nm slits in a screen, the light waves underwent diffraction in the metamaterial, forming a focal spot of the full width at half-maximum intensity FWHM$=\lambda/17$. However, the authors have failed to explain in which way, being so sharp, the focal spot could be output from the metamaterial. A nanoslit tens of nanometers wide can be used for confining and guiding light waves, similarly to a waveguide [252]. The nanoslits have been used in this way in near-field microscopy [36]. The light scattered by the surface of the specimen under study passes through a slit in the cantilever and arrives to a high NA objective lens.

On the other hand, it is possible to perform the tight focusing of light using graded-index microlenses, in particular, planar photonic-crystal lenses [36, 253]. In Reference [254] the focusing into a spot of size about $\lambda/4$ was performed with the aid of an array of holes in a taper. The diffraction of light from a nanoslit in metamaterial was analyzed in References [255, 256]. The obtaining of focal spots of size FWHM$=\lambda/10$ [255] and FWHM$=\lambda/5$ [256] was reported. The lens utilized in References [255, 256] contained dozens of 5-nm-thick silver layers on top of 10-nm-thick SiC. Note, however, that such a lens is difficult to realize in practice.

In this section, we combine the advantages of using a slit of a width of several dozen nanometers for light confinement [36] and a gradient lens for the sharp focusing of light [257]. The FDTD simulation has shown that a planar binary microlens in silicon with a 50-nm-wide slit can generate a near-surface focal spot of size FWHM $= \lambda/23$ with a 44% energy efficiency. This value is smaller than those reported in the above-quoted works. With the focus occurring near the lens surface, it may find use in various nanophotonics applications. The nanoslit placed near the lens output surface in the focus region serves a dual function. The portion of TM-waves that propagate at small angles to the optical axis is confined by the nanoslit so that the nanoslit size defines the focal spot size. In the meantime, the portion of radiation that comes to focus at larger-than-critical angles relative to the optical axis, generating a surface wave, are scattered by the nanoslit and output from the lens, thus increasing the focusing efficiency.

Figure 3.34 depicts a scheme of a planar graded-index slit lens. We consider a hyperbolic secant lens whose refractive index depends on the transverse coordinate x as follows [257]: $n(x) = n_0 \text{ch}^{-1}[\pi x / (2H)]$, where n_0 is the refractive index on the axis and H is the lens length.

The slit in the planar HS lens of width W_1 is located on the optical axis, stretching as far as the lens output plane. The slit can be extended through the entire lens ($W_2 = H$) or found in the rear section of the lens ($W_2 < H$). The light was propagated through the lens by the FDTD simulation implemented using the commercial software FullWAVE (by RSoft) and the software MEEP. The simulation parameters were as follows: computation domain, 8 μm × 4 μm, incidence wavelength, 1.55 μm, and computation domain step on the x- and y-axes, $\lambda/500$. Figure 3.35 depicts the intensity distribution profile in the focal plane of a slitless lens 10 nm after the lens output plane. The simulation parameters were $H = 1.95$ μm, $L = 4.8$ μm, the refractive index on the optical axis, $n_0 = 3.47$ (silicon), for a plane incident TE-wave. The lens

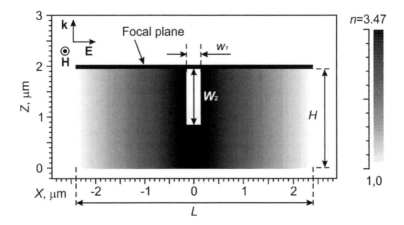

FIGURE 3.34 Gray-level refractive index distribution of an HS slit lens: k, wave number, $\mathbf{E}$ and $\mathbf{H}$, the electric and magnetic field strengths.

length H was fitted to obtain the minimal focal spot. For the TM-wave, the slitless graded-index lens would produce a focal spot 4 times wider.

The focal spot in Figure 3.35 has a size of FWHM $=0.117\lambda$. This value is smaller than the diffraction-limited spot size for silicon: FWHM $= 0.44\lambda/3.47 = 0.127\lambda$. The intensity peak in Figure 3.35 is 6 times larger than the incident intensity. We can show that using a slit on the lens optical axis with $n=1$ and the incident TM-wave, it is possible to obtain a sharper focal spot. Figure 3.36(a) shows the focal spot width FWHM as a function of the slit width W_1. The slit length is assumed to be equal to the lens length: $W_2=H=2.2$ µm, the remaining simulation parameters being the same as in Figure 3.35. Note that the simulation results obtained using the FullWAVE software are in good agreement with a similar simulation of the lens performance conducted using the MEEP software.

From Figure 3.36(a) the focal spot width FWHM is seen to be a linear function of the slit width W_1. The focal spot is slightly wider than the slit. Figure 3.36(b) shows the intensity profile in the lens focus at $W_1=50$ nm.

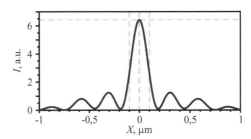

FIGURE 3.35 Intensity profile in the focal plane $|E_y|^2$ generated by a slitless lens with the incident TE-wave.

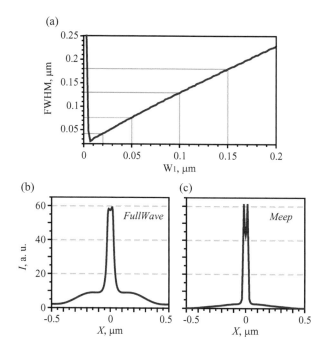

FIGURE 3.36 (a) The focal spot width FWHM as a function of the slit width W_1; and the intensity profile $I = |E_x|^2 + |E_z|^2$ in the focal plane (10 nm after the lens) at $W_1 = 50$ nm, simulated using (b) FullWAVE and (c) MEEP software.

The light confinement in the nanoslit can be explained in a similar way to generating a fundamental TM-mode in a planar slitted waveguide [252]. In this case, the field within the slit is described by the relation $E_x(x,z) = \exp(i\beta x)\mathrm{ch}\left[\left(\beta^2 - k^2\right)^{1/2}|x|\right]$, $|x| < a$,

where a is the nanoslit boundary coordinate, $2a = W_1$, β is the propagation constant of the fundamental TM-mode, and k is the wave number in vacuum. Figure 3.36(b) shows that there is an intensity dip in the focal spot center defined by a hyperbolic cosine. The field amplitude $E_x(x,z)$ undergoes a breakdown at the slit boundary $|x| = a$, so that near the slit at $|x| > a$ the amplitude is n^2 times decreased. Considering that Figure 3.36(b) shows the intensity profile calculated at distance 10 nm after the lens, the focal spot intensity peak is wider if compared to that produced by the lens. Besides, the intensity peak in the focus center is smaller compared to that observed in the slit.

The diffraction efficiency (DE) of focusing η_D depends on the slit width W_1. The DE is maximal at $W_1 \approx 40$ nm. With increasing slit width, the DE η_D is decreasing (Figure 3.37(a)) because the light intensity in the focus is decreasing with increasing W_1 (Figure 3.37(b)). The DE η_D was calculated as the ratio of the energy contained in the central lobe of the focal diffraction pattern (on the interval -75 nm $< x < 75$ nm) to the entire energy reaching the output plane of width L. It can also be seen from Figure 3.36(a) that for $W_1 < 5$ nm the focal spot size starts to grow. The minimal focal spot size FWHM is achieved at $W_1 = 5$ nm, being equal to FWHM $= \lambda/119$ at distance

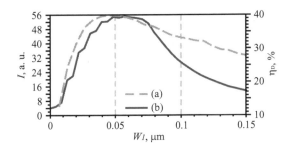

FIGURE 3.37 The (a) DE η_D and (b) intensity at the lens focus (10 nm away from the lens) as a function of the slit width W_1, at $W_2 = H$.

10 nm after the lens. For comparison, the focal spot in Figure 3.35 has FWHM $= \lambda/8$ and the DE $\eta_D = 60\%$.

From Figure 3.37 it is seen that the maximal DE is $\eta_D = 39.9\%$, whereas the focal spot size is FWHM $= \lambda/28$ (Figure 3.36(a)). In Reference [252] it was demonstrated that in the silicon waveguide less than 30% of the waveguide mode optical energy was confined in nanoslits of an arbitrary size (ranging from 10 nm to 150 nm).

Figure 3.38 depicts the DE η_D (a) and the focal intensity I (b) as a function of the slit length W_2, assuming the slit width $W_1 = 50$ nm. In calculating η_D, the focal spot size was again considered up to the nearest sidelobes, -75 nm $< x < 75$ nm. From Figure 3.38 it is seen that the maximal values of the DE $\eta_D = 43.4\%$ and intensity I are observed when the slit length is chosen so as to provide a $\lambda/2$ phase delay, i.e. at

$$W_2 = \lambda \left[2\left(n_0 - 1 \right) \right]^{-1} = 0.314 \, \mu m.$$ We note that the optical intensity in the lens focus is

about 20% larger than for $W_2 = H$.

Considering that the fabrication of a graded-index lens using the state-of-the-art nanolithography techniques presents a challenge, the above-described sharp focusing of the TM-wave can be implemented using a photonic-crystal lens with a slit on the optical axis, which would be analogous to the graded-index lens in terms of the average refractive index distribution. Figure 3.39(a) depicts the refractive index distribution in the xz-plane of a photonic-crystal lens similar to the graded-index lens of Figure 3.34.

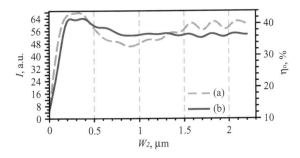

FIGURE 3.38 The (a) DE η_D and (b) intensity I in the lens focus as a function of the slit length W_2, at $W_1 = 50$ nm.

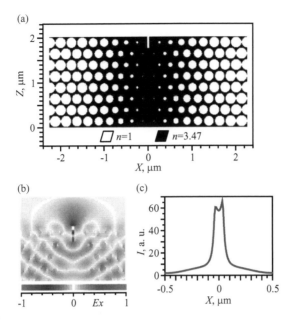

FIGURE 3.39   (a) Refractive index distribution in a photonic-crystal lens with a slit; (b) the instantaneous distribution of the field E_x in such a lens at instance $cT = 32$ µm; and (c) intensity profile in the transverse focal plane 10 nm after the lens.

The lens in Figure 3.39a consists of 8 rows of holes on the z-axis and 20 rows of holes on the x-axis arranged in a staggered order. The holes' centers are arranged with a period of 266 nm on the x-axis and 250 nm on the y-axis. The minimal hole diameter is 30 nm, maximal is 250 nm, the lens length is 2 µm, width is 4.8 µm, the lens material refractive index is $n = 3.47$, $W_1 = 50$ nm, and $W_2 = 0.25$ µm. For such a lens, the value of η_D depends on the slit length W_2 at $W_1 = 50$ nm in a similar way to the graded-index lens, as seen from Figure 3.39(b). The maximal DE is $\eta_D = 44.3\%$ at $W_2 = 250$ nm. In this case, the focal spot size is FWHM $= \lambda/23$, with the focal spot intensity being 60 times higher than the incident wave intensity. If such a lens is illuminated by a Gaussian beam with a waist radius of $\sigma = 2.4$ µm, the focal spot shape is preserved but the focal spot intensity is decreased by 1.7 times.

Summing up, using the 2D-FDTD simulation, we have shown that if a planar graded-index hyperbolic secant microlens of size 5 µm × 2 µm in silicon that contains a nanoslit of width 50 nm and length ~300 nm on the optical axis is illuminated by a plane TM-wave, a focal spot of width FWHM $= \lambda/28$ is generated at the lens output with the DE $\eta_D = 43\%$.

3.6   PROPAGATION OF HYPERGEOMETRIC LASER BEAMS IN A MEDIUM WITH THE PARABOLIC REFRACTIVE INDEX

Paraxial hypergeometric (HyG) modes were proposed for the first time in 2007 [258]. Hypergeometric Gaussian laser beams generated on the basis of the HyG modes followed soon after [259]. A large family of HyG beams which includes both particular cases of the HyG modes [258] and the HyG Gaussian beams [259] was discussed in Reference [86]. The HyG laser beams were experimentally generated using diffractive optical elements [260] and computer-synthesized holograms [261]. Recently analytical expressions that describe the propagation of the HyG modes in a hyperbolic-index medium [262] and a uniaxial crystal [263] have been derived.

On the other hand, a quadratic-index profile is widely used to implement graded-index waveguides, fibers, and lenses [264]. A new technique to fabricate the planar graded-index waveguide in the form of subwavelength binary grating was proposed in Reference [264]. The Gaussian beam [265], the Hermite–Gaussian beam [266], the Laguerre–Gaussian beam [267], the Ince–Gaussian beam [268] and the Airy–Gaussian beam [269, 270] are the exact solutions of the paraxial wave equation in a parabolic-index (PI) fiber.

In this section, a relationship that describes the propagation of the HyG beams in a three-dimensional (3D) parabolic-index waveguide is proposed. We have shown that the HyG beams are the exact solution of the paraxial wave equation in a parabolic-index fiber. The light field of the HyG beams is shown to be periodically varying, with the Fourier spectrum pattern emerging every half-period. According to geometric optics laws, the beams have been known to be periodically self-reproducing in a parabolic-index medium [271]. In this work, we demonstrate that an analogous effect occurs for the paraxial HyG beams. As a particular case, we study the propagation of a circular Gaussian optical vortex in the parabolic-index medium. A relationship to describe the maximum intensity on the optical ring is deduced.

Also, a new family of solutions of the non-paraxial Helmholtz equation for the mode propagation in the parabolic-index medium is derived. The solutions are proportional to Kummer's functions, becoming divergent at infinity. However, for some values of Kummer's function parameters, the solutions become finite, changing to the familiar non-paraxial Laguerre–Gaussian (LG) modes. This finding also shows that the LG modes in a parabolic-index medium obey both paraxial and non-paraxial Helmholtz equations of wave propagation.

Recent years have seen a notable increase of interest in the graded-index microoptical elements. For instance, a graded-index Luneburg microlens was studied in References [235, 236, 272]. Experiments with a planar Luneburg lens were conducted in References [235, 236], and simulation of a planar photonic-crystal Luneburg lens [272] has shown the lens to generate a focal spot size of intensity FWHM $= 0.44\lambda$ for $n_0 = 1.41$ (n_0 is the refractive index at the center of the lens). A planar [171, 246] and a 3D Mikaelian lens with the refractive index varying as a hyperbolic secant [49] were also reported. In Reference [49], the graded-index HS lens was shown to form a focal spot of size FWHM $= 0.40\lambda$. A planar subwavelength graded-index binary HS lens to generate a focal spot of size FWHM $= 159$ nm $= 0.102\lambda = 0.35\lambda/n$ (refractive index is $n = 3.47$) was discussed in Reference [273].

In this section, we study the propagation of the Gaussian beam in a parabolic-index fiber. A definite length of the fiber can be treated as a parabolic-index (PI) lens. We show by the numerical simulation that the well-known PI lens can find use for the subwavelength focusing of laser light. In terms of its focusing performance, the PI lens is on a par with unconventional lenses such as the Luneburg lens [235, 236, 272] and the hyperbolic secant lens [49, 273]. The numerical simulation has shown that such a lens produces a focal spot of size FWHM=0.42λ for the refractive index on the lens axis of n_0=1.5. An analytical relationship for the radii of jumps of a binary lens approximating the PI lens has also been derived. The elliptic focal spot's smaller size generated with the PI lens and calculated using the finite-difference time-domain (FDTD) method is FWHM=0.45λ. It is noteworthy that References [49, 171, 235, 236, 246, 272, 273] conducted the analysis of planar graded-index lenses whose diffraction limit was defined by the *sinc*-function, resulting in an FWHM=0.44λ, whereas in this work we study a 3D PI lens in which the diffraction limit is defined by the Airy disk size of $(2J_1(x)/x)$, producing a larger value of FWHM=0.51λ.

3.6.1 PARAXIAL HYPERGEOMETRIC BEAMS IN A PARABOLIC-INDEX MEDIUM

Assume a parabolic-index medium with the refractive index in the form:

$$n^2(r) = n_0^2\left[1 - \frac{\left(n_0^2 - n_1^2\right)}{n_0^2}\left(\frac{r}{r_0}\right)^2\right], \tag{3.57}$$

where r is the radial transverse coordinate, n_0 and n_1 are, respectively, the refractive indices on the optical axis (r=0) and at r=r_0. Shown in Figure 3.40 is the profile of the refractive index in Equation (3.57).

The paraxial equation of light propagation in a parabolic-index medium in Equation (3.57) is given by

$$\left[2ik\frac{\partial}{\partial z} + \frac{\partial^2}{\partial x^2} + \frac{\partial^2}{\partial y^2} - (\tau k)^2\left(x^2 + y^2\right)\right]E(x,y,z) = 0, \tag{3.58}$$

where $\tau = \left(n_0^2 - n_1^2\right)^{1/2}\big/\left(r_0 n_0\right)$, $k = \left(2\pi/\lambda\right)n_0$, and λ is the free-space wavelength. In Reference [274], the general solution to Equation (3.58) was shown to take the form:

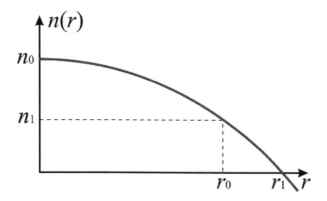

FIGURE 3.40   The parabolic refractive index against the radial coordinate.

$$E(x,y,z) = \frac{k\tau}{2\pi i \sin(\tau z)} \exp\left[\frac{i\tau k}{2\tan(\tau z)}\left(x^2+y^2\right)\right]$$

$$\times \int\limits_{-\infty}^{\infty}\int\limits_{-\infty}^{\infty} E_0(\xi,\eta) \tag{3.59}$$

$$\times \exp\left\{\frac{i\tau k}{2\sin(\tau z)}\left[\left(\xi^2+\eta^2\right)\cos(\tau z)-2\left(x\xi+y\eta\right)\right]\right\}d\xi\,d\eta$$

Note that the integral transform in Equation (3.59) coincides with the partial Fourier transform within designations [275, 276]. The refractive index $n(r)$ becomes imaginary at $r > f_0 = 1/\tau = r_0 n_0 / \left(n_0^2 - n_1^2\right)^{1/2}$ and therefore the medium becomes a perfect absorber. But nevertheless the solution (3.59) remains valid at $r < r_1$ and in further analysis we will ignore the finiteness of the waveguide. In the cylindrical coordinates and for the initial field $E_0(r,\varphi) = E_0(r)\exp(in\varphi)$, Equation (3.59) takes the form of an optical vortex with nth topological charge:

$$E(\rho,\theta,z) = (-i)^{n+1}\frac{k}{f_2}\exp\left(\frac{ik\rho^2}{2f_1}+in\theta\right)$$

$$\times \int\limits_0^{\infty} E_0(r)\exp\left(\frac{ikr^2}{2f_1}\right)J_n\left(\frac{kr\rho}{f_2}\right)r\,dr, \tag{3.60}$$

where n is an integer number and $J_n(x)$ is the Bessel function,

$$f_1 = f_0\tan\left(z/f_0\right),\ f_2 = f_0\sin\left(z/f_0\right). \tag{3.61}$$

Relationships that describe the transformations that the paraxial LG beams undergo in a parabolic-index fiber were derived in Reference [267, 277]. Let us consider a Gaussian beam with power apodization (HyG beam). Below, we study the propagation of such beams characterized by a complex amplitude in the initial plane in the PI fiber of Equation (3.57) [86]:

$$E_0(r) = \left(\frac{r}{\delta}\right)^{m+i\gamma}\exp\left(-\frac{r^2}{2\sigma^2}\right), \tag{3.62}$$

where m, γ are real numbers, δ and σ are, respectively, a scale factor of the amplitude power component and the Gaussian beam radius. We note that in Equation (3.62) the exponent $(m+i\gamma)$ is complex and different from the number n of the angular harmonic $\exp(in\varphi)$. This is the reason why the HyG beams are essentially different from the well-known LG beams. The real part m means power apodization while the imaginary part γ acts as a logarithmical axicon, having the transmittance function $\exp[i\gamma \ln(r/\sigma)]$. Substituting Equation (3.62) into Equation (3.60) and making use of the reference integral [278]:

$$
\int_0^\infty y^{\mu-1} \exp\left(-\beta y^2\right) J_n(cy) dy
$$

$$
= \frac{1}{2^{n+1} n!} c^n \beta^{-(n+\mu)/2} \Gamma\left(\frac{n+\mu}{2}\right) {}_1F_1\left(\frac{n+\mu}{2}, n+1, -\frac{c^2}{4\beta}\right),
$$

(3.63)

where $\Gamma(x)$ is the gamma function and ${}_1F_1(a, b, x)$ is Kummer's function [84], Equation (3.60) is rearranged to

$$
E_{m\gamma n}(\rho,\theta,z) = \frac{(-i)^{n+1}}{n!}\left(\frac{k\sigma^2}{f_2}\right)\left(\frac{\sqrt{2}\sigma}{\delta}\right)^{m+i\gamma} \exp\left(\frac{ik\rho^2}{2f_1} + in\theta\right)
$$

$$
\times \left(1 - \frac{ik\sigma^2}{f_1}\right)^{-(m+i\gamma+2)/2} \Gamma\left(\frac{n+m+i\gamma+2}{2}\right) x^{n/2}
$$

(3.64)

$$
\times {}_1F_1\left(\frac{n+m+i\gamma+2}{2}, n+1, -x\right),
$$

where

$$
x = \frac{\rho^2}{2\omega^2(z)} + \frac{ik\rho^2}{2R(z)},
$$

$$
\omega(z) = \sigma\left[\cos^2\left(\frac{z}{f_0}\right) + \left(\frac{f_0}{k\sigma^2}\right)^2 \sin^2\left(\frac{z}{f_0}\right)\right]^{\frac{1}{2}},
$$

(3.65)

$$
R(z) = \frac{1}{2} f_0 \sin\left(\frac{2z}{f_0}\right)\left[1 + \left(\frac{f_0}{k\sigma^2}\right)^2 \tan^2\left(\frac{z}{f_0}\right)\right].
$$

At $z \approx 0$, it follows from Equations (3.60) and (3.65) that $f_1 \approx f_2 \approx z$, $\omega(z) \approx \sigma$, $R(z) \approx z$, $x \approx \rho^2/(2\sigma^2) + ik\rho^2/(2z)$. Hence, the argument x of the hypergeometric function in Equation (3.64) tends to infinity, but the complex amplitude (3.64) tends to the boundary condition in Equation (3.62). We note that $R(z)$ does not describe the curvature radius of the HyG beam. At $\gamma=0$ and $m=n$ function $\omega(z)$ is the width of the Gaussian envelope. The light field in Equation (3.64) will be self-reproduced in modulus with a period of $L = \pi f_0$. The field amplitudes in planes separated by half-period $L_1 = (\pi f_0)/2$ are related through the Fourier transform:

$$E_{m\gamma n}\left(\rho,\theta,z=L_1\right) = \frac{(-i)^{n+1}}{n!}\left(\frac{k\sigma^2}{2f_2}\right)\left(\frac{\sqrt{2}\sigma}{\delta}\right)^{m+i\gamma}\exp(in\theta)$$

$$\times \Gamma\left(\frac{n+m+i\gamma+2}{2}\right)\left(\frac{\rho^2}{2\omega_1^2}\right)^{n/2} \tag{3.66}$$

$$\times {}_1F_1\left(\frac{n+m+i\gamma+2}{2},n+1,-\frac{\rho^2}{2\omega_1^2}\right),$$

where $\omega_1 = f_0/(k\sigma)$ is the effective radius of the light field in the Fourier plane. The functions $E_{m\gamma n}(\rho,\theta,z)$ of Equation (3.64) are orthogonal to each other at different values of the topological charge n. Note that at $f_0 = k\sigma^2$ the real part of Kummer's function argument x ceases to be z-dependent: $\omega(z) = \sigma = \text{const}$, whereas the imaginary argument's component remains to be z-dependent: $R(z) = f_0 \tan(z/f_0)$. As a result, the total amplitude of the field of Equation (3.64) appears to be z-dependent. Because of this, at no values of parameters and numbers m, n, γ can the HyG beams of Equation (3.64) be represented as modes of a parabolic-index medium, except for a particular case of $m = n$ and $\gamma = 0$.

Let us consider this particular case in more detail. From Equation (3.64) at $\gamma = 0$ and $m = n$, we obtain:

$$E_{n0n}\left(\rho,\theta,z\right) = (-i)^{n+1}\left(\frac{k\sigma^2}{f_2}\right)\left(\frac{\sqrt{2}\sigma}{\delta}\right)^n\left(1-\frac{ik\sigma^2}{f_1}\right)^{-(n+2)/2},$$

$$\times x^{n/2}\exp\left[-\frac{\rho^2}{2\omega^2(z)}+\frac{ik\rho^2}{2R_1(z)}+in\theta\right] \tag{3.67}$$

where

$$x = \frac{\rho^2}{2\omega^2(z)}+\frac{ik\rho^2}{2R(z)},$$

$$R_1(z) = 2k\left(\frac{f_0}{k\sigma^2}-\frac{k\sigma^2}{f_0}\right)^{-1}\omega^2(z)\sin^{-1}\left(\frac{2z}{f_0}\right). \tag{3.68}$$

From Equation (3.67), the wavefront curvature radius $R_1(z)$ is seen to be equal to infinity in the planes $z = \pi f_0 p/2$, $p = 0, 1, 2, \ldots$. In the initial plane ($z = 0$), the terms $(k\sigma^2/f_2)$ and $x^{n/2}$ tend to infinity in the absolute value. However, the term $(1-ik\sigma^2/f_1)^{-(n+2)/2}$ tends to zero. As a result, these two terms are mutually compensated, because at $z \to 0$, $f_0 \approx f_1 \approx R(z) \approx z$. The beams in Equation (3.67) can be referred to as Gaussian optical vortices that propagate in a parabolic-index medium. While propagating in the parabolic-index medium, these beams behave in a different way than the LG beams [274]. At $f_0 = k\sigma^2$, the beam of Equation (3.67) becomes a parabolic-index medium mode, with its radius remaining unchanged during propagation. Figure 3.41

depicts the intensity patterns of the beam in Equation (3.67) in the transverse planes separated by a half-period distance: $z = L/2$ and $z = L$, and in the longitudinal plane $y = 0$. The simulation was conducted for the incident wavelength $\lambda = 532$ nm, the waveguide parameter $\tau = 1/(20\lambda)$, the topological charge $n = 4$, the Gaussian beam waist radius $\sigma = \lambda$, the scale factor $\delta = 2\lambda$, and the dimensionless parameter $\gamma = 0$. The computational domain was $-10\lambda \leq x, y \leq 10\lambda$ in Figure 3.41(a) and (b), and $-25\lambda \leq x \leq 25\lambda$, $0 \leq z \leq 500\lambda$ in Figure 3.41(c).

From Equation (3.67) it can be derived that the intensity maximum is found on the ring $\rho = \sqrt{n}\,\omega(z)$ and equal to

$$I_{\max} = \left(\frac{k\sigma^2}{f_2}\right)^2 \left[1 + \left(\frac{k\sigma^2}{f_1}\right)^2\right]^{-(n+2)/2} \left[\frac{k\sigma^3 n\omega(z)}{\delta^2 f_2}\right]^n \exp(-n). \tag{3.69}$$

From Equation (3.69) it is seen that as the beam is alternatively narrowed and widened during propagation, its radius varies from σ to $f_0/(k\sigma)$, with the intensity maximum undergoing a $[f_0/(k\sigma^2)]^2$–fold change. In particular, for the above-specified parameters, the intensity maximum in Figure 3.41(b) is about 10 times higher than the intensity maximum in Figure 3.41(a).

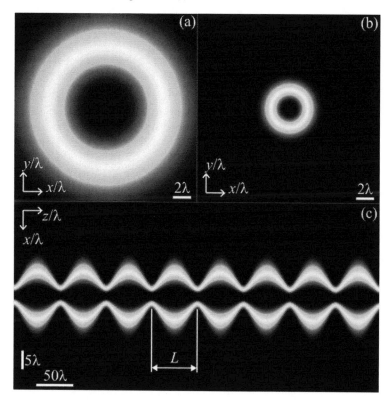

FIGURE 3.41 The intensity pattern from the beam of Equation (3.67) in the transverse planes (a) $z = L/2$ and (b) $z = L = 20\pi\lambda$, and (c) the longitudinal plane $y = 0$.

Thus, in this subsection we have derived a family of exact analytical solutions in Equation (3.64) of the paraxial equation for the parabolic-index medium in Equation (3.58). As a particular case, this solution describes the Gaussian optical vortices of Equation (3.67), which become the paraxial modes of the parabolic-index medium, different from the well-known LG modes under the condition $f_0 = k\sigma^2$. In the next subsection, an attempt is made to derive other non-paraxial modes that would propagate in a parabolic-index medium.

3.6.2   Non-Paraxial Modes of a Parabolic-Index Medium

Let there be a parabolic-index medium of Equation (3.57), $n^2(r) = n_0^2(1 - \tau^2 r^2)$. The solution to the Helmholtz equation for a parabolic-index medium

$$\left(\frac{\partial^2}{\partial r^2} + \frac{1}{r}\frac{\partial}{\partial r} + \frac{1}{r^2}\frac{\partial^2}{\partial \varphi^2} + \frac{\partial^2}{\partial z^2} + k_0^2 n^2(r) \right) E(r,\varphi,z) = 0 \tag{3.70}$$

with the refractive index of Equation (3.57) will be sought as a mode with a Gaussian envelope and an optical vortex:

$$E(r,\varphi,z) = r^p \exp\left(-\frac{r^2}{2\omega_2^2} \right) \exp(in\varphi) \exp(i\beta z) F(sr^q), \tag{3.71}$$

where F is a function, β is the mode propagation constant, ω_2 is the waist radius of the Gaussian envelope, n is the optical vortex topological charge, and s, p and q are constants to be defined later. Then the Helmholtz Equation (3.70) takes the form:

$$s^2 q^2 r^{2q-2} F''(sr^q) + \left[(2p+q)sqr^{q-2} - \frac{2sq}{\omega_2^2} r^q \right] F'(sr^q)$$

$$+ \left[\frac{1}{\omega_2^4} r^2 - \frac{2(p+1)}{\omega_2^2} + \frac{p^2 - n^2}{r^2} \right] \tag{3.72}$$

$$+ k_0^2 n_0^2 \left(1 - \tau^2 r^2 \right) - \beta^2 \bigg] F(sr^q) = 0.$$

Assuming $q = 2$ and $n = \pm p$, and dividing Equation (3.72) by $4s$ we obtain:

$$sr^2 F''(sr^2) + \left(p + 1 - \frac{r^2}{\omega_2^2} \right) F'(sr^2)$$

$$+ \frac{1}{s} \left[\frac{r^2}{4\omega_2^4} - \frac{p+1}{2\omega_2^2} + \frac{k_0^2 n_0^2 \left(1 - \tau^2 r^2 \right) - \beta^2}{4} \right] F(sr^2) = 0. \tag{3.73}$$

If we suppose that $k_0 n_0 \tau \omega_2^2 = 1$ and $s = 1/\omega_2^2$, then Equation (3.73) becomes Kummer's differential equation:

$$\xi F''(\xi) + (p+1-\xi)F'(\xi)$$

$$-\left[\frac{p+1}{2} + \frac{\omega_2^2}{4}\left(\beta^2 - k_0^2 n_0^2\right)\right]F(\xi) = 0. \tag{3.74}$$

where $\xi = r^2/\omega_2^2$. Therefore, the solution of Equation (3.70) can be shown to be given by a family of functions:

$$E(r,\varphi,z) = r^n \exp\left(-\frac{r^2}{2\omega_2^2}\right)\exp(\pm in\varphi)\exp(i\beta z)$$

$$\times {}_1F_1\left[\frac{n+1}{2} + \frac{\omega_2^2}{4}\left(\beta^2 - k_0^2 n_0^2\right), n+1, \frac{r^2}{\omega_2^2}\right], \tag{3.75}$$

where ${}_1F_1(a,b,x)$ is Kummer's function, as before, $\omega_2 = \left[1/(k_0 n_0 \tau)\right]^{1/2}$. Considering that at $\xi \to \infty$, Kummer's function asymptotics is given by [84]

$${}_1F_1(a,b,\xi) = \frac{\Gamma(b)}{\Gamma(a)}\exp(\xi)\xi^{a-b}\left(1 + O(1/\xi)\right), \tag{3.76}$$

where $O(x)$ tends to zero faster than x, the function in Equation (3.75) is diverging at $r \to \infty$, thus the beam has infinite energy. The set of solutions (3.75) also contains non-diverging solutions. Putting Kummer's function first parameter to be a negative integer, Kummer's function becomes equal to the polynomial, whereas the solution (3.75) converges to zero at $r \to \infty$. Thus, under the condition that

$$\frac{n+1}{2} + \frac{\omega_2^2}{4}\left(\beta^2 - k_0^2 n_0^2\right) = -s, \tag{3.77}$$

Equation (3.75) reduces to the familiar Laguerre–Gaussian modes:

$$E(r,\varphi,z) = r^n \exp\left(-\frac{r^2}{2\omega_2^2}\right)$$

$$\times \exp(\pm in\varphi)\exp(i\beta z){}_1F_1\left[-s, n+1, \frac{r^2}{\omega_2^2}\right]$$

$$= \frac{s!}{(n+1)^s}r^n \exp\left(-\frac{r^2}{2\omega_2^2}\right)$$

$$\times \exp(\pm in\varphi)\exp(i\beta z)L_s^n\left(\frac{r^2}{\omega_2^2}\right), \tag{3.78}$$

where

$$\beta = k_0 n_0 \sqrt{1 - \frac{2\tau}{k_0 n_0}(2s + n + 1)}. \tag{3.79}$$

If in the LG mode of Equation (3.78) the radius of the Gaussian beam does not meet the condition $\omega_2 = [1/(k_0 n_0 \tau)]^{1/2}$, the LG beam is no mode any more. In the unbounded waveguide of Equation (3.57) only Laguerre–Gaussian modes (3.78) can propagate, but if the waveguide is bounded by radius $r_0 < \infty$, then besides modes (3.78) there would be other modes described by Equation (3.75).

Thus, in this subsection we have shown that although there is a wide class of mode solutions of the Helmholtz equation in the cylindrical coordinates for a parabolic-index medium, only solutions coincident with the LG modes possess a finite energy, thus being physically realizable. In other words, in this subsection, we have shown that, except for the LG modes, no other non-paraxial modes with cylindrical symmetry propagate in the parabolic medium. This result also proves that the GL modes in the parabolic-index medium satisfy both equations: paraxial Equation (3.58) [265] and non-paraxial Equation (3.67). Besides, we can infer that both the Gaussian beam of Equation (3.64) at $f_0 = k\sigma^2$ and the Gaussian beam of Equation (3.78) at $\omega_2 = [1/(k_0 n_0 \tau)]^{1/2}$, have the same radius:

$$\omega_2 = \left(k_0 n_0 \tau\right)^{-1/2} = \sigma = \sqrt{f_0 / k} = \left(\frac{\lambda r_0}{2\pi \sqrt{n_0^2 - n_1^2}}\right)^{1/2}. \tag{3.80}$$

3.6.3 A PARABOLIC-INDEX MICROLENS

The parabolic-index microlens is well-known [271]. However, until now it has not yet been demonstrated that such a microlens can be utilized for the subwavelength focusing of light. In this subsection, based on the solution in Equation (3.64) we derive a paraxial estimate of the focal spot size. Considering that its amplitude is periodically self-reproducing during propagation in the parabolic-index fiber (3.57), the Gaussian beam enters into the family of solutions (3.64) as a principal beam. Assuming $n = m = \gamma = 0$, from Equation (3.64) we obtain:

$$E_0(\rho, \theta, z) = (-i)\left(\frac{2k\sigma^2}{f_2}\right)\left(1 - \frac{ik\sigma^2}{f_1}\right)^{-1}$$

$$\times \exp\left(-\frac{\rho^2}{2\omega^2(z)} - \frac{ik\rho^2}{2R_1(z)}\right), \tag{3.81}$$

where

$$R_1^{-1}(z) = \frac{\cos^2(z/f_0) + \left(\dfrac{f_0}{k\sigma^2}\right)^2 \sin^2(z/f_0) - 1}{f_0 \tan(z/f_0)\left[\cos^2(z/f_0) + \left(\dfrac{f_0}{k\sigma^2}\right)^2 \sin^2(z/f_0)\right]}. \tag{3.82}$$

From Equations (3.81) and (3.82), it follows that at $f_0 = k\sigma^2$ the Gaussian beam remains unchanged while propagating in the parabolic-index fiber, preserving its diameter. If $f_0 \neq k\sigma^2$, the Gaussian beam's radius changes (see Equation (3.65)):

$$\omega(z) = \sigma\left[\cos^2\left(\frac{z}{f_0}\right) + \left(\frac{f_0}{k\sigma^2}\right)^2 \sin^2\left(\frac{z}{f_0}\right)\right]^{\frac{1}{2}}, \tag{3.83}$$

suggesting that the minimal radius of $\omega_1 = f_0/(k\sigma)$ (if $f_0 < k\sigma^2$) is attained at distance $L_1 = \pi f_0/2$ from the initial point ($z = 0$). The Gaussian beam's diameter at intensity half-maximum equals

$$\text{FWHM} = \left(\frac{\sqrt{\ln 2}}{\pi}\right)\frac{\lambda r_0}{\sigma\sqrt{n_0^2 - n_1^2}}. \tag{3.84}$$

Thus, we can consider a PI lens in the form of a section of a parabolic-index fiber of radius r_0 and optical-axis length $L_1 = (\pi f_0)/2$. Such a PI lens is able to focus the incident plane Gaussian beam of radius σ into a focal spot of size in Equation (3.84), which is formed near the output surface of the lens (Figure 3.42).

From Equation (3.84), the NA of the PI lens ($n_1 = 1$) is given by

$$\text{NA} = \frac{\sigma}{r_0}\sqrt{n_0^2 - 1}. \tag{3.85}$$

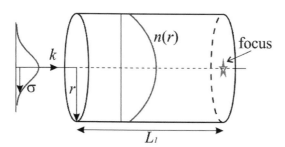

FIGURE 3.42   Focusing of the Gaussian beam with a PI lens.

When $\sigma = r_0$, the NA in Equation (3.85) equals that of a planar HS lens [246]. It should be noted that since Equation (3.85) has been derived as a paraxial approximation, for it to be used correctly one needs to choose the condition $\sigma \ll r_0$. Note, however, that the well-known relation that describes the diffraction-limited size of the focal spot in a medium with refractive index n has been also deduced in the paraxial approximation (the Airy disk diameter in terms of the intensity FWHM): $\mathrm{FWHM} = 0.51\lambda/(n\sin\theta)$. Nonetheless, it has been used for the larger-than-unity NAs: $\mathrm{FWHM} = 0.51\lambda/n$.

From Equation (3.84) it follows that at $n_0 = 1.5$, $n_1 = 1$, $\sigma = r_0$, the PI lens has the length of $L_1 = 2.1r_0$ and the focal spot size of $\mathrm{FWHM} = 0.24\lambda$. The latter value is 1.4 times smaller than the diffraction limit in the medium with $n_0 = 1.5$: $\mathrm{FWHM} = 0.34\lambda$. It should be noted, however, that in deducing Equation (3.83), the lens was supposed to be of infinite radius, with the refractive index of Equation (3.57) becoming $n(r) < 1$, thus the estimated focal spot size ($\mathrm{FWHM} = 0.24\lambda$) cannot be realized for a real lens of limited radius r_0, at which $n(r_0) = 1$. For comparison, we note that the diameter of the intensity FWHM of the minimal non-paraxial LG mode in a parabolic-index waveguide can be obtained from Equation (3.80) at $r_0 = \lambda$:

$$\mathrm{FWHM} = 2\sqrt{\ln 2}\left(\frac{\lambda r_0}{2\pi\sqrt{n_0^2 - n_1^2}}\right)^{1/2} = 0.63\lambda. \qquad (3.86)$$

Although fabrication techniques for synthesis of continuous-profile graded-index lenses have been widely known, including a chemical vapor deposition method [279] and an ion exchange method [280], they do not enable obtaining an arbitrarily specified refractive index profile. Because of this, in the next subsection we discuss a binary lens as an approximation of the parabolic-index lens. Such a lens can be fabricated by e-beam lithography.

3.6.4 BINARY PARABOLIC-INDEX LENS

The continuous parabolic-index lens can be approximated by a binary parabolic-index lens by the rule schematically illustrated in Figure 3.43.

According to Figure 3.43, the radius r_0 of the PI lens is broken down into N equal segments of length Δ: $r_0 = N\Delta$, with the segment's origin and end given by the radii $r_p = p\Delta$, $p = 0, 1, 2, \ldots N-1$.

The origin of the pth binary ring is at radius r_p, whereas the end $\bar{r}_p$ of the pth binary ring is found from the equation:

$$\pi\left(r_{p+1}^2 - \bar{r}_p^2\right) + n_0\pi\left(\bar{r}_p^2 - r_p^2\right) = 2\pi n_0 \int_{r_p}^{r_{p+1}} \sqrt{1 - \tau^2 r^2}\, r\, dr. \qquad (3.87)$$

This equation means that inside each ring $r_p \leq r \leq r_{p+1}$ the average refractive index is the same for both binary and gradient-index lenses. From Equation (3.87), an explicit relation for the end radius of the pth binary ring is

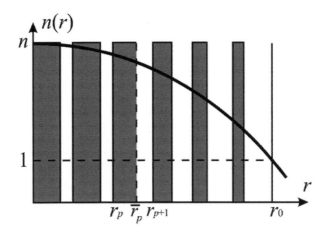

FIGURE 3.43 Schematic representation of the continuous PI lens approximated by a piece-wise-constant (binary) lens (dark gray, material, white, air).

$$\overline{r}_p^2 = \frac{n_0 r_p^2 - r_{p+1}^2}{n_0 - 1} + \frac{2n_0}{3(n_0 - 1)\tau^2}\left\{\left[1 - (\tau r_p)^2\right]^{3/2} - \left[1 - (\tau r_{p+1})^2\right]^{3/2}\right\}, \qquad (3.88)$$

where $p = 0, 1, 2, \ldots N-1$. Note that in Equation (3.87) it is possible to choose non-equidistant radii r_p of the refractive index jumps, also taking account of the fabrication capabilities with regard to a minimal feasible zone size.

3.6.5 SIMULATION RESULTS

3.6.5.1 A Planar Parabolic-Index Lens

First, we studied a 2D PI lens. The FDTD simulation of focusing an incident TE-wave was conducted using a commercial program FullWAVE (RSoft). The computational mesh samplings on the optical z-axis and the transverse x-axis were $\lambda/130$ (4.1 nm), the time step being $\Delta(cT) = 2.8$ nm.

Figure 3.44 shows a gray-level map of the refractive index in the PI lens. The lens radius is $r_0 = 1$ μm, $n_0 = 1.5$, $\alpha = 0.745$ μm^{-1}, and $\lambda = 0.532$ μm. Considering that $\alpha = 1/f_0$, then $f_0 = 1/\alpha = 1.342$ μm, the lens length being $L_1 = \pi f_0/2 = \pi/(2\alpha) = 2.1$ μm.

Figure 3.45 shows the intensity distribution of the E-field 10 nm behind the lens for a plane incident beam and a Gaussian beam of radius $\sigma = 1$ μm. At the same time, a part of the Gaussian beam passes by the lens.

The focal spot size is FWHM $= 0.388\lambda = 0.2$ μm for the plane wave and FWHM $= 0.5\lambda = 0.27$ μm for the Gaussian beam. It can be seen from Figure 3.45 that the intensity sidelobes amount to 30% of the major intensity peak, signifying that the lens length is not optimal.

A more tightly focused spot can be achieved by varying the lens length. Figure 3.46 shows the focal spot size at intensity FWHM as a function of the lens length L_1.

From Figure 3.46, the minimal focal spot size for the plane wave and the Gaussian beam is seen to be attained at different values of the lens length L_1. The plane wave

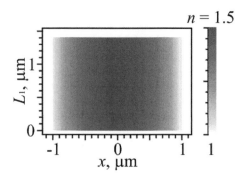

FIGURE 3.44   The refractive index (in gray level) in a planar gradient-index PI lens.

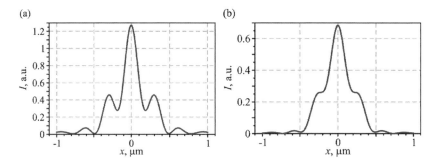

FIGURE 3.45   The intensity distribution ($|E^2|$) in the focal plane found 10 nm behind the lens for (a) a plane incident wave, and (b) a Gaussian incident beam.

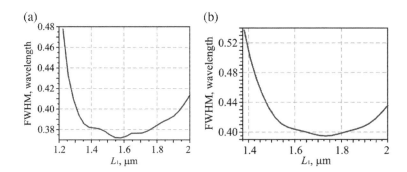

FIGURE 3.46   The focal spot size generated 10 nm away from the lens output as a function of the lens length L_1 for (a) a plane wave and (b) a Gaussian incident beam.

is most tightly focused at $L_1 = 1.56$ µm, producing FWHM$= 0.375\lambda$, whereas the Gaussian beam produces a minimal focal spot size of FWHM$= 0.394\lambda$ at lens length $L_1 = 1.73$ µm. This result reveals a general pattern of laser light focusing: an increase in the sidelobes (Figure 3.45(a)) is accompanied by a tighter focal spot (and increased focus depth).

3.6.5.2 3D Parabolic-Index Lens

The 3D simulation involved a lens with a graded-index profile in the xy-plane, which was dependent on the radius r. The graded refractive index profile of the 3D lens in the xy-plane is shown in Figure 3.47.

The 3D simulation was performed on a computational mesh of size $\lambda/40$ (13.3 nm) along all three axes. We simulated the propagation of the linearly polarized light through the lens (for the major incident field component E_y). The incident light was a plane Gaussian beam of radius $\sigma = 1$ µm. The plot in Figure 3.48(a) shows the focal spot size FWHM on the x-axis as a function of the lens length L_1.

The minimum is seen to be attained at the lens length of $L_1 = 1.6$ µm, with the focal spot size on the x-axis being FWHM $= 0.42\lambda$. The intensity profile ($I = |E_x^2| + |E_y^2| + |E_z^2|$) within the focal spot of the lens for the said parameters is shown in Figure 3.48(b). Because the incident light is linearly polarized, the focal spot is widened on the y-axis, amounting to FWHM $= 0.70\lambda$ (Figure 3.48(c)).

3.6.5.3 3D Binary Parabolic-Index Lens

In practice, it appears difficult to fabricate a lens with a continuous parabolic-index profile. Meanwhile, a binary microlens can be fabricated using techniques for

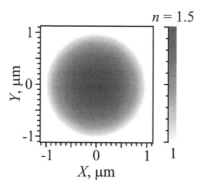

FIGURE 3.47 Lens gray-level graded-index profile in the plane xy.

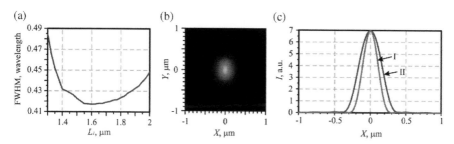

FIGURE 3.48 (a) The focal spot size on the x-axis formed 10 nm away from the lens output plane against the lens length L_1 for the linearly polarized incident Gaussian beam of radius $\sigma = 1$ µm; (b) the intensity pattern in the focal spot of the lens for an optimal length of $L_1 = 1.6$ µm; (c) the intensity cross-sections passing through the spot center along the x-axis (curve I) and y-axis (curve II).

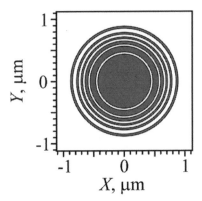

FIGURE 3.49   Binary refractive index distribution in the xy-plane, with dark rings denoting the refractive index $n = 1.5$ and white rings denoting $n = 1$.

fabricating 3D photonic-crystal waveguides or Bragg's waveguides [281]. Shown in Figure 3.49 is a binary refractive index distribution in the xy-plane for a binary lens of Equation (3.87), which approximates to the continuous-profile parabolic-index lens.

The focusing performance of such a lens is somewhat inferior to the continuous-profile parabolic-index lens of Equation (3.57). Figure 3.50 depicts the intensity pattern in the focal plane of the binary parabolic-index lens (Figure 3.49) and the distribution cross-sections (Figure 3.50(b)) drawn through the spot center along the x-axis (curve I) and y-axis (curve II). The distributions correspond to an optimal lens length of $L_1 \approx 1.9$ μm, whereas the paraxial theory predicts a somewhat larger value of $L_1 = 2.1$ μm. The linearly polarized incident wave has a plane wavefront and a Gaussian amplitude distribution of radius $\sigma = 1$ μm. In the binary lens, the minimal zone size (i.e. the difference between the radii of neighboring refractive index jumps) is 35 nm. Note that the implementation of such a lens requires a technology that would enable obtaining a binary microrelief with the aspect ratio of 60 (relief-depth/relief-width ratio). Such fabrication techniques have been utilized for

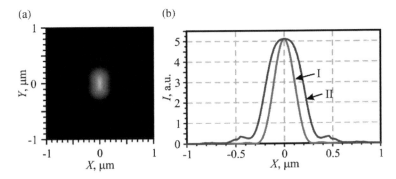

FIGURE 3.50   (a) The intensity distribution in the focal plane (10 nm behind the lens) of the binary PI lens of Figure 3.49; and (b) the intensity cross-sections drawn through the spot center along the x-axis (curve I) and y-axis (curve II).

the implementation of X-ray zone plates (ZP). A ZP with the minimal (outermost) zone of width 100 nm and height 1300 nm (aspect ratio = 13) was fabricated using the positive resist polymethyl methacrylate (PMMA) in Reference [282]. To prevent the collapse of the ZP walls, the resist was dried using a special technique [282].

The collapse of the binary lens in Figure 3.49 can be prevented by replacing the air in the between-wall spacing with a different material. Using a "jelly-roll" technology, a sputtered-slice X-ray ZP composed of two alternating materials with a 160-nm outermost zone and the aspect ratio of greater than 1000 was fabricated in Reference [283].

We conducted the simulation on a computational mesh of size $\lambda/70$ (7.6 nm) for all three axes. The initial polarization plane was parallel to the zy-plane.

For the distributions shown in Figure 3.50, the focal spot size is FWHM = 0.45λ along the x-axis and FWHM = 0.78λ along the y-axis. The focal spot is seen to be extended along the y-axis and nearly devoid of sidelobes. The extension of the focal spot along the y-axis can be explained by the polarization of the field. In the xz-plane, the field propagates as a TE-wave, i.e. only the E_y component is non-zero. In the yz-plane, the field propagates as a TM-wave, i.e. the E_z component is also present, contributing to the total intensity and making the focal spot wider along the y-axis.

The comparison of Figures 3.48(c) and 3.50(b) suggests that due to the binary structure of the lens in Figure 3.49 the efficiency of focusing has decreased, with 8% less optical energy coming to the focal spot in Figure 3.50 when compared with that in Figure 3.48. Besides, the focal spot in Figure 3.50(b) has acquired the sidelobes which the focal spot from the parabolic-index lens has been nearly devoid of (Figure 3.48(c)). However, these drawbacks are an acceptable cost to be paid for the feasibility of fabricating such a lens using the available fabrication techniques [282, 283].

The major results in this section are as follows. The relationship for the complex amplitude of a family of paraxial hypergeometric laser beams propagating in a parabolic-index fiber has been derived; also, a relationship that describes a simple particular case of the HyG beams, namely, Gaussian optical vortices in the parabolic-index medium, has been deduced. Under certain parameters, the said Gaussian optical vortices become modes of the parabolic-index medium. A wide class of solutions of the Helmholtz equation in the cylindrical coordinates to describe the propagation of modes in a parabolic-index medium has been derived; the solutions are proportional to Kummer's function; it has been shown that only solutions coincident with the non-paraxial Laguerre–Gaussian modes have a finite energy, thus being physically realizable; there have been no other non-paraxial modes with cylindrical symmetry in the parabolic-index waveguide similar to paraxial modes. A definite length of the parabolic-index fiber has been treated as a parabolic lens, for which relationships for the NA and the focal spot size FWHM have been derived. The FDTD simulation of focusing a linearly polarized Gaussian beam with a 3D parabolic-index lens has shown that an optimal lens length is smaller than that predicted by the scalar theory, with the smaller diameter of the elliptic focal spot being FWHM = 0.42λ. An explicit relation for the ring radii of a binary lens that approximates the parabolic-index lens has been derived. It has been estimated that the elliptic focal spot generated with the binary lens has the smaller diameter of FWHM = 0.45λ.

4 Sharp Focusing of Light by Metasurface

4.1 FOUR-ZONE REFLECTIVE POLARIZATION CONVERSION PLATE

The use of subwavelength diffractive gratings for manipulation with polarization of laser light was introduced in Reference [284] for the first time. In References [285, 286] subwavelength diffractive gratings were used for conversion of polarization. When linearly polarized light is reflected from this grating it rotates, depending on the angle between the incident light polarization and the directions of the grating grooves. Circularly polarized laser light with a wavelength of 10.6 μm was converted to azimuthally polarized light using a binary subwavelength diffractive grating with curved grooves and varying period in Reference [285]. The period of the grating varied from 2 μm to 3 μm, and the radius was 9.6 mm. In Reference [286] the phase grating converted left circularly polarized light to radially polarized light and right polarized light to azimuthally polarized light. The grating was manufactured in GaAs substrate (index of refraction $n = 3.13$); the grating had a diameter of 10 mm, the period varied from 2 mm to 3 mm, the relief height was 2.5 um, and the wavelength was 10.6 um. In Reference [48, 287] the grating proposed in Reference [286] was upgraded. This grating converted laser light with circular polarization to radially polarized light with high efficiency. The polarization converter for near infrared light (with a wavelength of 1.064 um) was investigated in Reference [288]. Reference [288] also used a binary subwavelength grating. The grating had a diameter of 1 mm, a period of 240 nm, and a relief depth of 470 nm. The grating converted linearly polarized laser light to radially polarized light. References [48, 285–288] proposed elements transmit light to convert it.

In this section we assume a reflective four-zone binary subwavelength micrograting for visible light conversion from linear polarization to radial polarization [289].

4.1.1 DESIGN OF CONVERSION PLATE

In our numerical simulation we used the finite-difference time-domain (FDTD) method implemented in FullWAVE (http://optics.synopsys.com/rsoft/). Gold was used as the material for the proposed diffractive optical element (DOE). Figure 4.1 shows the directions of polarization of incident and reflected DOE light.

In Figure 4.1 α is the angle between the polarization of incident light and the direction of the grating grooves, θ is the angle between the polarization of incident light and the light reflected from the DOE. $\mathbf{E}_1$ is the incident electric field vector; $\mathbf{E}_2$ is the reflected electric field vector. Figure 4.2 shows the polarization angle of reflected light θ as a function of the incident polarization angle α for a grating with

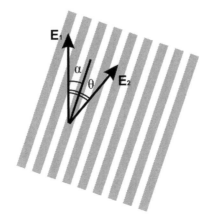

FIGURE 4.1   Directions of polarizations of incident $\mathbf{E}_1$ and reflected from DOE $\mathbf{E}_2$ light.

height $h = 110$ nm, produced in gold film (index of refraction for wavelength $\lambda = 633$ nm was $n = 0.312 + 3.17i$). The period of the DOE was 0.4 um.

Figure 4.2 shows that angle θ varies from 0° to 180° when angle α varies from 0° to 90°. In other words through the manipulation of the directions of the grating grooves it could be possible to form any polarization. Figure 4.3 shows the intensity of reflected light $|E|^2$ as a function of incident polarization angle α.

The intensity of refracted light strongly depends on the angle α. To decrease the difference between the intensities of light refracted from different parts of the element designed, the DOE is divided onto 4 parts. All the parts have a similar index of reflection. The polarization angles θ were equal to −135°, −45°, 45°, 135°, and the angles of the grating grooves α were equal to −70°, −40°, 40°, 70°. Figure 4.4 shows the sketch of the designed structure.

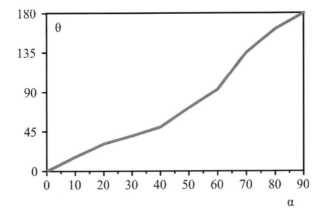

FIGURE 4.2   Polarization angle of reflected light θ as a function of incident polarization angle α.

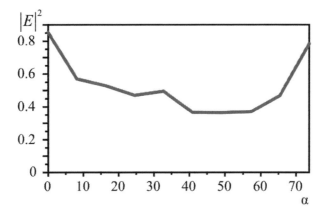

FIGURE 4.3   Intensity of reflected light $|E|^2$ as a function of incident polarization angle α.

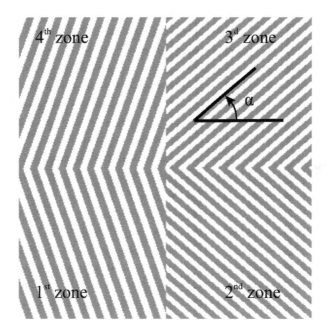

FIGURE 4.4   Sketch of designed DOE (grating-polarizer).

4.1.2   NUMERICAL MODELING

Figure 4.5 shows the intensity and polarization of refracted light on the different distances from the designed polarizer.

To estimate the effectiveness of focusing light we numerically simulate the focusing of light using a binary microaxicon with period of $T=\lambda$ at a distance of $\lambda/2$ from its surface. The refraction of light from the DOE was simulated using the FDTD method, then using Rayleigh–Sommerfeld integral we calculate a light field

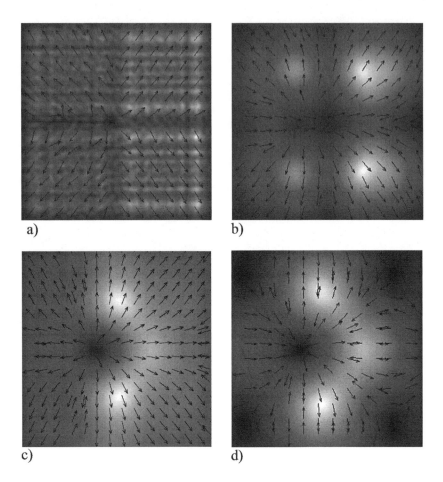

FIGURE 4.5 Intensity and polarization of reflected light at the distances z equal to (a) 5.2 μm, (b) 100 μm, (c) 200 μm, (d) 1000 μm. The image size is (a)–(c) 20×20 μm and (d) 800×800 μm.

distribution at a distance of $z = 200$ nm from the DOE. Propagation of light through the subwavelength axicon was calculated using the FDTD method. The axicon has refractive index of $n = 1.5$ and a relief height of $h = 1.266$ μm. The substrate was not taken into account (the axicon rings were located in free space). Figure 4.6 shows the intensity distribution at the focal point of the axicon.

The focused linearly polarized plane wave forms an elliptical focal spot with its diameters equal to full-width half-maximum $(\text{FWHM})_x = 0.64\lambda$ and $\text{FWHM}_y = 0.305\lambda$ along axes x and y. The radially polarized light refracted from the designed DOE forms a focal spot with its diameters equal to $\text{FWHM}_x = 0.407\lambda$ and $\text{FWHM}_y = 0.351\lambda$. Both diameters of this focal spot are below the diffraction limit of $\text{FWHM} = 0.51\lambda$. Despite the drawbacks of this design (different index of refraction from each zone) our DOE forms a beam that is different from the linearly polarized beam and could be used for tight focusing or superresolution.

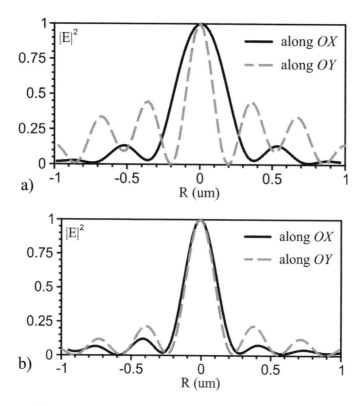

FIGURE 4.6 Focal point sections obtained by the focusing of (a) linearly polarized light and (b) radially polarized light reflected from conversion plate, at a distance of $\lambda/2$ from the axicon.

4.1.3 Manufacturing and Experiment

To manufacture a polarization converter a gold layer with thickness of 160–180 nm was coated onto a glass substrate. The gold layer was then covered with resist. A mask of a four-zone polarization converter was projected onto the resist using an electron beam. The element was etched in xylene, which dissolves the resist exposed by the electron beam. Using an ion etching mask this was transformed into a gold layer. Finally the resist was removed using oxygen plasma. The height of the manufactured DOE was about 110 nm. Figure 4.7 shows the SEM image of a central part of the designed grating-polarizer. The size of the manufactured polarizer was 100×100 um.

Using the optical scheme depicted in Figure 4.8 we experimentally obtained radially polarized light when we detected laser light with a wavelength of 633 nm reflected from the four-zone DOE. The light from an He-Ne laser (with wavelength of 633 nm) propagates through a neutral density filter (ND) and polarizer (P1), and is then focused by a lens with a long focal length (L) ($f=25$ cm) and a beam splitter cube on a substrate with four-zone polarization converters. The light reflected from the four-zone polarization converter propagates through a beam splitter cube (BS)

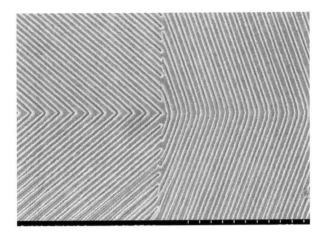

FIGURE 4.7 Image (13 × 13 um) of a four-zone polarization converter with period of 400 nm and relief depth of 110–120 nm.

and 10× objective (O), and was then registered by a charge-coupled device (CCD-) camera. The polarizer (P2) located between O and the CCD-camera could block the light from the incident beam reflected from the mirror surface surrounding the four-zone conversion plate.

Figure 4.9 shows an image of a four-zone conversion plate when the angle between the axis of the output polarizer the horizontal axis is equal to 45° (a), 135° (b), and 90° (c). The incident light was linearly polarized in a horizontal direction. This experiment confirmed that our four-zone plate converts the polarization of light.

A four-zone subwavelength binary diffraction optical microelement (size of 100 × 100 µm) for polarization conversion from linear-to-radial was calculated and designed. The grating period was equal to 400 nm, and the element was manufactured

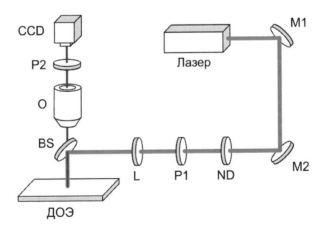

FIGURE 4.8 Optical scheme: M_1, M_2, mirrors, ND, neutral density filter, P_1, P_2, polarizers, L, lens with focal length of $f = 25$ cm, BS, beam splitter cube, DOE, substrate with four-zone conversion plate, O, 10× objective

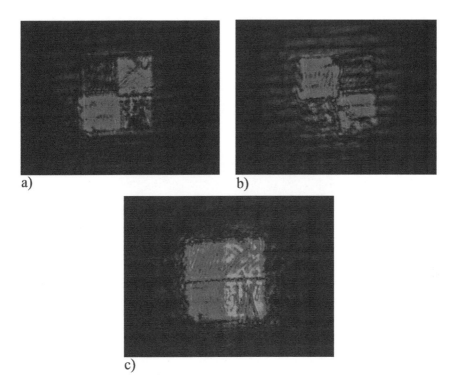

a)   b)

c)

FIGURE 4.9   Image of conversion plate when the angle between output polarizer and horizontal axis equals (a) 45°, (b) 135°, and (c) 90°.

for a wavelength of 633 nm in a gold film with a height of 110 nm. Simulation by the FDTD method and the Rayleigh–Zommerfeld integral and the experiment shows that there is a radial polarized light beam in reflected light.

4.2 TIGHT FOCUS OF LIGHT USING A MICROPOLARIZER AND A MICROLENS

The manipulation of polarization states of laser light and generation of a prescribed intensity distribution in a given plane by means of subwavelength diffraction gratings was first proposed in Reference [284]. Subwavelength diffraction gratings were utilized to convert the laser light polarization in References [47, 48, 285–288]. When a linearly polarized light is incident on such a grating, the magnitude of polarization rotation depends on the angle between the incident polarization vector and the direction of the grating lines/grooves. References [47, 48, 285–288] describe gratings operating in transmission and manufactured from a variety of materials for the infrared (IR) wavelength range. In particular, GaAs gratings for a wavelength of 10.6 μm were utilized in References [48, 285–287]. For the grating-polarizer, the etch depth was 2.5 μm. A GaAs grating operating at a wavelength of 1.06 μm and characterized by a 240-nm period and 470-nm etch depth was discussed in Reference [288]. In

Reference [47] an azimuthally polarized laser beam was generated using a subwavelength grating with birefringence properties at 1.55 μm. The structure was composed of alternating SiO_2 and SiN layers of the overall height 8 μm. It presents a challenging task to manufacture a subwavelength grating in transmission for the visible range due to high aspect ratio (of about 5). Considering this, a reflection four-sector binary grating in a gold film of period 400 nm and depth 110 nm was utilized in Reference [289] to generate a radially polarized light beam of wavelength 633 nm.

Note that no experiments on tight focusing of light with the aid of the grating-polarizers [47, 48, 285–289] have been conducted.

The present authors have reported [101, 102, 143] experiments on tightly focusing a linearly polarized laser beam by means of microoptic components. In Reference [101], the focusing was realized with a binary microaxicon with a circular grating of period 800 nm (numerical aperture (NA)=0.67). In Reference [102], a binary microlens of focus $f=0.532$ μm, diameter 14 μm, and microrelief depth 510 nm was utilized. Using a scanning near-field optical microscope (SNOM), we studied the propagation of a linearly polarized Gaussian beam of wavelength $\lambda = 532$ nm through the microlens. A focal spot of size at full-width of half-maximum (FWHM) intensity FWHM$=(0.44 \pm 0.02)\lambda$ was experimentally obtained. By replacing the 532-nm incident wavelength with $\lambda = 633$ nm, we have managed to obtain [143] a tighter focal spot of size FWHM$=(0.40 \pm 0.02)\lambda$ using the same microlens. However, the radially polarized light was not focused by the microlens.

Publications concerned with the subwavelength focusing of radially and azimuthally polarized laser beams have also been widely known [105–108]. Focal spots of size FWHM$=0.43\lambda$ (NA$=0.95$) [105, 107], FWHM$=0.30\lambda$ (NA$=1.4$) [106] have been obtained in the numerical and physical experiments. In Reference [106], the above-mentioned resolution was experimentally attained in recording of 3 terabytes of data on a conventional optical disk. Note that in References [105–108] microelements were not employed to obtain radially polarized beams. For instance, in Reference [106] radially polarized beams were generated using a twisted nematic liquid crystal device (Arcoptix SA).

This section is aimed at obtaining a subwavelength circular focal spot with the aid of a two-component microoptic design composed of a grating micropolarizer and a microlens. The former generates a radially polarized beam and the latter focuses it into a near-surface spot. To enhance the quality of the device, the micropolarizer had to be fabricated in reflection (thus, enabling a minimal microrelief depth) and comprising four sectors (thus, ensuring a better-quality grating). However, as a result of the fabrication restrictions, the light is only partially radially polarized, becoming a mixture of linear and radial polarizations, eventually leading to an elliptic focal spot. Nonetheless, the focal spot size is 40% smaller than that previously obtained with the same microlens for linearly polarized light [143].

In this section, a 633-nm laser beam was focused using a binary microlens of diameter 14 μm and focus 532 nm (NA$=0.997$) fabricated in resist. The beam composed of a mixture of linearly and radially polarized beams, generated by reflection of a linearly polarized Gaussian beam from a gold-coated four-sector subwavelength binary diffractive optical microelement (micropolarizer) of size 100×100 μm, was focused near the microlens surface into a near-circular focal spot measuring

FWHM = $(0.35 \pm 0.02)\lambda$ and FWHM = $(0.38 \pm 0.02)\lambda$. This value is smaller than the diffraction limit of FWHM = 0.51λ.

The measurement of the focal spot was conducted with a 20-nm step using an SNOM with a metal cantilever shaped as a hollow pyramid tip with a 70° vertex angle and a 100-nm pinhole. Such a cantilever is 3 times more sensitive to the transverse electric field component than to the longitudinal field component [290].

We are not the first to have harnessed a four-sector polarizer. In a classical work [291], a linear-to-radial polarization conversion was implemented using a four-sector half-wave plate. It is worth noting that a defocused intensity pattern was devoid of radial symmetry. Four is the minimal number of sectors required to obtain radial polarization, irrespective of whether the said sectors are in the form of half-wave plates or subwavelength diffraction gratings. For a larger number of sectors, the fabrication quality of diffraction gratings deteriorates in the near-center region, where the sectors meet. The focal spot reported in Reference [291] had the size FWHM = $0.16\lambda^2$, whereas in this work we attained a tighter focus of FWHM = $0.105\lambda^2$.

The radially polarized light is known to be focused into a circular spot, as distinct from the linearly polarized light generating an elliptic focal spot. In Reference [146], it was numerically shown that when focusing a radially polarized beam the focal spot size was FWHM = 0.37λ (based on the Richards–Wolf formula) using an aplanatic lens of NA = 0.99, and FWHM = 0.39λ using a zone plate of focus $f = \lambda$ (based on the FDTD method). In this section these simulation results are confirmed experimentally.

4.2.1 Reflection Binary Micropolarizer

A gold layer of thickness 160–180 nm was deposited on a glass substrate by electron beam evaporation. Then, the gold layer was coated with a layer of electron beam resist (ZEON ZEP520A), and the pattern of the four-sector grating-polarizer projected into the resist using a 30-kV electron beam. The sample was then developed in xylene to dissolve resist fragments that had been exposed to the electron beam. The pattern of the grating-polarizer was then transferred into the gold layer by sputtering with an argon plasma. In the final stage, the remaining resist was removed with an oxygen plasma, resulting in a grating-polarizer pattern engraved in gold. The duration of the reactive ion etching process was optimized so as to achieve the etch depth of the gold of about 110 nm.

Figure 4.10(a) depicts an SEM image of the grating-polarizer central fragment. Note that the overall size of the polarizer is 100 × 100 μm and the fill factor was equal to 0.5. The quality of the micropolarizer is superior to that of Reference [289]. The grating grooves in the sectors are tilted by 40°, −40°, 70°, and −70°. Each sector of the micropolarizer in Figure 4.10(a) acts as a half-wave plate, with the linearly polarized electric field component incident in parallel with the grooves remaining unchanged and the perpendicular component acquiring a half-wave delay after reflection. Thus, following the reflection at a grating sector, the polarization is rotated by a definite angle.

When the linearly polarized white light is reflected at the micropolarizer under study (Figure 4.10(a)) just two of the four sections of the polarizer appear as bright

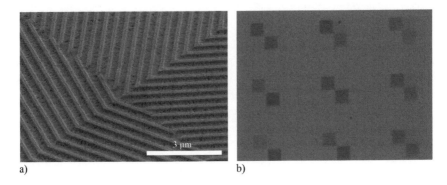

a)   b)

FIGURE 4.10 (a) SEM image of the central 13×13-µm fragment of a golden binary four-sector grating-polarizer: period, 400 nm, groove depth, 110 ± 10 nm; and (b) image of a golden substrate comprising nine micropolarizer prototypes obtained in linearly polarized white light and observed through a polarizer rotated by 45°. Each dark square measures 50×50-µm.

(or dark) in the image plane (Figure 4.10(b)) if observed through the polarizer rotated by +45° or −45° about the incident light polarization plane. From Figure 4.10(b), the light reflected at each micropolarizer sector is seen to be devoid of circular symmetry, whereas the average intensities across two diagonal squares are seen to be different.

Figure 4.11(a) shows an image of the four-sector micropolarizer of Figure 4.10(a) obtained using an output polarizer with its axis directed vertically, considering that the incident light is linearly polarized in the horizontal plane. In this case, all four sectors of the micropolarizer turn out to be bright (Figure 4.10(a)). Note that the two right sectors of the micropolarizer have a higher (10%) reflectance when compared with the two left ones. Although it is possible to achieve equal reflection coefficients in the four sectors, the price to be paid is different periods and fill factors of the gratings in each sector. Figure 4.11(b) shows a micropolarizer image obtained when the axis of the polarizer placed in reflected light is rotated by 45° about the horizontal axis. Figure 4.11(c) shows the same pattern but with the output polarizer axis rotated by −45° about the horizontal axis. In this case, two of four micropolarizer's sectors found on the diagonals are observed as bright.

(a)   (b)   (c)

FIGURE 4.11 Image of a 100×100-µm four-sector micropolarizer (Figure 4.10(a)) obtained in reflection under illumination by a 633-nm laser beam for the differently directed axis of the output polarizer (put before a CCD-camera): (a) 0°, (b) 45°, (c) −45°.

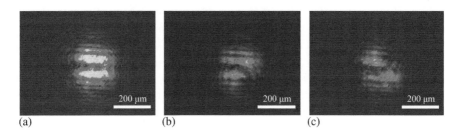

(a) (b) (c)

FIGURE 4.12 Far-field intensity patterns for the laser light (633 nm) reflected at the micro-polarizer (Figure 4.10): (a) without an output polarizer, and with a polarizer rotated by (b) 45° and (c) −45°.

Unlike a perfect radially polarized beam that generates a ring-like intensity pattern, in our case, the lack of circular symmetry in the far-field results in the micropolarizer generating a beam with a square-shaped transverse intensity pattern (Figure 4.12(a)) with its intensity maxima (Figures 4.12(b) (c)) found on the corresponding diagonals (as in Figures 4.11(b) and (c)). When the output polarizer is rotated by +45° and −45°, the resulting beam is a mixture of linearly and radially polarized beams.

4.2.2 BINARY MICROLENS IN A RESIST

A high quality binary microlens [102, 143] was fabricated in ZEP520A resist (refractive index $n = 1.52$) by electron beam lithography. Figure 4.13 depicts a SEM image of the microlens: mirorelief depth 510 nm, diameter 14 μm, and outermost zone $0.5\lambda = 266$ nm. The microlens is composed of 12 rings and a central disk.

The radii of the microlens rings were calculated using the familiar formula $r_m = (m\lambda f + m^2\lambda^2/4)^{1/2}$, where $f = \lambda = 532$ nm is the focal length and m is the radius number. The microlens has NA = 0.997. This formula suggests that when the focal length is small, $f = \lambda$, the microlens radii are increasing near linearly with increasing

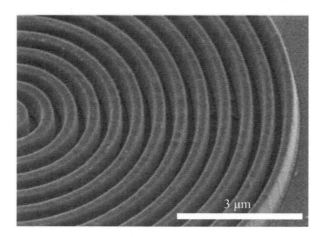

FIGURE 4.13 SEM image of the microlens with at 18000× magnification.

number m. Because of this, the microlens in Figure 4.13 looks very much like a diffractive axicon or a radial diffraction grating. When the microlens in Figure 4.13 is illuminated by a linearly polarized Gaussian beam of wavelength 633 nm and radius approximately equal to that of the microlens (7 μm), the focal spot is formed nearer than is suggested by the calculations (532 nm), being found 230 nm away from the microlens, as shown in Reference [143]. Using the FDTD method [143], the numerically calculated minimum and maximum sizes of the focal spot were shown to equal $FWHM_{min} = (0.40 \pm 0.02)\lambda$ and $FWHM_{max} = (0.87 \pm 0.02)\lambda$. In the meantime, the minimum and maximum sizes of the focal spot experimentally measured with an SNOM were $FWHM_{min} = (0.40 \pm 0.02)\lambda$ and $FWHM_{max} = (0.60 \pm 0.02)\lambda$. In the next section, we demonstrate that the microlens under study (Figure 4.13) is able to focus a radially polarized beam and a mixture of linear-radial polarizations into a tighter focal spot.

4.2.3 SIMULATION

Using the FDTD method, the focusing of a radially polarized $R\text{-}TEM_{01}$ mode of wavelength $\lambda = 633$ nm and mode parameter $R = 10\lambda$ by means of a microlens of focal length 532 nm intended to operate at a wavelength of 532 nm was simulated. In Figure 4.14 are: (a) the intensity profile in the focal point; and (b) the expected experimental profile that would be measured using an SNOM with a pyramid metal cantilever [290], In Figure 4.14(a), the focal spot size is $FWHM = 0.37\lambda$.

Figure 4.14 suggests that should there be a perfect radially polarized binary-microlens-aided focus (Figure 4.13), the SNOM image of the focal spot would appear as a circle (Figure 4.14(b)) of size $FWHM = 0.47\lambda$. That is to say, the image of a tight focus of a perfect radially polarized beam obtained with an SNOM with a hollow metal pyramid tip will be larger than the real spot. In the case under study, instead of the real size of $FWHM = 0.37\lambda$ the SNOM image is expected to be of size $FWHM = 0.47\lambda$.

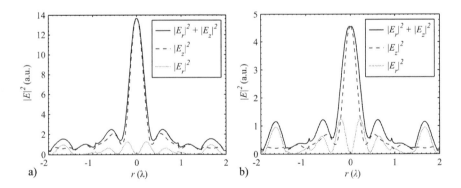

FIGURE 4.14 (a) Intensity profile in the focus near the microlens surface when focusing the $R\text{-}TEM_{01}$ mode with the parameter $R = 10\lambda$ (a) u and (b) the same NSOM-aided pattern 3 times less sensitive to the longitudinal component.

However, the four-sector micropolarizer in Figure 4.10(a) generates a mixture of linearly and radially polarized beams [289]. Below, we simulate in which way this type of beam is focused by the microlens of Figure 4.13 and in which way the focal spot is to be imaged by the SNOM. Figure 4.15 shows the intensity pattern generated 200 μm away from the micropolarizer of Figure 4.10(a), with the arrows showing the polarization directions. Figure 4.16 shows the intensity profiles in the focus of the microlens in Figure 4.13, which is illuminated by a mixed linearly and radially polarized beam (Figure 4.15) from the micropolarizer in Figure 4.10(a).

From Figure 4.16, the transverse intensity component is seen to be characteristic of the linearly polarized light, with the longitudinal component corresponding to the radially polarized light. This has prompted us to define the beam under study as having a mixed linear-radial polarization. In this case, instead of being a perfect circle (see a perfect radial polarization in Figure 4.14), the focal spot is elliptical with the axial dimensions $FWHM_{max} = (0.40 \pm 0.02)\lambda$ (Figure 4.16a) and $FWHM_{min} = (0.35 \pm 0.02)\lambda$ (Figure 4.16(b)). Let us recall that a circular focal spot for perfect radial polarization has been found to have an interim size of $FWHM = 0.37\lambda$ (Figure 4.14a). In terms of their area, the difference between these focal spots is as low as 3%. From Figure 4.16, the transverse intensity E_x^2 is seen not to take a zero value on the optical axis, taking a maximum value instead, although the longitudinal intensity E_z^2 is 6 times the transverse intensity component E_x^2. Figure 4.17 depicts intensity profiles similar to those shown in Figure 4.16(b) for the 3-times decreased longitudinal intensity. In this case, the focal spot is seen to be somewhat larger in size: $FWHM_{min} = (0.36 \pm 0.02)\lambda$ (Figure 4.17). A comparison of Figures 4.14 and 4.17 suggests that for both radial and mixed linear-radial polarization, SNOM-aided measurements with a metal hollow pyramid cantilever [290] (3 times more sensitive to the transverse than to the longitudinal electric field component), give an intensity maximum at the focal spot center.

In this subsection, we show that a mixed linearly and radially polarized beam from the micropolarizer of Figure 4.10(a) can be focused by means of the microlens of Figure 4.13 into a near-circular focal spot, which is tighter than that reported in

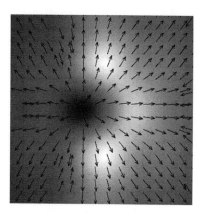

FIGURE 4.15 Numerically modeled intensity pattern |E|², 200 μm away from the micropolarizer of Figure 4.10(a). The frame size is (5×5) μm.

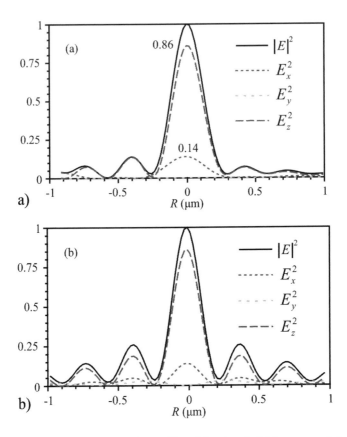

FIGURE 4.16 FDTD method: intensity profiles of the electric field components in the focus of the microlens of Figure 4.13 along the (a) *x*- and (b) *y*-axes.

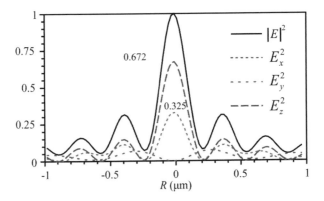

FIGURE 4.17 FDTD method: intensity profiles for the electric field components in the focus of the microlens in Figure 4.13 on the *y*-axis. The longitudinal intensity E_z^2 has been decreased by a factor of 3.

Reference [143], which is in agreement with the simulation results (Figures 4.14 and 4.16).

4.2.4 EXPERIMENTAL FOCUSING OF A MIXED LINEAR-RADIALLY POLARIZED BEAM WITH A BINARY MICROLENS

This experiment studied the tight focusing of an optical beam reflected from a micro-polarizer (Figure 4.10(a)), in which the linear polarization was converted into a radial one. The experimental optical setup is shown in Figure 4.18.

The light from a He-Ne laser ($\lambda = 633$ nm) was focused with an objective O_1 onto a four-sector micropolarizer (Figure 4.10(a)), which converted the linear polarization into the mixed linear-radial one. The polarizers P_1 and P_2 served the double purpose of verifying that the light incident on the micropolarizer was linearly polarized and attenuating the light power. After reflecting at the micropolarizer, the beam was focused onto the bottom of the microlens of Figure 4.13 of focal length 532 nm, producing a subwavelength focal spot after passing it. The intensity pattern of the focal spot was measured with an SNOM NTEGRA Spectra (NT-MDT) (shown in a dashed-line rectangle in Figure 4.18). Figure 4.19(a) illustrates an SNOM-aided intensity pattern, with the minimal and maximal size of the focal spot measuring $FWHM_{max} = (0.38 \pm 0.02)\lambda$ (Figure 4.19(b)) and $FWHM_{min} = (0.35 \pm 0.02)\lambda$ (Figure 4.19(c)). The simulation results for this case are shown in Figures 4.16 and 4.17: with the focal spot measuring $FWHM_{max} = (0.40 \pm 0.02)\lambda$ (Figure 4.16(a)) and $FWHM_{min} = (0.36 \pm 0.02)\lambda$ (Figure 4.17). Thus, the discrepancy between the experiment and simulation falls within the root mean square (rms) error 5%. In the focal spot, the sidelobes of the diffraction pattern amount to 30% of the major intensity peak (Figure 4.19). Thus, one can infer that a high-NA binary microlens (Figure 4.13) operates as a binary axicon, generating a focal spot with the intensity profile described by the squared zero-order Bessel function. For a Bessel beam, the diffraction limit is known to equal $FWHM = 0.36\lambda$.

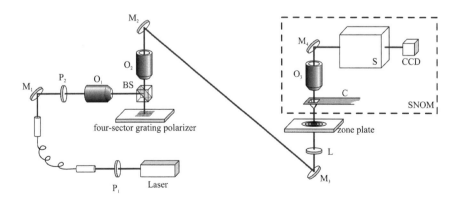

FIGURE 4.18 Experimental setup. Laser, He-Ne laser ($\lambda = 633$ nm), P_1, P_2, polarizers, M_1, M_2, M_3, M_4, mirrors, O_1, 3.7× objective, BS, light-splitting cube, O_2, 20× objective, O_3, 100× objective, C, cantilever, L, lens, S, spectrometer, CCD, CCD-camera.

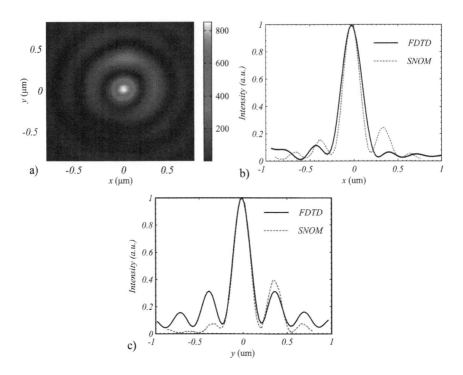

FIGURE 4.19 Illustration of an SNOM-aided intensity distribution when a beam reflected at the micropolarizer is focused by the microlens: (a) two-dimensional intensity pattern; and (b), (c) intensity profiles along the perpendicular axes (dashed lines). FDTD simulation from Figures 4.16 and 4.17 (solid lines).

To evaluate the effect of the micropolarizer, the substrate containing the micro-polarizer was shifted in the transverse direction to allow the light to be reflected at the relief-free golden surface. This resulted in a microlens-aided focal spot of dimensions $FWHM_{max} = (0.50 \pm 0.02)\lambda$ and $FWHM_{min} = (0.40 \pm 0.02)\lambda$, similar to that reported in Reference [143].

Using a binary microlens of diameter 14 μm and focal length 532 nm (NA = 0.997) we have focused a 633-nm laser beam into a near-circular focal spot with dimensions $(0.35 \pm 0.02)\lambda$ and $(0.38 \pm 0.02)\lambda$ at full-width of half-maximum intensity. The focal spot area is $0.105\lambda^2$. The incident light is a mixture of linearly and radially polarized beams generated by reflecting a linearly polarized Gaussian beam at a gold 100 μm × 100 μm four-sector subwavelength diffractive optical microelement, with the sector gratings of period 400 nm and microrelief depth 110 nm. The focusing of a linearly polarized laser beam (other conditions being the same) has been found to produce a larger focal ellipse measuring $(0.40 \pm 0.02)\lambda$ and $(0.50 \pm 0.02)\lambda$, similar to that reported in Reference [143]. The subwavelength focusing of light using a pair of microoptic elements (a binary microlens and a micropolarizer) has been implemented for the first time. It is notable that recently [292] a focal spot of size 0.25λ has been attained by focusing an azimuthally polarized first-order vortex beam by means of a conventional immersion microlens with NA = 1.4. When reduced to the common NA, the focal spot in Reference [292] and the one reported in this work, equal to 0.35λ, have nearly the same size, namely,

$$\text{FWHM} = 0.25\lambda = 0.25\lambda * 1.4 / \text{NA} = 0.35\lambda / \text{NA},$$

$$\text{FWHM} = 0.35\lambda = 0.35\lambda * 0.997 / \text{NA} = 0.35\lambda / \text{NA}.$$

From the formulae above it follows that the immersion microlens (Figure 4.13) gives the same result as in Reference [292].

4.3   A FOUR-SECTOR REFLECTING AZIMUTHAL MICROPOLARIZER

Recent years have seen the publication of a large number of articles handling techniques for generating cylindrical vector beams. One way to obtain cylindrical vector beams is through the use of subwavelength diffraction gratings. By properly choosing the feature height in such a grating, it may serve as a half- (or quarter-) wave plate, meanwhile the groove direction defines the spatial orientation of the plate.

For the first time, the use of the subwavelength gratings as polarization converters was reported in Reference [284]. As the first publications to report on the fabrication of the gratings, we may denote References [285, 286], where a circularly polarized 10.6-μm light beam was converted into an azimuthally polarized beam. Generation of a radially polarized light beam for a subwavelength grating operating at a 1064-nm wavelength was discussed in Reference [288]. A technique for fabricating a polarization converter in silicon for wavelengths ranging from 1030 to 1060 nm was proposed in Reference [293].

It is worth noting that the majority of papers concerned with the use of subwavelength gratings for polarization conversion analyze the IR range of wavelengths because the fabrication of diffraction gratings for the visible light is a technologically challenging task. We have only managed to find one paper, [294], in which a subwavelength annular diffraction grating fabricated in aluminum performed the conversion of a circularly polarized beam into a radially polarized one for a visible wavelength of 633 nm. Note, however, that such a grating is unable to form an azimuthally polarized beam. Earlier, we discussed a four-zone grating-polarizer intended to form a radially polarized beam [289]. In a follow-up study, we used the resulting radially polarized beam to form a subwavelength focal spot of area $(0.35 \times 0.38)\lambda$ by means of a Fresnel zone plate [295]. The radially polarized light has been primarily used for the sharp focusing of light. In recent years, the interest of researchers has been attracted to the analysis of optical vortices with radial and azimuthal polarization.

To the best of our knowledge, the first work to study the focusing of non-uniformly polarized optical vortices was reported in Reference [296]. In particular, using the Richards–Wolf integrals, the researchers have shown that a radially or azimuthally polarized optical vortex is able to form a focal spot with central intensity peak only for the topological charge equal to $n = 1$. Focusing radially and azimuthally polarized optical vortices with the topological charges $n = 1$ and $n = 2$ in a birefringent crystal was numerically studied in Reference [297]. The vortex with $n = 1$ was shown to generate an intensity peak in the focus of a lens with NA = 0.9, whereas the vortex with the topological charge $n = 2$ generated an annular intensity pattern. The propagation of an optical vortex initially characterized by radial polarization was analyzed in Reference [298]. The effect of a spiral plate introduced into an azimuthally polarized laser beam to obtain a tighter focal spot was studied in Reference [299]. Such a beam was shown to generate a focal spot of the area $0.147\lambda^2$, which is 13.5% smaller when compared to a similar focal spot from a radially polarized beam ($0.17\lambda^2$). In Reference [300], an azimuthally polarized beam was transmitted through concentric annular filters, with the rings acting as spiral plates arranged so as to ensure a phase shift of π between the adjacent rings. The beam was focused with a wide-aperture lens (NA = 0.95) within a magneto-optical material, forming a needle-shaped focus of diameter FWHM = 0.38λ and depth 7.48λ. In a similar way, a radially polarized optical vortex was reported in Reference [301] to be focused in the magneto-optical material using a 4π-system. A radially polarized optical vortex transmitted through an annular mask and focused by a 4π-system has been reported to generate foci chains over 30λ long [302]. It should be noted that in the above-mentioned papers, the focusing of radially and azimuthally polarized beams was studied numerically, mainly with the use of the Richards–Wolf formulae. The only experimental research we have managed to discover, [292], reported on the focusing of an azimuthally polarized optical vortex by means of a lens with NA = 1.4. The resulting focal spot had an area of $0.089\lambda^2$, which is 31% smaller than a focal spot from a circularly polarized beam.

In this section, we fabricated and studied the performance of a 100×100-µm four-sector binary subwavelength grating-polarizer in a golden film. It was experimentally demonstrated that a linearly polarized 532-nm Gaussian beam reflected at the polarizer was converted to an azimuthally polarized beam. Putting a spiral phase plate (SPP) with the topological charge $n = 1$ into the azimuthally polarized beam from the micropolarizer was experimentally shown to enable the conversion of the annular intensity pattern into a central intensity peak.

4.3.1 NUMERICAL SIMULATION OF THE MICROPOLARIZER PERFORMANCE

The proposed micropolarizer consists of four sectors (Figure 4.20(a)): in the two sectors on the right, the microrelief features have a 0.46-µm period and make angles of $70°$ and $-70°$ with the y-axis, whereas in the sectors on the left, the features have a 0.4-µm period and make angles of $40°$ and $-40°$ with the y-axis. The micropolarizer has a size of 100×100 µm and the microrelief height of 110 nm.

Figures 4.20(b) and (c) depict the arrangement of gratings in a four-sector micropolarizer to convert linear polarization (vertical arrows) into azimuthal (Figure 4.20(b)) and radial (Figure 4.20(c)) polarization. The arrows within the enumerated squares (1–4) mark the polarization direction of the reflected light. The angles of the grating grooves with the vertical axis are shown in degrees. Figures 4.20 (b) and (c) suggest that by merely rotating the polarizer in Figure 4.20(b) by 90 degrees it is not possible to obtain the polarizer in Figure 4.20(c). To accomplish this requires rearranging the sub-gratings anticlockwise: 4 to 1, 1 to 2, 2 to 3, and 3 to 4. Also, note that the sub-gratings have different periods.

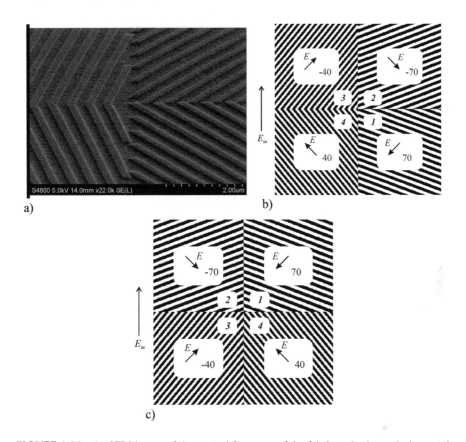

FIGURE 4.20 (a) SEM image of the central fragment of the fabricated micropolarizer, and the arrangement of gratings of the four-sector polarizer to form (b) azimuthal. and (c) radial polarization.

The performance of the micro-polarizer in Figures 4.20(a) and (b) was simulated as follows: first, the FDTD-aided simulation of the complex amplitude of the field reflected at the polarizer was implemented using the FullWAVE software. A linearly polarized plane wave of wavelength 532 nm was assumed to hit the polarizer normally. The mesh of the FDTD method had a $\lambda/30$ step. The refractive index of the grating grooves and the substrate was $n = 0.312 + 3.17i$ (gold). The grating feature height was put at 110 nm, with the substrate thickness being 150 nm. The field distribution at a significant distance from the polarizer was calculated using the Rayleigh–Sommerfeld integral, with the FDTD-aided complex amplitude calculated 200 nm away from the surface taken as an initial field guess. Figure 4.21 depicts the intensity patterns calculated at distances of 5 μm, 300 μm, and 500 μm from the element, the arrows marking the polarization directions.

The arrows in Figure 4.21 mark the polarization directions. The fact that a number of polarization arrows are oriented in the opposite direction is due to an error of π in determining the polarization direction at this point. Following

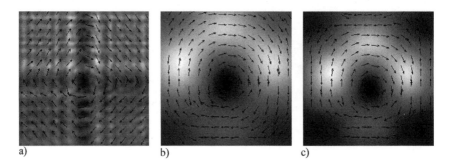

FIGURE 4.21   Intensity patterns and polarization directions for the reflected beam at distances (a) 5.2 µm, (b) 300.2 µm, and (c) 500.2 µm from the polarizer surface. The image size is (a), (b) 20×20 µm and (c) 40×40 µm.

the reflection at the polarizer, the plane wavefront turns into a spherical one at a large distance, whereas the observation remains in a plane. As a result, there occurs an extra phase shift that introduces an error of π when determining the polarization direction. It is worth noting that such a "failure" in polarization determination occurs locally, not having an impact on the general pattern of beam polarization.

4.3.2   Fabrication of a Micropolarizer and Generation of Azimuthal Polarization

The micropolarizer in Figure 4.20(a) was fabricated by electron beam lithography. A golden layer of thickness 160–180 nm was coated onto a glass substrate, followed by applying a resist layer, on which a four-sector grating-polarizer pattern was generated by electron beam lithography (at 30 kV). Then the sample was developed by etching with xylene, which dissolved the fragments exposed to the electron beam. Next, using reactive ion etching, the grating-polarizer pattern was transferred into the golden layer, which was etched in places with no resist. Using argon plasma, gold particles were spattered from areas unprotected with the resist. At the final stage, the resist residue was eliminated using oxygen plasma, producing a micropolarizer template "engraved" in gold. The reactive ion etching time was optimized so as to achieve an etch depth of about 110 nm.

The performance of the fabricated micropolarizer was experimentally tested using a linearly polarized beam of 1-mm width from a 532-nm laser. The beam was focused with a 10× lens O_1 onto the substrate containing the grating micropolarizer (Figure 4.22). The size of the spot focused on the micropolarizer was controlled by varying the distance from the lens O_1 to the micropolarizer surface. Although in this case, the micropolarizer was not found in the beam waist and the incident wave was spherical, the experimental results we discuss below corroborate that the micropolarizer still operated in a proper way. This may be explained by the fact that while changing from a plane to a spherical wavefront the beam polarization does not acquire an azimuthal component, only acquiring a longitudinal component along the propagation axis,

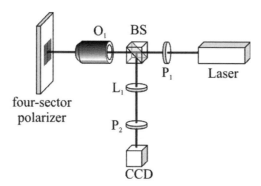

FIGURE 4.22 Experimental setup: P_1, P_2 are linear polarizers, BS is a non-polarizing light-splitting cube, O_1 is a 10× lens (NA=0.25), L_1 is a lens ($f \approx 1.5$ cm), and a CCD-camera.

meaning that the angle between the polarization direction in the polarizer plane and the microrelief grooves remains unchanged. The image of the four-sector grating-polarizer was displayed in a CCD-camera using a lens L_1 ($f \approx 1.5$ cm, NA=0.01). The polarization of the input beam was determined using a polarizer/analyzer P_2.

Figure 4.23 depicts the central fragment of a four-sector polarization converter illuminated by a linearly polarized Gaussian beam from a 532-nm laser. The illuminated area is somewhat smaller than the entire micropolarizer of size 100×100 µm. The linear polarization of the light outgoing from the splitting cube and incident on the micropolarizer was checked using a standard polarizer. Following reflection at the micropolarizer, the polarization state of light was changed. Figure 4.24 depicts images of the micropolarizer in Figure 4.20(a) with a polarizer P_2 placed in front of the CCD-camera and making angles 0°, 90°, −45°, and 45° with the incident light polarization axis. Figure 4.25 depicts the far-field intensity pattern for a beam

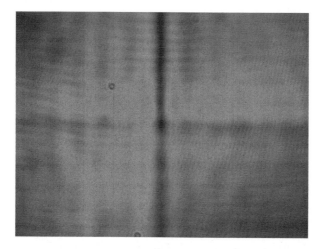

FIGURE 4.23 Image of the micropolarizer in Figure 4.22 obtained in laser light without an output polarizer.

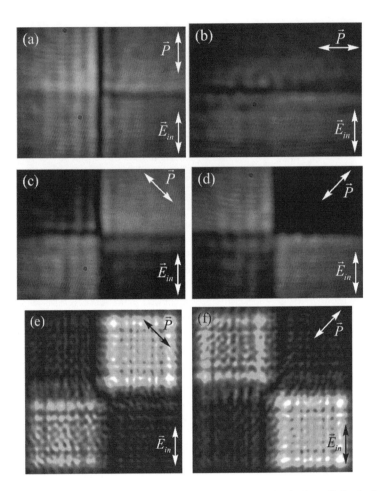

FIGURE 4.24 Image of the micropolarizer in Figure 4.20(a) in the laser light. In front of the camera was an output polarizer/analyzer P_2, making an angle of (a) 0°, (b) 90°, (c) −45°, and (d) 45° with the incident beam. The frame size was 4 mm×3 mm. Numerically simulated intensity patterns of the beam in Figure 4.21(a), which was reflected at the four-sector polarizer and then transmitted through the analyzer P_2, with the analyzer in Figures 4.24(e) and 4.24(f) being, respectively, rotated by −45° and 45° with respect to the incident light polarization axis.

reflected at the micropolarizer in Figure 4.20(a), which occurs in the focal plane of lens L_1. The CCD-camera image was obtained using a 20×lens.

From Figures 4.24(c) and (d) it is possible to obtain a quantitative estimate of the micropolarizer efficiency by calculating the ratio of the maximal to the minimal energy in the corresponding quadrants. In particular, in Quadrant I (first quarter of a circle), the ratio of maximal (Figure 4.24(c)) to minimal (Figure 4.24(d)) energy was found to be 185:1, in Quadrant II (second quarter of a circle), 17:1, and in the bottom quadrants, 6:1.

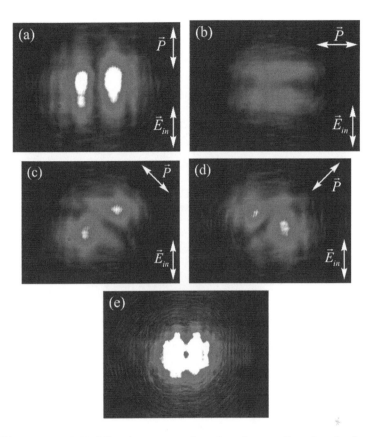

FIGURE 4.25 Far-field diffraction patterns for a laser beam reflected at the four-sector micropolarizer. In front of the camera was an output polarizer/analyzer P_2 making the angles of (a) $0°$, (b) $90°$, (c) $-45°$, and (d) $45°$ with the incident beam polarization axis; (e) the annular intensity with no polarizer. The frame size is $4\,\text{mm} \times 3\,\text{mm}$.

Based on Figure 4.25, we can infer that the four-sector micropolarizer in Figure 4.20(a) converts an incident linearly polarized beam into an azimuthally polarized light beam. However, the intensity of the reflected beam is not distributed uniformly over the ring. The reason is that the four-sector polarizer in Figure 4.20(a) is devoid of radial symmetry. It should be noted that a classical work on the sharp focus of radially polarized laser light [291] also utilized a four-sector polarizer, which was made up of mutually rotated half-wave plates. Hence, we can infer that four sectors would suffice to form a radially [291] or azimuthally polarized beam. Note that the use of a four-sector micropolarizer in our previous work [295] made it possible to focus the laser beam into a focal spot of area $(0.35 \times 0.38)\lambda$, meanwhile using the four-sector polarizer made up of half-wave plates [291] enabled a focal spot of size 0.45λ. Let us note that both four- and eight-sector polarizers were discussed in Reference [303]. However, the resulting near-field diffraction patterns were of inferior quality (see Figure 1(c) in Reference [303]). It is also worth noting that the micropolarizers under study are simpler to fabricate compared to the set of half-wave plates [303].

4.3.3 Optical Vortices with Different Polarization of the Initial Beam

In this subsection, using the Rayleigh–Sommerfeld integrals we derive conditions for an on-axis intensity maximum for optical vortices with different initial polarization. The propagation of coherent monochromatic light in free space is known to be described by the vector Rayleigh–Sommerfeld integrals [284]:

$$E_{x,y}(\rho,\theta,z) = -\frac{1}{2\pi}\int_0^\infty\int_0^{2\pi} E_{x,y}(r,\varphi,0)\frac{\partial}{\partial z}\left(\frac{e^{ikR}}{R}\right)rdrd\varphi, \tag{4.1}$$

$$E_z(\rho,\theta,z) = \frac{1}{2\pi}\int_0^\infty\int_0^{2\pi}\left[E_x(r,\varphi,0)\frac{\partial}{\partial x}\left(\frac{e^{ikR}}{R}\right)\right.$$

$$\left. +E_y(r,\varphi,0)\frac{\partial}{\partial y}\left(\frac{e^{ikR}}{R}\right)\right]rdrd\varphi, \tag{4.2}$$

where

$$R = \left(z^2 + r^2 + \rho^2 - 2r\rho\cos(\varphi-\theta)\right)^{1/2}$$

$$\approx z + \frac{r^2}{2z} + \frac{\rho^2}{2z} - \frac{r\rho}{z}\cos(\varphi-\theta), \tag{4.3}$$

k is the wave number, (ρ, θ, z) are the cylindrical coordinates, and (r, φ) are the polar coordinates in the initial plane $z=0$. The approximate relation in Equation (4.3) is valid in the paraxial case, when $z \gg r, \rho$.

In the initial plane let a uniformly (linearly, circularly, or elliptically) polarized optical vortex be described by the projection of the E-field:

$$E_{x,y} = (r,\varphi,0) = A_{x,y}(r)e^{in\varphi}, \tag{4.4}$$

where n is an integer number referred to as the topological charge of an optical vortex. Then, substituting Equation (4.4) into Equations (4.1) and (4.2), and making use of the paraxial approximation Equation (4.3), we obtain the light field projections at any z in the form:

$$E_{x,y}(\rho,\theta,z) = \frac{(-i)^{n+1}k}{z}\exp\left(ikz+in\theta+\frac{ik\rho^2}{2z}\right)$$

$$\times\int_0^\infty A_{x,y}(r)e^{\frac{ikr^2}{2z}}J_n\left(\frac{kr\rho}{z}\right)rdr, \tag{4.5}$$

$$E_z(\rho,\theta,z) = \frac{(-i)^n k}{2z^2} \exp\left(ikz + in\theta + \frac{ik\rho^2}{2z} \right)$$

$$\times \int_0^\infty e^{\frac{ikr^2}{2z}} \left\{ re^{i\theta} \left[A_x(r) - iA_y(r) \right] J_{n+1}\left(\frac{kr\rho}{z} \right) \right.$$

$$\left. - re^{-i\theta} \left[A_x(r) + iA_y(r) \right] J_{n-1}\left(\frac{kr\rho}{z} \right) \right.$$

$$\left. - 2i\rho \left[A_x(r)\cos\theta + A_y(r)\sin\theta \right] J_n\left(\frac{kr\rho}{z} \right) \right\} r dr.$$

(4.6)

From Equations (4.5) and (4.6), the intensity distribution of transverse E-field components of a uniformly polarized field is seen to be radially symmetric, with the intensity distribution of the axial E-field component being devoid of radial symmetry. Equations (4.5) and (4.6) also suggest that on the optical axis ($\rho=0$) the amplitude of all three E-field projections takes a zero value at $n>1$ and $n<-1$. At $n=0$, the axial E-field component equals zero, whereas the transverse components can take non-zero values. On the contrary, at $n=\pm1$, the transverse E-field components are equal to zero, whereas the axial component is non-zero. However, an exception takes place if the initial field in Equation (4.4) is circularly polarized clockwise, $A_x(r) = iA_y(r)$, and the topological charge is $n=-1$, then the on-axis component of field (4.6) takes a zero value: $E_z(\rho=0,\theta,z)=0$. Similarly, if the initial field is polarized circularly anticlockwise, $A_x(r) = -iA_y(r)$, and $n=1$, we again have $E_z(\rho=0,\theta,z)=0$. Summing up, there are three possible combinations of the initial polarization type and the topological charge of an optical vortex that can result in an on-axis focal spot (Table 4.1).

With a view of deriving relations similar to (4.5) and (4.6) for non-uniformly (radially and azimuthally) polarized light fields, let us write down formulae to change from the transverse E-field projections written in the Cartesian coordinates (E_x, E_y) to the polar coordinates (E_r, E_φ):

TABLE 4.1
Conditions for Generating an On-Axis Focal Spot for Uniformly Polarized Light

No.	Topological Charge	Type of Uniform Polarization
1	$n=0$	$A_x(r)+\alpha A_y(r)$ any
2	$n=1$	$A_x(r)=iA_y(r)$ circular clockwise
3	$n=-1$	$A_x(r)=-iA_y(r)$ circular anticlockwise

$$\begin{pmatrix} E_r \\ E_\varphi \end{pmatrix} = \begin{pmatrix} \cos\varphi & \sin\varphi \\ -\sin\varphi & \cos\varphi \end{pmatrix} \begin{pmatrix} E_x \\ E_y \end{pmatrix},$$

$$\begin{pmatrix} E_x \\ E_y \end{pmatrix} = \begin{pmatrix} \cos\varphi & -\sin\varphi \\ \sin\varphi & \cos\varphi \end{pmatrix} \begin{pmatrix} E_r \\ E_\varphi \end{pmatrix}.$$

(4.7)

In view of Equation (4.7), the Rayleigh–Sommerfeld integrals in Equations (4.1)–(4.3) for non-uniformly polarized light fields take the form:

$$E_r(\rho,\theta,z) = -\frac{1}{2\pi} \int_0^\infty \int_0^{2\pi} \big[E_r(r,\varphi,0)\cos(\varphi-\theta)$$

(4.8)

$$-E_\varphi(r,\varphi,0)\sin(\varphi-\theta)\big] \frac{\partial}{\partial z}\left(\frac{e^{ikR}}{R}\right) r\,dr\,d\varphi,$$

$$E_\varphi(\rho,\theta,z) = -\frac{1}{2\pi} \int_0^\infty \int_0^{2\pi} \big[E_r(r,\varphi,0)\sin(\varphi-\theta)$$

(4.9)

$$+E_\varphi(r,\varphi,0)\cos(\varphi-\theta)\big] \frac{\partial}{\partial z}\left(\frac{e^{ikR}}{R}\right) r\,dr\,d\varphi,$$

$$E_z(\rho,\theta,z) = \frac{1}{2\pi} \int_0^\infty \int_0^{2\pi} \big[E_r(r,\varphi,0)\big(r-\rho\cos(\varphi-\theta)\big)$$

(4.10)

$$+E_\varphi(r,\varphi,0)\rho\sin(\varphi-\theta)\big] \frac{\partial}{R\partial R}\left(\frac{e^{ikR}}{R}\right) r\,dr\,d\varphi.$$

Equations (4.8)–(4.10) suggest that in the general case, the initial radially polarized, $E_\varphi(r,\varphi,0)=0$, and azimuthally polarized, $E_r(r,\varphi,0)=0$, light fields do not preserve their respective radial and azimuthal polarization as they propagate. From Equation (4.10), the contribution to the on-axis intensity ($\rho=0$) is seen to be only made by the radially polarized component $E_r(r,\varphi,0)$. Thus, we can infer that if the initial light field has only an azimuthal component $E_\varphi(r,\varphi,0)$, the axial component ($\rho=0$) of the E-field equals zero: $E_z(\rho=0,\theta,z)=0$.

Let the initial light field be in the form of a radially polarized optical vortex with the complex amplitude:

$$\begin{cases} E_r(r,\varphi,0) = A_r(r)e^{in\varphi}, \\ E_\varphi(r,\varphi,0) = 0. \end{cases}$$

(4.11)

Then, integrals (4.8)–(4.10) in the paraxial approximation are given by

$$E_r(\rho,\theta,z) = \frac{(-i)^n k}{2z} \exp\left(ikz + in\theta + \frac{ik\rho^2}{2z}\right)$$

$$\times \int_0^\infty A_r(r) e^{\frac{ikr^2}{2z}} \left[J_{n-1}\left(\frac{kr\rho}{z}\right) - J_{n+1}\left(\frac{kr\rho}{z}\right) \right] r dr,$$

(4.12)

$$E_\varphi(\rho,\theta,z) = \frac{(-i)^{n-1} k}{2z} \exp\left(ikz + in\theta + \frac{ik\rho^2}{2z}\right)$$

$$\times \int_0^\infty A_r(r) e^{\frac{ikr^2}{2z}} \left[J_{n-1}\left(\frac{kr\rho}{z}\right) + J_{n+1}\left(\frac{kr\rho}{z}\right) \right] r dr,$$

(4.13)

$$E_z(\rho,\theta,z) = \frac{(-i)^{n-1} k}{2z^2} \exp\left(ikz + in\theta + \frac{ik\rho^2}{2z}\right)$$

$$\times \int_0^\infty A_r(r) e^{\frac{ikr^2}{2z}} \left\{ \begin{array}{l} 2r J_n\left(\frac{kr\rho}{z}\right) \\[2mm] + i\rho\left[J_{n+1}\left(\frac{kr\rho}{z}\right) - J_{n-1}\left(\frac{kr\rho}{z}\right) \right] \end{array} \right\} r dr.$$

(4.14)

From Equations (4.12)–(4.14) we can infer that (1) the initial radial polarization of a vortex field does not remain unchanged during propagation, turning into a mixture of radially and azimuthally polarized components; and (2) the initially radially polarized vortex field with a radially symmetric intensity distribution retains radial symmetry of the intensity during propagation, which means that such a beam can be focused into a radially symmetric on-axis focal spot. Equations (4.12) and (4.13) also suggest that taking the initial radially polarized non-vortex beam ($n=0$) and, considering that $J_{-n}(x) = (-1)^n J_n(x)$, we obtain the azimuthal E-field component with zero amplitude: $E_\varphi(\rho=0,\theta,z) = 0$. Thus, the radially polarized and radially symmetric light field $E_r(r,\varphi,0) = A_r(r)$ retains its radial polarization during propagation.

Note that at $n>1$ and $n<-1$, all E-field components in Equations (4.12)–(4.14) equal zero on the optical axis ($\rho=0$). If $n=0$, only longitudinal components take non-zero values on the optical axis ($\rho=0$): $E_z(\rho=0,\theta,z) \neq 0$, whereas at $n=\pm 1$, only transverse E-field components are non-zero on the optical axis: $E_r(\rho=0,\theta,z) = \mp i\, E_\varphi(\rho=0,\theta,z) \neq 0$. Thus, there are two ways to form an on-axis focal spot: (1) by means of a radially polarized non-vortex ($n=0$) initial field; and (2) using an initial radially polarized vortex field at $n=\pm 1$. The latter approach is preferable because the annular on-axis focal spot is contributed to by the transverse E-field components. The transverse E-field components propagate along the optical axis and can be used by the observer. With the former approach, when the focal spot is obtained using a non-vortex radially symmetric field, light does not propagate along the optical axis.

Let the initial field be defined by an azimuthally polarized optical vortex with a complex amplitude:

$$\begin{cases} E_\varphi(r,\varphi,0) = A_\varphi(r)e^{in\varphi}, \\ E_r(r,\varphi,0) = 0. \end{cases} \tag{4.15}$$

Then, integrals (4.8)–(4.10) in the paraxial approximation take the form:

$$E_r(\rho,\theta,z) = \frac{(-i)^{n+1}k}{2z} \exp\left(ikz + in\theta + \frac{ik\rho^2}{2z} \right)$$

$$\times \int_0^\infty A_\varphi(r)e^{\frac{ikr^2}{2z}} \left[J_{n-1}\left(\frac{kr\rho}{z}\right) + J_{n+1}\left(\frac{kr\rho}{z}\right) \right] rdr, \tag{4.16}$$

$$E_\varphi(\rho,\theta,z) = \frac{(-i)^n k}{2z} \exp\left(ikz + in\theta + \frac{ik\rho^2}{2z} \right)$$

$$\times \int_0^\infty A_\varphi(r)e^{\frac{ikr^2}{2z}} \left[J_{n-1}\left(\frac{kr\rho}{z}\right) - J_{n+1}\left(\frac{kr\rho}{z}\right) \right] rdr, \tag{4.17}$$

$$E_z(\rho,\theta,z) = \frac{(-i)^{n+1}k\rho}{2z^2} \exp\left(ikz + in\theta + \frac{ik\rho^2}{2z} \right)$$

$$\times \int_0^\infty A_\varphi(r)e^{\frac{ikr^2}{2z}} \left[J_{n-1}\left(\frac{kr\rho}{z}\right) + J_{n+1}\left(\frac{kr\rho}{z}\right) \right] rdr, \tag{4.18}$$

From comparison of (4.16) and (4.18), the radial and longitudinal components of the E-field are seen to be related as

$$E_z(\rho,\theta,z) = \frac{\rho}{z} E_r(\rho,\theta,z). \tag{4.19}$$

From Equations (4.16)–(4.18) it follows that (1) the initial field defined by an azimuthally polarized vortex does not retain azimuthal polarization during propagation, becoming a mixture of radial and azimuthal polarization; and (2) the initial field defined by an azimuthally polarized vortex with a radially symmetric intensity distribution retains radial symmetry of the intensity upon propagation; i.e. such a field forms a radially symmetric on-axis focal spot. From Equations (4.16) and (4.17) it also follows that given a non-vortex azimuthally polarized initial field ($n=0$) and considering that $J_{-n}(x) = (-1)^n J_n(x)$, the amplitude of the radial component of the E-field is zero: $E_r(\rho=0,\theta,z) = 0$. Hence, an azimuthally polarized and radially symmetric field $E_\varphi(r,\varphi,0) = A_\varphi(r)$ remains azimuthally polarized during propagation. It can

TABLE 4.2

Conditions for Generating an On-Axis Focal Spot by a Non-Uniformly Polarized Light Beam

No.	Topological Charge	Initial Non-Uniform Polarization
1	$n=0$	$A_r(r)$ radial
2	$n=\pm1$	$A_r(r)$ radial or $A_\phi(r)$ azimuthal

be noted that the above conclusions relating to the field in Equations (4.16)–(4.18) coincide with the conclusions relating to the field in Equations (4.12)–(4.14), with the term "azimuthal" being replaced with "radial." However, the subsequent conclusions are going to be different. At $n \neq \pm1$, all E-field components in Equations (4.16)–(4.18) are zero on the optical axis ($\rho=0$). The longitudinal component (4.19) always equals zero on the optical axis: $E_z(\rho=0,\theta,z)=0$. If $n=\pm1$, only transverse E-field components take non-zero values on the optical axis: $E_r(\rho=0,\theta,z)=\mp iE_\varphi(\rho=0,\theta,z)\neq0$. Thus, using the initial azimuthally polarized vortex field with the topological charge $n=\pm1$, the on-axis focal spot can be generated in a unique way. The other conditions being the same, the resulting annular focal spot should be of the same size as for the radially polarized initial vortex field with ±1. This can be inferred from the fact that Equations (4.12) and (4.14) are functionally equivalent to Equations (4.16) and (4.18) (see Table 4.2).

4.3.4 A Spiral Phase Plate in an Azimuthally Polarized Laser Beam

In the experimental setup in Figure 4.26, a linearly polarized beam from a 532-nm laser was focused using a 10× objective O_1 onto a substrate coated with a four-zone grating-polarizer. Following the reflection at the micropolarizer, the

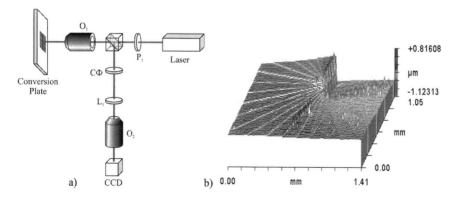

FIGURE 4.26 (a) Experimental setup: P_1 is a linear polarizer, O_1 is a 10× objective (NA=0.25), L_1 is a lens (f $\approx$ 1.5 cm), O_2 is a 20× objective (NA=0.4), CCD is a CCD-camera, and (b) SPP is a spiral phase plate with $n=1$.

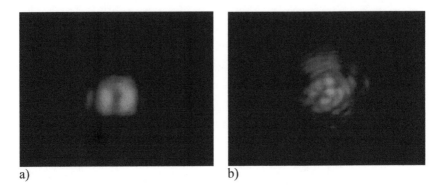

a)   b)

FIGURE 4.27 Lens-aided focal spots from an azimuthally polarized beam (a) without and (b) with an SPP with n = 1 in the beam's path. Frame size is (2.7 × 2) mm.

beam passed through a spiral phase plate with $n = 1$ and lens L_1 ($f \approx 1.5$ cm), forming a focal spot, with the intensity distribution observed using a 20× objective O_2 (Figure 4.26).

Figure 4.27 depicts the intensity pattern in the focus without Figure 4.27(a) and with Figure 4.27(b) a spiral phase plate put in the beam's path.

From Figure 4.27 the azimuthally polarized beam is seen to be focused by the lens into a near-annular focal spot. It is also seen that by introducing an SPP into the beam's path the focal intensity pattern is changed to a central intensity peak with sidelobes. This can be seen clearly in Figure 4.27, which shows the profiles of the intensity patterns in Figure 4.28.

To verify the experimental results, we additionally simulated the focusing of an azimuthally polarized beam with $n = 1$ using the Richards–Wolf formulae. In the first run of simulation, the lens (NA = 0.3) was assumed to focus a plane wave in which polarization changed sector-wise, depending on the value of the azimuthal angle φ. In the second run of simulation, the plane wave from the first run was complemented with an optical vortex. The resulting intensity patterns are shown in Figure 4.29.

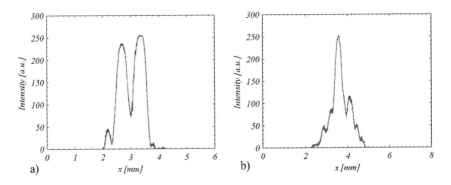

FIGURE 4.28 Lens-focused intensity profiles of Figure 4.27 (a) without and (b) with an SPP with n = 1 in the beam's path.

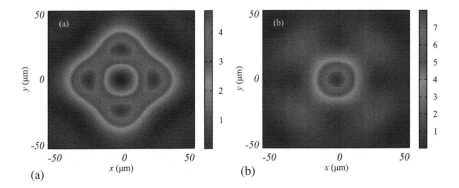

FIGURE 4.29 Intensity patterns (negative) in the lens focus for an azimuthally polarized beam (a) without and (b) with an SPP placed in the beam's path.

The simulation results in Figure 4.29 show that the focal spot size (b) coincides with the size of a low-intensity region in the center of the four-sector ring (a). A similar pattern can be seen in the experiment in Figures 4.27 and 4.28.

Summing up, we have for the first time designed and manufactured a four-sector binary subwavelength reflecting micropolarizer in a gold film to convert a linearly polarized light beam into an azimuthally polarized beam. It should be noted that the four-sector polarizer composed of subwavelength gratings to convert linear polarization into the radial one [289] is different from the four-sector micropolarizer in Figure 4.20 to convert linear polarization into azimuthal polarization. Using FDTD-aided simulation and physical experiments, we have shown the manufactured four-sector micropolarizer to form an azimuthally polarized beam in the near- and far-field diffraction regions. Using vector Rayleigh–Sommerfeld integrals, conditions for the on-axis intensity peak have been deduced for circularly, azimuthally, and radially polarized vortices. It has been numerically and experimentally shown that by introducing an SPP with $n=1$ into an azimuthally polarized beam the focal intensity pattern can be changed from "annular" to a central intensity peak for low numerical apertures (NA <0.3).

4.4 TIGHT FOCUSING OF A QUASI-CYLINDRICAL OPTICAL VORTEX

Cylindrical vector beams (beams with polarization with a radial direction of symmetry) are currently an active topic of research [304]. Recent years have also seen an increased interest in the study of azimuthally and radially polarized optical vortices. It should be noted that a radially polarized beam forms a sharp peak in a focal spot, whereas azimuthally polarized light forms a ring in a focal spot. Thus, an azimuthally polarized beam needs a phase singularity to produce a peak in the focal spot.

In Reference [299], it was shown that an azimuthally polarized optical vortex produces a focal spot whose area ($0.147\lambda^2$) is 13.5% smaller than a focal spot from a radially polarized beam ($0.17\lambda^2$). Optical needles generated by azimuthally polarized vortices were investigated in Reference [305]. These needles have a depth of

12λ and a subwavelength width which varies from 0.42λ to 0.49λ. In Reference [105], an azimuthally polarized beam propagated through a multibelt phase hologram and high-NA lens (NA = 0.95) was used to generate a focal spot with a depth of focus (DOF) of 4.84λ and a subwavelength width of 0.53λ. In Reference [306], a similar multibelt phase hologram combined with an axicon lens was used to generate an optical needle with a large DOF of 11λ and a small width of 0.38λ. An optical needle with a subwavelength diameter of 0.38λ and a longitudinal depth of 7.48λ was obtained in Reference [300]. A focal spot limited by subdiffraction was obtained in Reference [307].

The authors of [302] used 4π focusing to focus a radially polarized optical vortex into a spot with a width of 0.43λ and a depth of 0.45λ. This type of focusing was also used in Reference [301] to produce spherical and subwavelength longitudinal magnetization. A solid immersion lens (SIL) was used in Reference [104] to produce a focal spot with a diameter of 0.305λ. The effect of coma on tightly focused cylindrically polarized vortex beams was investigated in Reference [308]. A beam quality measuring technique was introduced in Reference [309]. The conversion of cylindrically polarized laser beams from radial to azimuthal polarization was demonstrated in Reference [310] by introducing a higher-order vortex phase singularity.

There are several ways to obtain beams with sectoral azimuthal or radial polarization (or quasi-cylindrical vector beams), including the use of half-wave plates [291, 303, 311, 312], nonlinear optical crystals [313], polarizing films [314], and subwavelength gratings [289, 315, 316]. In addition sectoral binary elements could be added to a lens to obtain a smaller focal spot [317, 318].

The tight focusing of quasi-cylindrically polarized beams was previously investigated in detail in Reference [319] using numerical analysis. It was shown that a deviation of an eight-sector beam does not exceed 5.3% from the ideal beam. However, azimuthally polarized optical vortices have not been previously investigated.

In this section, we numerically investigate the tight focusing of a quasi-azimuthally polarized optical vortex with a wavelength of 532 nm using a Fresnel zone plate with NA = 0.95. It is shown that the focal spot produced by a beam with six sectors does not differ from the ideally azimuthally polarized optical vortex; the difference in the focal spot diameter does not exceed 0.001λ. For a four-sectoral beam, the difference does not exceed 0.03λ.

4.4.1 NUMERICAL SIMULATION

Our numerical simulation was performed using the Richards–Wolf formula [6]:

$$\mathbf{E}\left(\rho,\psi,z\right) = -\frac{if}{\lambda}\int_{0}^{a}\int_{0}^{2\pi} B(\theta,\varphi)T(\theta)\mathbf{P}(\theta,\varphi)$$

$$\times \exp\left\{ik\left[\rho\sin\theta\cos(\varphi-\psi)+z\cos\theta\right]\right\}\sin\theta d\theta d\varphi$$

(4.20)

where $B(\theta, \varphi)$ is the electrical field of focused light (θ is the polar angle and φ is the azimuthal angle), $T(\theta)$ is the apodization function, f is the focal length, $k=2\pi/\lambda$ is the wave number, and $\mathbf{P}(\theta, \varphi)$ is the polarization matrix:

$$\mathbf{P}(\theta,\varphi) = \begin{bmatrix} \left[1+\cos^2\varphi\left(\cos\theta-1\right)\right]a(\theta,\varphi) + \sin\varphi\cos\varphi\left(\cos\theta-1\right)b(\theta,\varphi) \\ \sin\varphi\cos\varphi\left(\cos\theta-1\right)a(\theta,\varphi) + \left[1+\sin^2\varphi\left(\cos\theta-1\right)\right]b(\theta,\varphi) \\ -\sin\theta\cos\varphi a(\theta,\varphi) - \sin\theta\sin\varphi b(\theta,\varphi) \end{bmatrix} \quad (4.21)$$

where $a(\theta, \varphi)$ and $b(\theta, \varphi)$ are polarization functions for the x- and y-components of the focused beam. In the simulation, we assume that a Fresnel zone plate ($T(\theta)=\cos^{-3/2}(\theta)$, NA$=0.95$ is the same as in References [105, 300, 305–307]) and is illuminated using a plane wave that has a different polarization and phase in each sector. In this case, a four-sector beam, for example, will have $a(\theta, \varphi)$, $b(\theta, \varphi)$ and $B(\theta, \varphi)$ as follows:

$$a(\theta,\varphi) = \begin{cases} -1, & 0 \le \varphi < \dfrac{\pi}{2} \\ -1, & \dfrac{\pi}{2} \le \varphi < \pi \\ 1, & \pi \le \varphi < \dfrac{3\pi}{2} \\ 1, & \dfrac{3\pi}{2} \le \varphi < 2\pi \end{cases} \quad (4.22)$$

$$b(\theta,\varphi) = \begin{cases} 1, & 0 \le \varphi < \dfrac{\pi}{2} \\ -1, & \dfrac{\pi}{2} \le \varphi < \pi \\ -1, & \pi \le \varphi < \dfrac{3\pi}{2} \\ 1, & \dfrac{3\pi}{2} \le \varphi < 2\pi \end{cases} \quad (4.23)$$

$$B(\theta,\varphi) = \begin{cases} 1, & 0 \le \varphi < \dfrac{\pi}{2} \\ i, & \dfrac{\pi}{2} \le \varphi < \pi \\ -1, & \pi \le \varphi < \dfrac{3\pi}{2} \\ -i, & \dfrac{3\pi}{2} \le \varphi < 2\pi \end{cases} \quad (4.24)$$

Figure 4.30 shows a sketch of the simulation. The light is polarized azimuthally in four sectors; it propagates through a sectoral spiral phase plate with topological

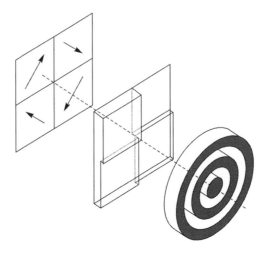

FIGURE 4.30 Sketch of the simulation: the four-sector azimuthally polarized beam and four-sector SPP.

charge equal to unity, and it is then focused by the phase zone plate. The sectoral polarizer and sectoral SPP could be manufactured as a single element [289, 315, 316].

4.4.2 FOCUSING OF SECTOR-POLARIZED BEAM TRANSMITTED THROUGH THE SECTOR SPP

Figures 4.31 through 4.33 show the simulation results. Figure 4.31 shows a focal spot produced by the beam consisting of four sectors, and Figure 4.32 shows the focal spot produced by the beam, consisting of six sectors. Figure 4.33 shows the focus of the ideal azimuthally polarized optical vortex. The intensity was calculated as $I = I_r + I_z = |E_r|^2 + |E_z|^2$.

Figures 4.31 through 4.33 show that the sector-polarized beams transmitted through the sectoral SPP have a longitudinal component of the electric field, unlike the ideal azimuthally polarized optical vortex. Although its contribution to the formation of focus is small, the maximum transverse component of the four-sector beam is approximately 21 times the maximum longitudinal component, and that of the six-sector beam about 80 times. Minimal and maximal focal spot diameters at

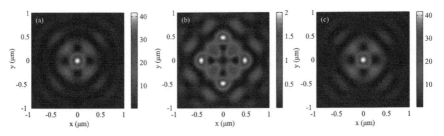

FIGURE 4.31 Intensity in the focal plane for (a) I_r, (b) I_z, and (c) I. Focusing of a four-sector-polarized beam transmitted through the four-sector SPP.

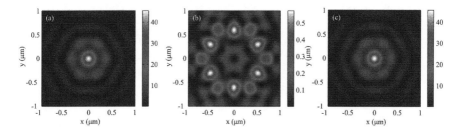

FIGURE 4.32 Intensity in the focal plane for (a) I_r, (b) I_z, and (c) I. Focusing of a six-sector-polarized beam transmitted through the six-sector SPP.

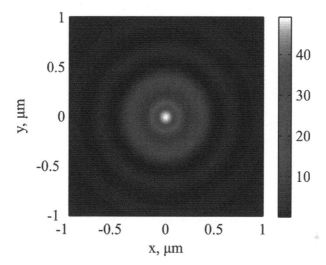

FIGURE 4.33 Intensity in the focal plane. Focusing of ideally polarized optical vortex.

half-maximum for the six-sector beam (Figure 4.32(c)) give $\text{FWHM}_{min} = 0.460\lambda$ and $\text{FWHM}_{max} = 0.461\lambda$ consequently. The difference in the focal spot diameters with the ideal beam (Figure 4.33) does not exceed 0.001λ. Additionally we simulated the focusing if there was a shift between the sectors of the six-sector polarizer and the sectors of the six-sector SPP. In this case the difference also does not exceed 0.001λ.

Figure 4.34 shows the relative error in the intensity differences in the focus of an ideal beam (I_{ideal}) and a sectoral polarized beam (I_{quasi}), calculated as $|I_{ideal} - I_{quasi}|/\max(I_{ideal})$.

The maximum relative error in Figure 4.34(a) does not exceed 18%, and that in Figure 4.34(b) does not exceed 9%.

4.4.3 FOCUSING OF SECTOR-POLARIZED BEAM TRANSMITTED THROUGH THE CONTINUOUS SPP

In the second stage of numerical simulation, it was assumed that the sectoral azimuthally polarized light passes through a continuous (helical) SPP ($B(\theta, \varphi) = e^{i\varphi}$) and

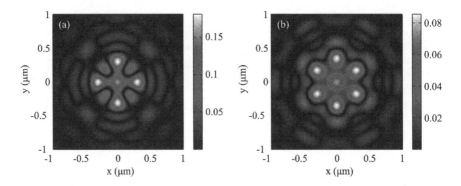

FIGURE 4.34 Relative error in focus, calculated as $|I_{ideal} - I_{quasi}|/\max (I_{ideal})$ for (a) a four-sector and (b) a six-sector azimuthally polarized beam.

is then focused by the zone plate (Figure 4.35). The intensity distribution in the focus in this case is shown in Figure 4.36. Figure 4.37 shows the relative error.

4.4.4 FOCUSING OF IDEALLY POLARIZED BEAM TRANSMITTED THROUGH SECTOR SPP

The third stage of numerical simulation modeled an ideal azimuthally polarized beam propagating through the sector SPP. The intensity distribution in the focus in this case is shown in Figure 4.38. Figure 4.39 shows the relative error. It should

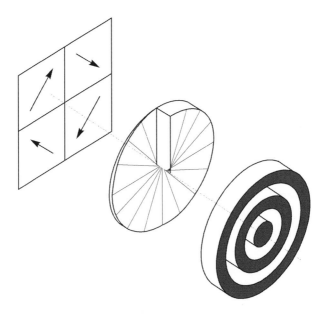

FIGURE 4.35 Sketch of the simulation: four-sector azimuthally polarized beam and continuous SPP.

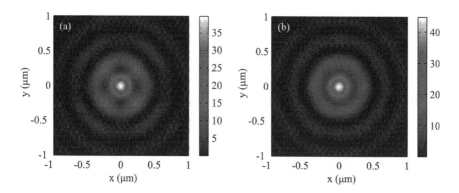

FIGURE 4.36   Intensity in the focal plane for (a) a four-sector and (b) a six-sector azimuthally polarized beam transmitted through the continuous SPP.

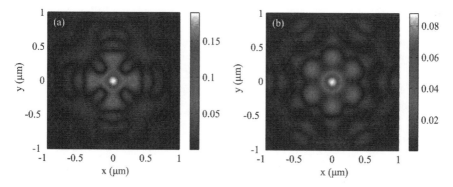

FIGURE 4.37   Relative error in the focus, calculated as $|I_{ideal} - I_{quasi}|/\max(I_{ideal})$ for (a) a four-sector and (b) a six-sector azimuthally polarized beam.

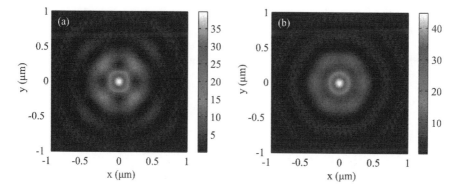

FIGURE 4.38   Intensity in the focal plane. Focusing of ideal azimuthally polarized beam transmitted through (a) a four-sector and (b) a six-sector SPP.

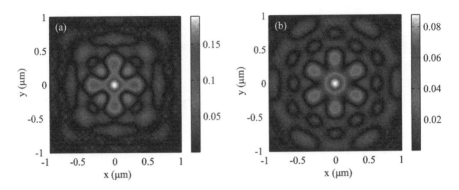

FIGURE 4.39   Relative error in the focus, calculated as $|I_{ideal} - I_{quasi}|/\max(I_{ideal})$ for (a) a four-sector and (b) a six-sector SPP.

be noted that in this case, due to the lack of a longitudinal component in the initial electric field, the focal spot does not contain the longitudinal component.

Tables 4.3 and 4.4 show the results of these simulations for comparison. The error is shown in Table 4.3, and the diameters of spots in Table 4.4.

Table 4.3 shows that the sector beam and sector SPP have approximately the same effect on the error in the focus. However, a comparison of Figures 4.34, 4.37 and 4.39 indicates that the simultaneous use of the sector beam and sector SPP lead to a greater discrepancy in the sidelobes.

Table 4.4 shows that the focal spot produced by the beam with six sectors does not differ from the ideally azimuthally polarized optical vortex; the difference in the focal spot diameter does not exceed 0.001λ. For a four-sectoral beam, the divergence does not exceed 0.03λ.

In this section, we numerically investigated the tight focusing of a quasi-cylindrical optical vortex with azimuthal polarization and a wavelength of 532 nm using a Fresnel zone plate with a numerical aperture of NA = 0.95. It was shown that the focal spot produced by a beam with six sectors does not differ from the ideally azimuthally polarized optical vortex; the difference in the focal spot diameter does not exceed 0.001λ. For a four-sectoral beam, the difference does not exceed 0.03λ. The obtained results could be used in subwavelength focusing of laser light [105, 291, 299, 305, 306].

TABLE 4.3
Maximum Error in the Intensity Distribution of the Focus

	Number of Sectors	Maximum Relative Error (%)
Sector SPP, sector polarization	4	18.0
	6	8.6
Continuous SPP, sector polarization	4	18.9
	6	8.8
Sector SPP, continuous polarization	4	18.9
	6	8.8

TABLE 4.4
Focal Spot Diameters at Half-Maximum

	Number of Sectors	FWHM$_{min}$ (λ)	FWHM$_{max}$ (λ)
Sector SPP, sector polarization	4	0.456	0.489
	6	0.460	0.461
Continuous SPP, sector polarization	4	0.460	0.468
	6	0.460	0.460
Sector SPP, continuous polarization	4	0.452	0.474
	6	0.460	0.460
Ideal azimuthally polarized optical vortex		0.460	

4.5 SUBWAVELENGTH FOCUSING OF LASER LIGHT OF A MIXTURE OF LINEARLY AND AZIMUTHALLY POLARIZED BEAMS

Recent years have seen an increase in interest by the research community in studying metasurfaces – elements enabling the amplitude, phase, and polarization of the laser beam to be manipulated (e.g. see review articles [320, 321]). As a particular case of the metasurface, we can mention diffraction gratings with subwavelength features [284], which are suitable for polarization control of a passing light beam.

The first subwavelength binary micropolarizers to be manufactured operated in the IR range [285, 286], implementing conversion of circularly polarized light of wavelength 10.6 µm into an azimuthally polarized beam. Generation of a radially polarized laser beam using a subwavelength diffraction grating operating at 1064 nm has also been reported [288]. A radial polarization interferometer that employs a subwavelength grating described in Reference [288] has also been reported [322].

A concentric-ring metal grating for generating radially polarized light of wavelength 633 nm from a circularly polarized beam has been discussed [294]. Using a similar approach, a terahertz vector radially polarized beam was generated in Reference [323]. However, it is worthwhile noting that the beam discussed in Reference [294] cannot be defined as being radially polarized in a strict sense. While being characterized by the radial polarization vector and a doughnut transverse intensity pattern, the beam's transverse phase distribution was not uniform. As a remedy for the said drawback, a combined holographic element, composed of the concentric metal ring of Reference [294] and a "fork" hologram, was introduced in References [324, 325].

The above-listed articles can be divided in two groups. The first group [285, 286, 294, 323–325] discusses the conversion of circularly polarized light into cylindrically symmetric beams. Being an analog of quarter-wave plates, such optical elements are easier to fabricate, as they have a smaller aspect ratio of grating microrelief features. The second group contains elements intended to transform linearly polarized light into the radially or azimuthally polarized beams [48, 288, 322]. Such elements can be looked upon as analogs of a half-wave plate.

In earlier publications, we reported on the characterization of four-sector subwavelength gratings for incident light polarization manipulation. A reflective 4-SPC proposed in Reference [295] was intended to convert linearly polarized incident light of wavelength λ into a radially polarized beam, which was then focused into an elongated subwavelength focal spot with smaller and larger sizes, respectively, measuring at full-width half-maximum of intensity FWHM $= 0.40\lambda$ and FWHM $= 0.50\lambda$. In a similar way, reflective [316] and transmission azimuthal micropolarizers [326] have been described. Being a follow-up of our previous study [326], in this work we discuss obtaining a tight focal spot by focusing a laser beam transmitted through a 4-SPC.

Below, we discuss a four-sector transmission micropolarizer that enables converting linearly polarized incident light into an azimuthally polarized beam. The resulting azimuthally polarized beam is characterized by a phase shift of π between the diametrically opposite beam points. We show that by placing a Fresnel zone plate (ZP) of focus 532 nm immediately behind the four-sector micropolarizer, light can be focused onto a subwavelength focal spot with smaller and larger sizes measuring $\text{FWHM}_x = 0.28\lambda$ and $\text{FWHM}_y = 0.45\lambda$, which is below the diffraction limit. It is also shown that if after passing through the micropolarizer the beam is allowed to propagate in free space over a distance of 300 μm before being focused by the ZP, the focal spot has respective sizes $\text{FWHM}_x = 0.42\lambda$ and $\text{FWHM}_y = 0.81\lambda$ (with the contribution of the transverse E-field component being $\text{FWHM}_x = 0.42\lambda$ and $\text{FWHM}_y = 0.59\lambda$). Note that the experimental verification of the last numerical result has given a focal spot measuring FWHM $= 0.46\lambda$ and FWHM $= 0.57\lambda$.

4.5.1 Simulation of the Far-Field Diffraction from the Micropolarizer

At the first stage of our study, we numerically simulated the focusing of laser light of wavelength $\lambda = 633$ nm, having passed through a four-sector micropolarizer, by means of a Fresnel ZP of focus $f = 532$ nm. The reason for using the Fresnel ZP as a focusing element was that given identical numerical apertures it enables obtaining a tighter focal spot when compared with an aplanatic lens [87]. The incident wavelength was chosen to be greater than the focal length of the ZP to reduce the focal length.

The FDTD-based simulation was implemented in the FullWAVE software. Parameters of a ZP used in simulation (Figure 4.40) were the same as those of a real Fresnel ZP that was later utilized in the experiment. The ZP of diameter 14 μm had a microrelief depth of 510 nm. Being composed of 12 rings and a central disk, the ZP was assumed to be fabricated in a resist with refractive index 1.52. The simulation step in the FDTD method was 0.02 μm.

The beam incident on the ZP (Figure 4.41) was assumed to have the complex amplitude distribution we discussed earlier [326]. The image size is 10×10 μm.

Figures 4.42 through 4.44 depict the simulation results. An elliptic focal spot is seen to have smaller and larger sizes measuring $\text{FWHM}_x = 0.42\lambda$ and $\text{FWHM}_y = 0.81\lambda$. Note that while taking into account only the transverse E-field component, the larger size of the spot is 0.59λ (dashed line in Figure 4.44(a)).

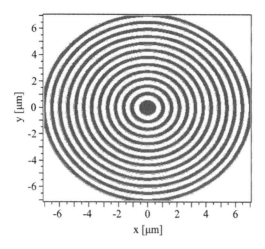

FIGURE 4.40 The Fresnel ZP under simulation as seen in a FullWAVE window.

4.5.2 Experiment

In the experiment, a laser beam of wavelength $\lambda = 633$ nm was focused with a ZP of focal length $f = 532$ nm after having passed through a 4-SPC. An optical arrangement for the experimental measurements is shown in Figure 4.45(a). Linearly polarized laser light from a He-Ne laser of wavelength 633 nm was delivered via an optical fiber to a substrate containing a 4-SPC on its surface. The 4-SPC under study was rigidly attached to the other substrate containing an array of ZPs (Figure 4.45(b)) [295]. A microscope image of the central part of the synthesized 4-SPC is shown in Figure 4.46.

Note that while, due to peculiarities of the fabrication process, the substrate contained many zone plates, just one of them was used in each run of the experiments. Shown in Figure 4.47(b) is an optical microscope image of the ZP against the background of a 4-SPC.

The location of the focal spot on the ZP and its size were controlled by shifting the mirror M_1. After passing the 4-SPC, the beam was focused by a ZP, with the focal spot intensity distribution being measured with a near-field scanning optical microscope (NSOM) NTEGRA Spectra (within a dashed-line contour in Figure 4.45(a)) using a tetrahedral pyramid tip C.

To ascertain that the beam has passed through the 4-SPC, a conventional polarizer can be placed after the objective O_1. Shown in Figure 4.48 are images of the transmitted beam for different positions of the polarizer. The images are identical to those reported earlier in Reference [326].

From the similarity of the images in Figures 4.41 and 4.48, we can assume that prior to hitting the ZP, radiation passes through the 4-SPC.

Measurements with a near-field scanning optical microscope have shown a focal spot to be formed at a distance of 200–250 nm from the ZP surface, with the spot size being FWHM $= 0.46\lambda$ and FWHM $= 0.57\lambda$ (Figure 4.49). Earlier, it was numerically

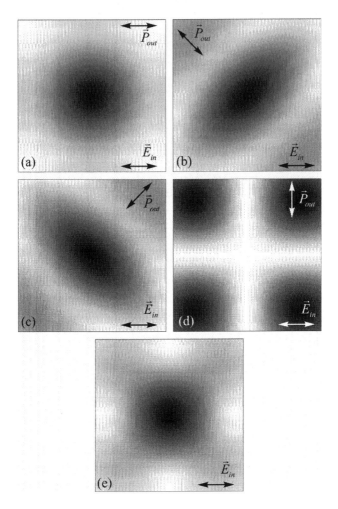

FIGURE 4.41 Intensity distribution (negative) for a beam $\vec{E}_{in}$ having passed through a micropolarizer, measured 300 μm from the micropolarizer surface. The analyzer $\vec{P}_{out}$ is rotated by (a) 0°, (b) 90°, (c) −45°, and (d) 45° with respect to the polarization vector $\vec{E}_{in}$.

shown that the focal spot size contributed to by the transverse E-field component measured FWHM = 0.42λ and FWHM = 0.59λ (Figure 4.44).

4.5.3 SIMULATION OF THE NEAR-FIELD DIFFRACTION FROM THE MICROPOLARIZER

In this subsection, the focusing of light of wavelength $\lambda = 633$ that was passed through a 4-SPC before being focused by a ZP of focus $f = 532$ nm was numerically simulated. The 4-SPC was assumed to be placed directly in front of the ZP.

As was the case in far-field simulation (Figure 4.50), the ZP parameters were the same as those of a real ZP utilized in the physical experiment. The FDTD-aided simulation was conducted with a 0.02-μm step. The ZP was assumed to be

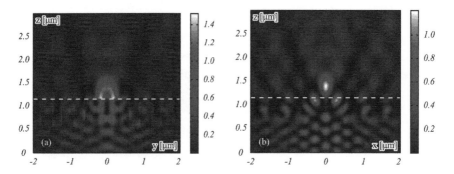

FIGURE 4.42 Intensity patterns (negative) within the calculated domain in the (a) yz- and (b) xz-planes. ZP boundary is marked with a black dashed line.

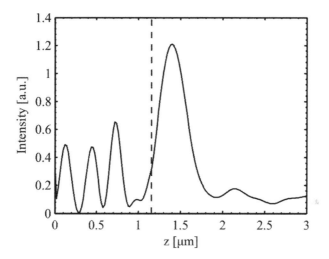

FIGURE 4.43 Intensity profile along the z-axis. ZP boundary is marked with a black dashed line.

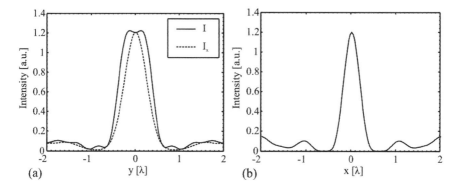

FIGURE 4.44 Intensity profiles in the focal spot along the (a) y- and (b) x-axes. The total intensity is marked with a solid line and the transverse intensity component is marked with a dashed line.

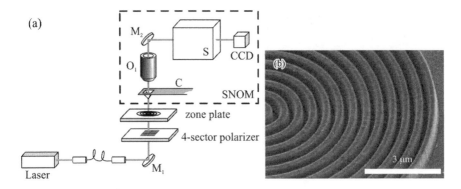

FIGURE 4.45 Experimental optical arrangement (a): M_1, M_2 are mirrors, O_1 is a 100× objective, C is a probe, S is a spectrometer, and CCD is a video-camera. (b) SEM image of ZP [295].

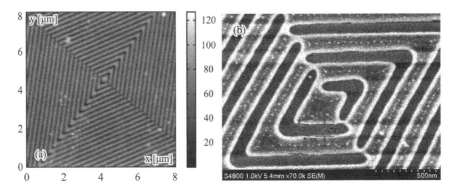

FIGURE 4.46 Images of the central part of the transmission 4-SPC using (a) an atomic force microscope and (b) a scanning electron microscope. The scale shows the microrelief depth in nm.

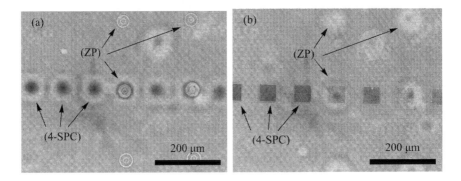

FIGURE 4.47 An optical microscope 20× image of (a) a ZP on the 4-SPC background, and (b) a 4-SPC on the ZP background.

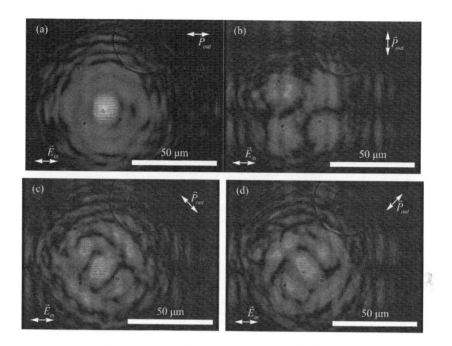

FIGURE 4.48 Images of the transmitted beam $\vec{E}_{\mathrm{in}}$ obtained at different polarizer positions $\vec{P}_{\mathrm{out}}$: (a) 0°, (b) 90°, (c) −45°, and (d) 45°.

illuminated by a laser beam whose complex amplitude distribution was reported earlier in Reference [326] and depicted in Figure 4.50.

The simulation results are presented in Figures 4.51 through 4.53. From Figure 4.51, an intensity peak (focal spot) is seen to be generated just behind the ZP surface. The focal spot is elliptic, measuring FWHM$_x$=0.28λ and FWHM$_y$=0.45λ at a distance of 40 nm from the surface. Note that the larger size contributed to solely by the transverse E-field component equals 0.42λ (dashed line in Figure 4.53(a)), implying that the longitudinal component just insignificantly widens the focal spot.

With increasing distance from the ZP surface, the focal spot gets larger and closer to a circle, e.g. 200 nm away from its surface the spot size is FWHM$_x$=0.37λ and FWHM$_y$=0.48λ (Figure 4.54).

From Figure 4.41, the central part of the beam transmitted through the 4-SPC is seen to be nearly completely linearly polarized along the y-axis (Figure 4.41(e)), in the meantime it was an azimuthally polarized beam that was observed at the 4-SPC exit (Figure 4.50). As a result, as the 4-SPC-to-ZP distance increases to 300 μm the focal spot increases in size from FWHM$_x$=0.28λ and FWHM$_y$=0.45λ to FWHM$_x$=0.42λ and FWHM$_y$=0.81λ.

We would like to note that with the ZP being synthesized in a substrate hundreds of micrometers thick, it appeared impossible to place the 4-SPC right up against the ZP. However, this difficulty can be obviated by obtaining a 4-SPC image in front of the ZP relief. In this case, the polarization vector in the image would be the same

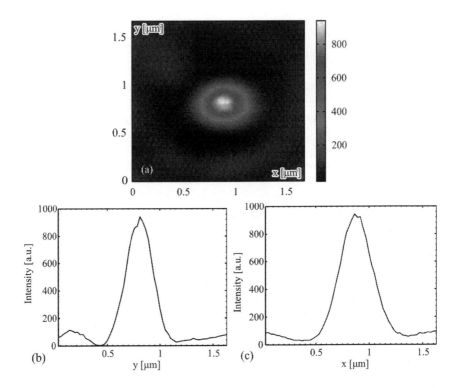

FIGURE 4.49 Intensity of the focal spot measured using an NSOM NTEGRA Spectra: 2D intensity distribution (a) and intensity profiles along the (b) *y*- and (c) *x*-axis.

as that at the 4-SPC exit (Figure 4.50). In this way, it will be possible to verify the simulation results discussed above. This is going to be the objective of our future study.

We have studied a four-sector transmission polarization convertor intended to convert linearly polarized laser light into an azimuthally polarized beam with its phase

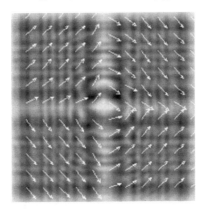

FIGURE 4.50 Intensity pattern (negative) and polarization vectors for a beam transmitted through the micropolarizer, generated 5.1 μm from its surface.

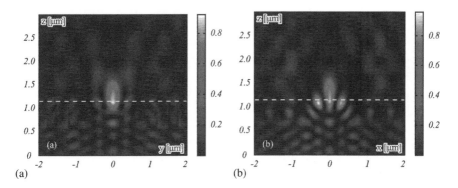

(a)

(b)

FIGURE 4.51 Intensity pattern (negative) in the central part of the calculated domain in the (a) *yz*- and (b) *xz*-planes. Black dotted line marks the ZP surface boundary.

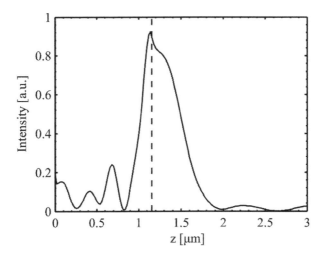

FIGURE 4.52 Intensity profile along the *z*-axis. Black dashed line marks the ZP relief boundary.

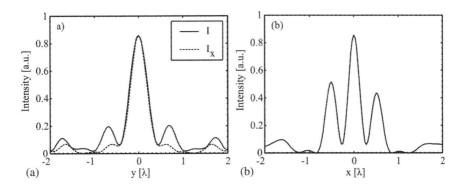

FIGURE 4.53 Intensity profile 40 nm from the ZP surface along the (a) *y*- and (b) *x*-axes. Solid line depicts the total intensity and the dotted line is for the transverse component of the intensity.

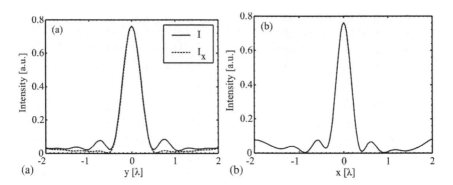

FIGURE 4.54 Intensity profile 200 nm from the ZP along the (a) y- and (b) x-axes. Solid line depicts the total intensity and the dotted line is for the transverse intensity component.

shifted by π in the diametrically opposed points, applying the proposed micropolarizer for tight focus of light. The results obtained are as follows. FDTD-aided simulation has shown that by illuminating the ZP with a laser beam transmitted through a 4-SPC it is possible to obtain an elliptic subwavelength focal spot whose size is smaller than the diffraction limit, measuring $FWHM_x = 0.28\lambda$ and $FWHM_y = 0.45\lambda$. If after passing through the 4-SPC the beam was allowed to propagate in free space over a distance of 300 μm, before being focused by the ZP, the resulting focal spot measured $FWHM_x = 0.42\lambda$ and $FWHM_y = 0.81\lambda$. Notably, the spot size contributed to only by the transverse E-field component was $FWHM_x = 0.42\lambda$ and $FWHM_y = 0.59\lambda$. The last numerical result above was verified experimentally using NSOM measurements. Experimentally, the measured size of the focal spot was $FWHM_x = 0.46\lambda$ and $FWHM_y = 0.57\lambda$.

5 Metalenses

5.1 A THIN HIGH-NUMERICAL-APERTURE METALENS

In recent years, optical researchers have been studying subwavelength-thick planar binary microoptics components in the form of metallic or semiconductor subwavelength element arrays (rods, slits, strips, and gratings [327]) that can simultaneously modify the polarization, amplitude, and phase of an incident electromagnetic wave. Such photonic components are referred to as metasurface optical elements (MOE). A review of MOE can be found in Reference [320]. By means of MOE it is possible to generate optical vortices [328]; synthesize sawtooth reflection gratings capable of reflecting 80% of light into a desired angle for the near-infrared (IR) spectrum (750 to 900 nm) [329]; and focus light into a ring [330] or transverse line [331]. Of particular interest is the use of the MOE as ultrathin lenses [143, 332–337].

Note that while the lenses in References [143, 332–336] operated in the infrared (IR) spectrum, only a single lens for the visible wavelengths (550 nm) has been demonstrated [337]. Lenses based on nanometallic antennae [332, 335, 336] have lower efficiencies compared to amorphous-silicon-based lenses [143, 333, 334, 337]. The best optical characteristics to have been achieved thus far are for a metalens constructed of silicon nanorod arrays of diameter 200 nm and height ~1 μm [333]. The lens has been reported to focus an incident linearly polarized beam to a minimal focal spot of diameter 0.57λ with a 40% efficiency. A disadvantage of the lens [333] is its high aspect ratio (5:1), which is necessary to realize high-quality silicon rods.

The design demonstrated in this work is closest to that of [337], which consisted of a binary microlens in amorphous silicon with a focal length of 100 μm (and a numerical aperture (NA) of NA = 0.43) for a wavelength of 550 nm. The metalens converted incident right-circular polarized light into a left-circularly polarized focal spot that measured 670 nm in full-width at half-maximum (FWHM). The metalens [337] was designed based on the Pancharatnam–Berry (PB) phase and was capable of operating only when illuminated with circularly polarized light, which is a shortcoming as an extra quarter-wave plate needs to be introduced to generate circular polarization. Additionally, the metalens of [337] features low NA of 0.43.

In this section, we propose an alternative approach to designing binary ultrathin metalenses capable of focusing linearly polarized laser light into a focal spot below the diffraction-limited size. With our method, subwavelength diffraction gratings (four would suffice) are synthesized in each annular zone of a binary Fresnel zone plate to convert linearly polarized incident light into a radially polarized wave. For instance, if in a certain zone of the Fresnel lens the polarization directions of transmitted light are given by the angles $+45°$, $+135°$, $-135°$, and $-45°$, the adjacent zone needs to produce polarization directions defined by the angles $-135°$, $-45°$, $45°$, and $135°$. Such an arrangement of gratings in the Fresnel lens zones produces a π-phase delay between the adjacent zones. Smaller heights of subwavelength grating features can be achieved by choosing a high-index material, namely, amorphous silicon. An amorphous-silicon film of width ranging from 50 to 120 nm deposited on a fused-silica substrate needs to be etched as far as the substrate so that light can pass through silicon only where protruding features of the diffraction gratings are found. The performance of the designed metalens was simulated using a finite-difference time-domain (FDTD) method in FullWAVE software.

5.1.1   DESIGN AND NUMERICAL SIMULATION OF A METALENS

Linearly polarized light sharply focused by means of microoptics components (a binary axicon [143] or a binary zone plate [146]) has been known [143, 146] to generate an elliptic subwavelength focal spot. By way of illustration, Figure 5.1(a) shows the arrangement of microrelief rings in a binary zone plate (ZP) fabricated from glass (refractive index $n=1.5$) for wavelength $\lambda=532$ nm, that features a subwavelength focal length of $f=200$ nm and microrelief depth $h=0.9$ μm. When illuminated with a linearly polarized Gaussian beam with waist radius $w=4\lambda$, such a ZP generates at distance $z=200$ nm behind its surface an elliptic ("dumb-bell") focal spot extended along the polarization axis (Figure 5.1(b)). Using the FDTD-based simulation in FullWAVE, the size of the focal spot was found to equal $\text{FWHM}_x=0.85\lambda$ and $\text{FWHM}_y=0.37\lambda$ (2.3:1 ellipticity) in full-width at half-maximum. Considering that its size on the y-axis is below the diffraction limit of $\text{FWHM}=0.51\lambda$, the spot is termed subwavelength.

It has also been known that by converting a laser beam from linear to radial polarization it is possible to obtain a circular subwavelength focal spot [291]. For instance, a four-sector micropolarizer composed of subwavelength gratings fabricated in a golden film proposed in Reference [289] converted a linearly polarized incident laser beam into a radially polarized beam. Figure 5.2(a) shows the layout of a four-sector polarizer composed of four binary diffraction gratings (period $T=460$ nm, wavelength $\lambda=633$ nm, depth $h=110$ nm) fabricated in a golden film ($n=0.312+i3.17$). Figure 5.2(b) depicts the near-surface intensity distribution of the light reflected at

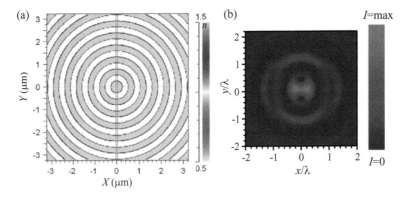

FIGURE 5.1   (a) The arrangement of rings of a binary zone plate that was used to calculate tightly focusing a linearly polarized Gaussian beam; (b) intensity distribution in the focal plane (at a distance of 200 nm).

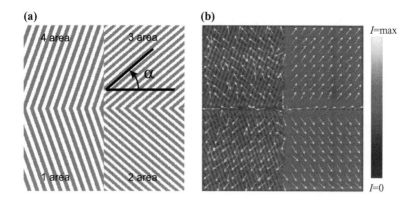

FIGURE 5.2 (a) The layout of a four-sector micropolarizer composed of four subwavelength binary diffraction gratings with period 460 nm (for the incident wavelength of 633 nm) in a golden film; and (b) the near-surface intensity distribution of the reflected light. Arrows show the polarization direction in each sector.

the micropolarizer. Arrows show the polarization direction for each zone. The light field was calculated using the FDTD method in FullWAVE. Simulations have shown that it will suffice to utilize four sectors in order to form a near-radially polarized light field [289, 291] that can be focused in a subwavelength focal spot [291, 295]. Note that a four-sector micropolarizer in the transmission mode can be realized in an amorphous-silicon film deposited on a transparent substrate [326]. In such a design, the gratings featured a period of $T = 230$ nm and a microrelief depth of $h = 130$ nm ($\lambda = 633$ nm). The refractive index of silicon used in the simulation was $n = 3.87 - i0.016$.

However, the use of two different elements in the scheme – a reflective/transmissive polarization converter and a zone plate – calls for its high-precision alignment, also leading to extra loss of energy due to reflection at additional surfaces. Thus, designing a microoptics element capable of simultaneously converting the polarization of laser light and generating a tight focal spot is challenging. Such a binary subwavelength optical element can be designed by combining two above-discussed optical elements, a zone plate or axicon (Figure 5.1(a)) and a four-sector micropolarizer (Figure 5.2(a)). The aim is to enable a phase jump by π in passing from ring to ring, in this way changing the polarization direction to the opposite one. To this end, gratings in the adjacent plate zones should be taken from diagonal sectors of the micropolarizer in Figure 5.2(a): 1 and 3 or 2 and 4. Figure 5.3 depicts the binary microrelief pattern of a metalens that combines the properties of a micropolarizer (Figure 5.2(a)) and a zone plate with high NA (similar to the axicon in Figure 5.3(a)).

The operation of the metalens was simulated using the following parameters: wavelength $\lambda = 633$ nm, focal length $f = 633$ nm (NA = 1), designed microrelief height $h = 0.24$ μm, pixel size 22 nm, grating period $T = 220$ nm, diffraction grating groove 110 nm (5 pixels), and step width 110 nm (5 pixels). The entire simulation domain 5×5 μm (Figure 5.3), refined refractive index of amorphous silicon $n = 4.35 + i0.486$ (measured using ellipsometry), and glass substrate ($n = 1.5$). In the FDTD method, the sampling grid was taken to be $\lambda/30$ for all three coordinates. The incident plane wave

FIGURE 5.3 The layout of the grooves of a transmission binary metalens that is capable of simultaneously converting the linear polarization into the radial one and generating a tight focal spot.

of the diameter equal to that of the metalens was polarized along the y-axis (vertical axis). It is noteworthy that within the proposed approach to designing the metasurface optical elements, the microrelief height does not critically affect the element's operation. The metalens (Figure 5.3) remains able to generate a subwavelength focal spot within a certain range (0.3–1.2 μm) of heights by converting the incident light from linear-to-radial polarization. The parameter affected by the microrelief height is the throughput efficiency of the lens.

Figure 5.4 depicts the focal spot size along the x- and y-axes and the on-axis intensity against the distance to the metalens. The metalens relief height is 70 nm, meaning that a 70-nm thick silicon film has been etched through as far as the substrate. Figure 5.4 also suggests that within the on-axis distance from the metalens ranging from 300 nm to 1.2 μm the focal spot retains its circular shape, with its diameter varying insignificantly. The metalens in Figure 5.4 with 70-nm-high relief features generated at distance $z = 391$ nm a focal spot of size $FWHM_x = 0.434\lambda$ and $FWHM_y = 0.432\lambda$. Figure 5.5 shows a pseudo-color intensity distribution of the electric field at distance $z = 391$ nm from the metalens (Figure 5.5(a)) and intensity profiles in the focal plane along the x- and y-axes (Figure 5.5(b)).

The focusing efficiency was found to depend on the diameter of the incident beam. Simulation was conducted for a plane wave limited by a circular aperture of radius R incident on a metalens with a 70-nm-high microrelief. With decreasing aperture of the beam, the focusing efficiency increased, while the lens focusing capability deteriorated, resulting in a larger focal spot measured in FWHM. For instance,

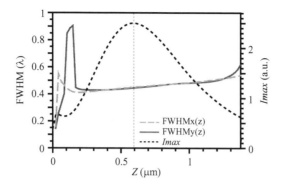

FIGURE 5.4 FWHM$_x$ (dashed curve), the focal spot size in full-width at half-maximum along the x-axis, generated at distance z from the microrelief's upper surface; FWHM$_y$ (solid curve), a similar estimate for the y-axis. I_{max} (dotted curve), the on-axis intensity as a function of distance z. The metalens relief height, 70 nm.

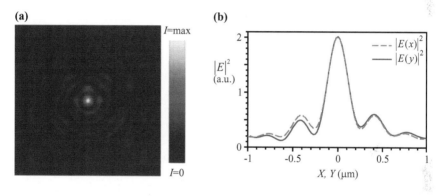

FIGURE 5.5 Intensity distribution $|E|^2$ at distance $z = 0.391$ µm. Dashed line, along the x-axis, solid line, along the y-axis.

by reducing the aperture radius to $R = 1.4$ µm, one can attain a focusing efficiency of $\eta = 5.6\%$, meanwhile the focal spot gets larger, measuring FWHM$_x = 0.61\lambda$ and FWHM$_y = 0.65\lambda$. Efficiency η is equal to the ratio of power in the focal spot, calculated till the first intensity minima, to all power incident on the element. At the maximum aperture radius analyzed, $R = 2.5$ µm, the efficiency was found to be as low as 2.5%. The increase of the aperture resulted in the increase of the intensity maximum at the focal spot center.

5.1.2 Fabrication of the Metalens and Measuring the Surface Relief

A metalens with the relief depicted in Figure 5.3 was fabricated using electron beam lithography. A 130-nm thick amorphous-silicon (a-Si) film deposited on a transparent pyrex substrate (with refractive index $n = 1.5$) was coated with a 320-nm thick polymethyl methacrylate (PMMA) resist, which was baked at a temperature of 180°C. The resist thickness of 320 nm was chosen to give both good etch resistance and high resolution patterning. To prevent charging, the surface was sputtered with a 15-nm thick gold layer. A binary template (Figure 5.3) was transferred onto the resist surface using a 30-kV electron beam. The specimen was developed in the water blended with isopropanol in the ratio 3:7

The template was transferred from the resist into the a-Si by reactive ion etching in the gaseous atmosphere of CHF_3 and SF_6. The aspect ratio of the etch rate of the material and the photomask was found to be 1:2.5. An electron microscope image of the metalens is shown in Figure 5.6. The entire metalens of diameter 30 µm is shown in Figure 5.6(a) and its magnified central part is shown in Figure 5.6(b).

The metalens microrelief was also characterized using an atomic force microscope. Figure 5.7(a) depicts the central fragment of the metalens microrelief and Figure 5.7(b) depicts a characteristic profile of the metalens microrelief. The depth of microrelief features varies from 80 nm to 160 nm, with the average depth being 120 nm. A probe tip of radius 10 nm was utilized in the microscope. The microrelief measurement error was 5%, with the transverse coordinate measurement error being 2.5%.

5.1.3 Modeling the Metalens with Account for Fabrication Errors

To account for fabrication errors when modeling the metalens, the microrelief characteristics measured by an atomic force microscope (Figure 5.7) were used in a FullWAVE simulation. Figure 5.8(a) shows the simulated microrelief obtained based on the one in Figure 5.7(a). The simulation parameters were as follows. The metalens size (Figure 5.8(a)) is 6.22×6.22 µm, which is equivalent to 256×256 samplings. The maximum relief height difference is 189 nm (from minimum to maximum points according to measurements), the plane linearly polarized incident wave has a wavelength of $\lambda = 633$ nm, the sampling grid is $\lambda/30$ on all three coordinates, the a-Si has the refractive index $n = 4.35 + 0.486i$, and the transparent substrate has the refractive index $n = 1.5$. Figure 5.8(b) depicts the simulated intensity distribution at distance $z = 600$ nm from the metalens. The focal spot size is $FWHM_x = 0.46\lambda$ and $FWHM_y = 0.52\lambda$ (along the polarization direction). In the focal spot found at distance $z = 600$ nm the intensity was maximal, being twice of that of the incident

FIGURE 5.6   (a) An electron microscope image of a metalens in an a-Si film of diameter 30 µm; and (b) the magnified 3×2-µm central fragment.

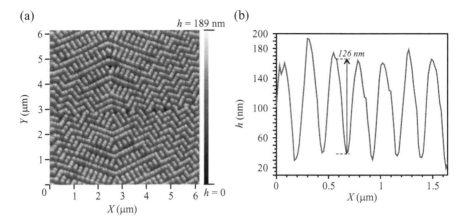

FIGURE 5.7 (a) An AFM image of the central part of the metalens microrelief, obtained by Solver Pro microscope; and (b) an example of the metalens relief profile.

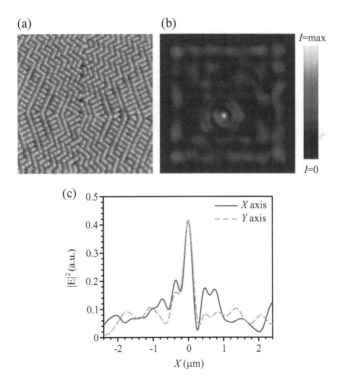

FIGURE 5.8 (a) Gray-level microrelief, with black corresponding to zero height and white to a height of 189 nm, which exactly corresponds to the image of a real metalens relief in Figure 5.7(a) incorporated in FullWAVE for modeling; (b) the simulated intensity pattern generated by the metalens of Figure 5.8(a); the pattern in Figure 5.8(b) has a size of 6×6 μm; (c) the focal intensity profile for $|E|^2$ along the x- and y-axes.

light. However, it is worth noting that the diameter of the focal spot remained nearly unchanged in the range from $z=200$ nm to $z=900$ nm. Figure 5.8(c) shows profiles of the E-field intensity along the x- and y-axes. The lack of central symmetry of the profiles is due to fabrication errors.

When scanning the microrelief with an atomic-force microscope (AFM) it is very hard to ensure that the center of the scanned image be coincident with the center of the element under scanning. This is the reason why the centers of the focal spot and scanned region are different.

5.1.4 Experiment on Focusing the Laser Light with a Metalens

The focusing properties of the metalens were experimentally studied by means of near-field scanning optical microscopy (NSOM). An experimental optical arrangement is shown in Figure 5.9.

In the experiment, a light beam from an He-Ne laser (wavelength 633 nm, power 50 mW) was fed via an optical fiber to the metalens under study, generating a subwavelength focal spot. The full-width of the incident beam was 30 μm. The intensity in the focal spot was measured using a hollow metallized pyramid-shaped tip C having a 100-nm pinhole in the vertex. Having passed through the pinhole the light was collected by a 100× objective O_1, before travelling through the spectrometer S (Solar TII, Nanofinder 30) to the CCD-camera (Andor, DV401-BV).

The experimentally measured focal length of the metalens was $z=0.6$ μm. Figure 5.10 depicts the focal intensity pattern experimentally measured by the NSOM. Figure 5.11 shows the intensity profiles in the focal spot (Figure 5.10) along the x- and y-axes. The maximal intensity in the focus was found to be 11 times that of the incident beam.

The experimentally measured size of the focal spot was $FWHM_x=0.49\lambda$ and $FWHM_y=0.55\lambda$. These values are just 8% different from those obtained via simulation ($FWHM_x=0.46\lambda$, $FWHM_y=0.52\lambda$) taking into account the metalens fabrication errors, also being 15% different from a focal spot generated by a perfect metalens ($FWHM_x=0.434\lambda$, $FWHM_y=0.432\lambda$), which has a regular microrelief of feature height 70 nm.

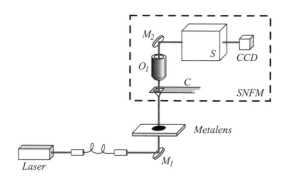

FIGURE 5.9 Experimental optical arrangement: M_1, M_2, mirrors, O_1, a 100× objective, C, a probe, S, a spectrometer, and CCD, a CCD-camera.

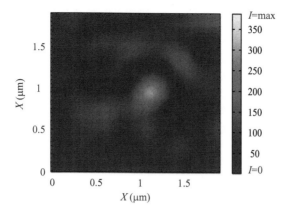

FIGURE 5.10 Intensity pattern at distance $z = 0.6$ μm from the metalens (Figure 5.6).

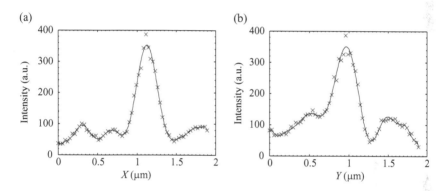

FIGURE 5.11 Measured intensity profiles of the focal spot (Figure 5.10) along the (a) x- and (b) y-axis. The crosses mark experimental values and the black curve presents a polynomial-based approximation.

A simple approach to synthesizing a binary microrelief of a subwavelength microlens in a thin a-Si film has been proposed. The rings of a Fresnel zone plate with a specified focal length are filled with subwavelength gratings, with each grating rotating the electric vector of the normally incident linearly polarized beam by a pre-given angle. The period, relief feature depth, and fill factor are fitted to be optimal to ensure that the intensity of light transmitted through each individual grating is approximately the same. At different angles of rotation of the polarization vector, the transmission of metalware will also be different. In this section, the linear polarization of light was converted into the radial polarization using just four different diffraction gratings, which rotated the polarization vector by four different angles, thus generating a circular subwavelength focal spot. The phase shift of π between adjacent zones was realized by providing that two local gratings abutting on the zone boundary rotated the polarization vectors by angles whose difference equaled π. The simulation has shown a thin-film silicon metalens, of diameter 5 μm and focal length 633 nm, composed of four differently oriented diffraction gratings of period 220 nm, will focus 2.5% of an incident linearly polarized beam of wavelength 633 nm and diameter 5 μm onto a circular focal spot below the diffraction limit at a distance varying from 200 nm to 1 μm. Notably, with the lens (silicon film) thickness varying in the range from 50 to 120 nm, the focal spot diameter has been found to change insignificantly, remaining smaller than the diffraction limit, in the range from 0.37 to 0.45 of the wavelength. When illuminated with a linearly polarized Gaussian beam, the lens generated a focal spot whose size along the Cartesian coordinates was $FWHM_x = 0.49\lambda$ and $FWHM_y = 0.55\lambda$. The experimental results are in good agreement with the simulation, measuring $FWHM_x = 0.46\lambda$ and $FWHM_y = 0.52\lambda$.

5.2 ENGINEERING A SECTOR-VARIANT HIGH-NUMERICAL-APERTURE MICROMETALENS

These days, metasurface photonic components have been actively studied. Their advantages include the possibility of simultaneously controlling the polarization and phase of laser light, while the microrelief height of dozens of nanometers for the visible and IR spectrum make them easy to fabricate. A review of metasurface photonic components can be found in Reference [320]. The metasurface components may find applications for generating optical vortices [328]; in sawtooth gratings capable of reflecting 80% of the incident light into a desired angle for a wide range of near-IR wavelengths [329]; and focusing light into a doughnut intensity distribution [330] or a transverse line [331]. Of special interest is the use of the metasurface components as ultrathin microlenses [331–333].

In a number of earlier published works, the present authors also studied some metasurface components [315, 316, 334]. In particular, microoptic components (micropolarizers and microlenses) to perform the linear-to-radial and linear-to-azimuthal polarization conversion and subwavelength focusing of the incident light beam have been considered. Such microoptic components are peculiar to the surface space-variant subwavelength diffraction gratings. Similarly to a half-wave plate, each grating can rotate the polarization plane by a definite angle. In total, four different space-variant gratings have been analyzed [315, 316, 334], which were able to rotate the polarization angle by four different angles. This was found to be sufficient to perform an approximate conversion of linearly polarized light into radial or azimuthal polarization.

In this section, we conduct the numerical simulation of focusing light with the aid of high numerical aperture (NA) micrometalenses composed of a varying number of local subwavelength diffraction gratings. We show that the linearly polarized light is best focused by an element with 16 local space-variant gratings, generating a near-circular-shape focal spot. For the metalens with 16 different azimuthal angles of the polarization plane, the focal spot size at full-width of half-maximum intensity was found to be $FWHM_x = 0.435\lambda$ and $FWHM_y = 0.457\lambda$ (corresponding to the eccentricity parameter, hereafter termed as ellipticity, of 1.05).

5.2.1 MODELING A SUBWAVELENGTH GRATING-POLARIZER

The metalens under study is synthesized based on the algorithm proposed in Reference [315], in which the binary lens is composed of a zone plate with desired focal length and numerical aperture. However, rather than considering the microrelief in a transparent substrate resulting in a half-wavelength inter-zone shift, the subwavelength binary microrelief of interest has local space-variant diffraction gratings in each zone. Each space-variant grating is supposed to rotate the polarization plane by a certain angle. Here, questions arise as to whether the angle φ of polarization plane rotation should depend in a linear way on the tilt α of the diffraction grating grooves and whether the light transmittance should be the same for different gratings. The first question can be answered by analyzing the plot for the rotation angle φ of the incident field polarization vector after passing through the subwavelength grating as a function of the angle α between the incident light polarization vector and the grating grooves (Figure 5.12(a)). The plot is shown in Figure 5.12(b) (solid curve), describing the numerically simulated propagation of the linearly polarized light (with the incident plane wave defined by the vertical polarization vector E_1 (Figure 5.12(a)) through a 120-nm high diffraction grating fabricated in an amorphous silicon with the refractive index $n = 4.35 + i0.486$ [315]. The second question can be clarified from the plot for the transmitted light intensity as a function of the angle α (Figure 5.12(b), dashed curve). The numerical simulation we conducted employed the FDTD-aided finite-difference solution of Maxwell's equations using the FullWAVE software. The grating under study had a 220-nm period, a 110-nm groove width, a 110-nm ridge width, and a 120-nm relief depth. The incident wavelength was $\lambda = 633$ nm.

From Figure 5.12(b), the output angle φ is seen to depend in a near-linear way on the input angle α for the material used, with the dependence being farthest from linear for the angles α of 40°, 50°, and 70°. However, the dependence does not need to be perfectly linear. Deviations of the curve plotted in Figure 5.12 from linearity just need to be accounted for when designing a metalens composed of local space-variant subwavelength gratings. The transmitted intensity I (incident intensity is equal to unity at each point) plotted in Figure 5.12(b) against the grating groove tilt angle α takes the form of a quasiperiodic function ranging in value from 0.7 to 1. Hence, rather than operating as a purely phase element, the metalens modulates the uniform incident intensity, in this way decreasing the efficiency.

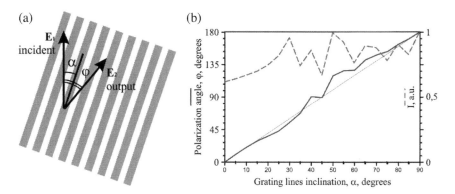

FIGURE 5.12 (a) Vectors of the incident and transmitted electric field intensity, E_1 and E_2; (b) the angle φ as a function of the transmitted electric field intensity E_2 (solid line) and the relative intensity I as a function of the grating groove tilt α (dashed line).

5.2.2   METALENSES WITH A VARYING NUMBER OF LOCAL GRATINGS

In this subsection, we conduct the numerical simulation of tightly focusing a linearly polarized flat-top laser beam by a binary metalens in amorphous silicon, composed of local diffraction gratings varying in number from 3 to 121. Figure 5.13(a) depicts a metalens composed of 16 space-variant diffraction gratings, which are seen as 16 sectors originating from the common center. The gratings in each sector are arranged so as to produce an azimuthally polarized output beam. Similar to the radially polarized light [299], the azimuthally polarized light with phase singularity has been known to enable focusing into a circular focal spot of subwavelength size [338]. To realize the phase singularity, we replaced a conventional zone plate with a spiral zone plate with the unit topological charge whose adjacent zones were composed of diffraction gratings with oppositely disposed sectors. In this way, a 180° angle between the polarization vectors of the adjacent zones was achieved, producing a phase shift of π. Thus, we removed the need to illuminate the focusing element by a wave with phase singularity. Functionally speaking, such a metalens is composed of three elements: a zone plate to focus light, a spiral phase plate to produce a wavefront singularity with the unit topological charge, and sectored half-wave plates in the form of sectored binary subwavelength gratings. Such a metalens is able to convert an incident flat-top or Gaussian beam with uniform phase into an azimuthally polarized beam with phase singularity, before focusing it at a designed distance. Figure 5.13(b) shows a spiral zone plate employed for constructing the light focusing metasurface of Figure 5.13(a). The metalens of Figure 5.13(a) has the focal length equal to the incident wavelength, $f = \lambda = 633$ nm. This lens was fabricated in an amorphous-silicon film using electron beam lithography and ion etching (Figure 5.13(c)). The relief of the film, measured with the help of the atomic force microscope Solver Pro P6, is seen to be very similar to the mask of Figure 5.13(a). Below is a simulation of the light propagation through such a metalens.

The metalens of a 8-μm diameter was illuminated by a linearly polarized plane wave. In building the sectored space-variant diffraction gratings, a 22-nm (0.03λ) mask step was utilized (the pixel size of a file containing the metalens structure).

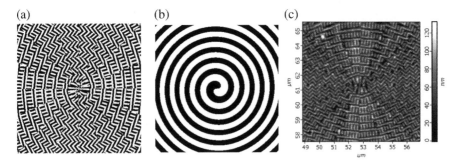

FIGURE 5.13 (a) A general view of the metalens of diameter 8 μm (364 × 364 pixels); (b) a spiral phase plate employed in constructing the metalens; (c) microrelief of a fabricated metalens measured using Solver Pro P6 scanning probe microscope (NT-MDT, Zelenograd, Russia).

Since the intensity peak along the z-axis was detected at a distance of $z=600$ nm above the upper edge of the metalens microrelief, hereafter the plots are given for that distance. Shown in Figure 5.14 are intensity $|E|^2$ in the focal spot 600 nm away from the metalens (inset) and intensity profiles along the x- and y-axes.

From Figure 5.14, the focal spot is seen to have a near-circular shape. The size of the focal spot at full-width of half-maximum intensity along the x- and y-axes was found to be FWHM$_x=0.435\lambda$ and FWHM$_y=0.457\lambda$. The focal spot size was calculated with an average error of 0.01λ equal to the calculation step size. By varying the number of metalens constituent diffraction gratings, we change both the focal spot shape and central intensity. Table 5.1 gives the results of the numerical simulation for metalenses with the number of sectors and respective polarization vectors taking from 3 to 121 different values. The last column of Table 5.1 shows the ellipticity of the focal spot in the Cartesian coordinates.

From Table 5.1, a three-sector metalens is seen to be able to focus light onto a near-circular subwavelength focal spot, although the resulting central intensity peak is 2.14 times lower than that from a 16-sector metalens. The maximum intensity and least ellipticity are seen to be generated by metalenses composed of 12 and 16 sectors. A further increase in the number of sectors is seen to result in decreasing central intensity peak, also making the focal spot more elliptical, with the semidiameter ratio reaching 1:2. This can be ascribed to the degradation of the sectored gratings, which become too narrow.

It is worth noting that the area of the focal spot produced by the 16-sector metalens (Figure 5.13(a)), measured in terms of the FWHM intensity, is $0.156\lambda^2$. In the meantime, when using a perfect aplanatic lens with NA = 1, the focal spot area is $0.147\lambda^2$ [338] for an azimuthally polarized beam with phase singularity and $0.17\lambda^2$ for a radially polarized beam. Hence, the metalens-aided focal spot (Figure 5.13(a)) is 10% smaller in area than the aplanatic-lens-aided minimal focal spot from a radially polarized beam.

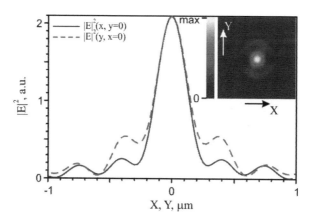

FIGURE 5.14 Intensity $|E|^2$ of the light field 600 nm from the metalens (inset) from Figure 5.13(a) and profiles of intensity $|E|^2$ along the x- and y-axes drawn through the focal spot center. The incident wave is E_y.

TABLE 5.1

Parameters of Focal Spots from Metalenses with a Varying Number of Sectored Gratings

Number of sectors	FWHM$_x$ (λ)	FWHM$_y$ (λ)	I_{max} (r.u.)	Ellipticity
3	0.443	0.464	0.997	1.05
4	0.428	0.460	1.84	1.07
6	0.424	0.461	1.67	1.09
8	0.436	0.46	1.8	1.06
12	0.427	0.465	2.13	1.09
16	0.435	0.457	1.94	1.05
24	0.43	0.466	1.95	1.08
32	0.4	0.504	1.68	1.26
54	0.375	0.534	1.68	1.42
81	0.411	0.575	0.919	1.40
121	0.387	0.764	2.13	1.97

Interestingly, unlike focusing radially polarized light, with an azimuthally polarized beam the focal spot is mainly contributed to by the transverse electric field components, E_x and E_y, with the longitudinal component being negligibly small. The intensities of the individual electric field components in the focal spot are depicted in Figure 5.15. When focusing light based on just the transverse electric field components, a major advantage stems from the fact that after coming to the focus almost the entire light then travels toward the on-axis observer (or photoreceiver).

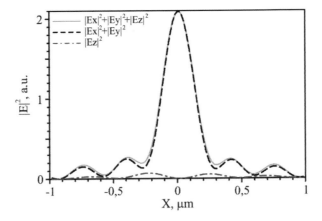

FIGURE 5.15 Contribution of the transverse and longitudinal components to the focal spot produced by a metalens focusing an azimuthally polarized beam with phase singularity (Figure 15.3(a)).

5.2.3 FIELD MEASUREMENTS USING A PROBE TIP: NUMERICAL SIMULATION

An extra benefit that the use of an azimuthally polarized beam with singular phase offers is that measurements of the focal spot can be conducted with a microscope tip with an on-axis nanohole, resulting in a lower measurement error thanks to a higher signal-to-noise ratio. When measuring the near-surface intensity of a light field by a scanning near-field optical microscope (SNOM), hollow aluminum probes with a square-shaped 100-nm nanohole in the pyramid tip are usually utilized. Such probes are sensitive to the transverse electric field components. Figure 5.16(a) depicts an SNOM cantilever and its characteristics (cantilever SNOM-NC by NT-MDT). Figure 5.16(b) presents the simulated measurement results for the focal spot of Figure 5.15 derived using the cantilever of Figure 5.16(a).

In the calculations, the refractive index of the aluminum probe was assumed to be $n = 1.27 + 7.3i$. The probe tip was placed in the focal plane 600 nm away from the metalens (Figure 5.13(a)). With the probe tip assumed to be found at $(x, y=0)$, the propagation of light through the metalens and the probe tip was numerically simulated. Having passed through the probe nanohole, light was detected by a virtual monitor placed inside the probe. The measurement was then conducted for a new probe position on the x-axis 50 nm apart from the previous point. The intensity profile in Figure 5.16(b) gives the focal spot size of $FWHM_x = 0.44\lambda$, which is just 5 nm larger than the $FWHM_x = 0.435\lambda$ for the focal spot of Figures 5.14 and 5.15. Thus, we can conclude that using the probe of Figure 5.16(a) with a 100-nm nanohole and a 50-nm scan-step, we can measure the metalens-aided near-surface focal spot from an azimuthally polarized beam with a negligibly small error of 2%.

Thus, the following numerical results have been obtained. When using a 8-μm metalens with unit NA and 633-nm focal length synthesized in a thin amorphous-silicon film (120 nm), the 633-nm linearly polarized plane incident beam is converted into an azimuthally polarized beam with phase singularity, producing a near-circular subwavelength focal spot of size $FWHM_x = 0.435\lambda$ and $FWHM_y = 0.457\lambda$. The maximum intensity in the focus is 2 times the incident intensity. Note that while the number of differently oriented subwavelength local gratings causing the polarization vector to turn by a different angle can vary from 3 to 16, this has been found not to

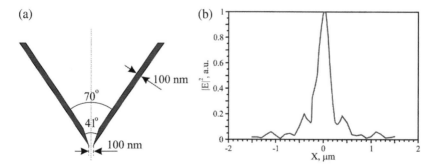

FIGURE 5.16 (a) Parameters of an aluminum SNOM probe with a square-shaped 100-nm nanohole in the tip; and (b) light intensity at the cantilever output as a function of the x-coordinate when scanning the focal spot of Figures 5.14 and 5.15.

cause essential changes in the focal spot size and shape, with the ellipticity being under 1.1. The smallest and most circular focal spot has been found to be produced by the 16-sectored metalens with space-variant gratings. We have shown that the metalens-aided subwavelength focus is only contributed to by the transverse components of the electric field intensity, enabling nearly error-free measurements to be done by a hollow pyramid aluminum probe with a 100-nm nanohole and a 50-nm scan-step.

5.3 SHARP FOCUSING OF VECTOR OPTICAL VORTICES USING A METALENS

The study of optical components with metasurfaces was prompted by References [321, 339], which demonstrated that using nanostructures synthesized in thin metal or dielectric films, the amplitude, phase, and polarization of laser light can be simultaneously controlled in a local manner at each point of the beam cross-section. Vector optical vortices have been generated using subwavelength rectangular apertures arranged along two concentric circles in a thin golden film for an incident laser wavelength of 1500 nm [340]. Cylindrical optical vortices with pre-set inhomogeneous polarization have been generated using two metasurfaces with space-variant diffraction gratings computed by a Pancharatnam–Berry-phase method and fabricated by femtosecond (fs) laser alblation in a fused silica for the incident 632-nm light [341]. A thin-film golden lens with a 10-μm focal length and elliptic apertures has been shown to simultaneously generate a sharp focus for a 405-nm wavelength and an optical vortex ring for a 532-nm wavelength in the same plane [342]. Such metalens may find uses in stimulated emission depletion microscopy (STED) microscopy.

It has been reported [343–345] that thin spiral plates based on a plasmonic metasurface can be utilized to convert a circularly polarized beam into an optical vortex with the topological charge $m = 1$ or $m = 2$. The said studies differ from each other only in the form of the local antennae (resonators) used. In Reference [343], an 800-nm silver film with 76–132-nm nanoholes produced with a tight ion beam and then filled with a PMMA photoresist was reported to operate at 532 nm, with the nanohole size defining the phase shift. In Reference [344], a 27-nm golden film was etched to produce an array of V-shaped nanoantennae of 209-nm side and 375-nm period. The phase shift of a circularly polarized 780-nm incident beam was controlled by the nanoantenna tilt angle. In Reference [345], a 100-nm golden film was etched with a 250-nm period to produce antenna arrays shaped as broken squares of 150-nm side. The scale and location of the square contour discontinuity defined the phase shift of a circularly polarized incident beam of wavelengths 710–900 nm.

In Reference [346], silicon nanodisks of 590-nm diameter and 243-nm height periodically arrayed in four squares of 10-μm side were used to control the phase of light by varying only a period of the nanodisk array in the 700–1000-nm range for a 1490-nm incident beam. The generation of an optical vortex with the topological charge $m = 1$ was demonstrated experimentally. A review of photonic components with plasmonic and dielectric metasurfaces designed using both the PB-phase method and the Huygens method [347] was made in Reference [348]. A metalens in transmission (10%) with a 10-μm focal length, composed of elliptic 90×180-nm nanoholes in a 120-nm golden film was fabricated to convert a 632.8-nm left-handed circularly polarized incident Gaussian beam into vector optical vortices with topological charges $m=\pm3$ and $m=\pm5$. The phase shift of the transmitted light was found to vary with the nanohole tilt angle in a near-linear manner. In Reference [349], a plasmonic metasurface was synthesized in the form of $30\times90\times300$-nm nanostrips fabricated on a golden surface using a PB-phase method. The metasurface was demonstrated to reflect a 700-nm incident Gaussian beam, producing a vector optical vortex with the topological charge $m=+1$ for right-handed circular polarization and $m=-1$ for left-handed circular polarization. The phase shift was shown to depend on the strip tilt, with 50% of the incident light being reflected. In Reference [350], various plasmonic nanosieves to generate optical vortices and Bessel beams were studied.

An array of 75×150-nm nanoholes located with a 500-nm period was generated by e-beam lithography in a 60-nm-thick golden film for a 632-nm incident wave. With the phase computed using a PB-phase method and a circularly polarized incident beam, the efficiency was as low as 3%. In Reference [351], half- and quarter-wave plates and q-plates were fabricated on a silicon surface using resonance elliptic nanorods of size $450\times285\times860$ nm located with a 750-nm period to convert linear incident polarization into the radial and azimuthal ones. The plates were shown to have a 90% transmittance and near 100% polarization conversion for incident wavelengths of 1400–1700 nm. With the nanorod parameters selected to be resonant, the nanorod-based metasurface was constructed based on a Huygens principle so that the destructive interference of the transverse electric (TE)- and transverse magnetic (TM)-modes from each nanorod, acting as a meta-atom, served as an anti-reflection coating. In a similar way, transmission meta-holograms composed of 750-nm-high circular silicon nanorods have been shown to produce a desired image at a 10-mm distance for a 1600-nm incident beam [352]. The phase was encoded in variations of the nanorod diameter, ranging from 79 to 212 nm with a 750-nm period. A review of recent advances in the design of plasmonic and dielectric metasurfaces can be found in Reference [353].

In Reference [354], an array of $410\times175\times466$ nm nanoplates arranged with a 600-nm period and fabricated in amorphous silicon (refractive index $n=3.9231+i0.1306$ for a 780-nm wavelength) was used as a quarter-wave plate, simultaneously converting the transmitted beam into several optical vortices with the topological charges $m=2$, 4, and 6 in different diffraction orders. In Reference [355], three phase elements with metasurfaces (a spiral plate, an axicon, and a zone plate), composed of diffraction gratings in an amorphous-silicon film computed using the PB-phase method and synthesized with an fs-laser were reported to generate perfect vector optical vortices with the topological charge ranging from 1 to 3, with the vortex ring diameter being independent of the topological charge. The phase elements were 6 mm in size, the incident wavelength was 633 nm, and the lens focal length was 200 mm. An immersion metalens with NA = 1.1 composed of elliptic TiO_2 nanorods of size $80\times220\times600$ nm arranged with a 240-nm period, operating as a half-wave plate was reported in Reference [356]. When illuminated by a circularly polarized 532-nm Gaussian beam, the metalens generated an immersion 240-nm focal spot with a 50% efficiency. In References [357, 358] the elliptic TiO_2 nanorods were used to synthesize spiral metalenses intended to generate optical vortices with the arbitrary orbital angular momentum.

In this section, using a FDTD-based approach implemented in the FullWAVE software, spiral metalenses capable of simultaneously generating vector optical vortices with the topological charges $m = 1$, 2, and 4 and a subwavelength focal spot or a doughnut intensity pattern are numerically simulated. The metalenses are illuminated by plane linearly polarized TE- or TM- waves and a left-handed and right-handed circularly polarized beam. Locally, the metalenses operate as half-wave plates. The novelty of our approach consists in the use of an original method proposed in our earlier work [338], in place of using the Pancharatnam–Berry-phase or Huygens methods when designing a metalens.

A question arises: why does one need to employ a metasurface when generating an optical vortex? There are several reasons for this. A metasurface-aided optical element has multiple functionalities, being capable of simultaneously manipulating the state of polarization, amplitude, and phase of the transmitted beam, as well as focusing it. In addition, depending on the incident beam polarization, an element with a metasurface is capable of changing the properties of the transmitted beam. Using differently polarized incident beams, it becomes possible to focus transverse or axial optical field components and generate circular or annular subwavelength focal spots, also choosing whether or not a reverse energy flux from focus needs to be formed. As distinct from the present study, a similar optical element with a metalens fabricated in Reference [338] did not generate an optical vortex.

5.3.1 COMPUTING AN OPTICAL SURFACE OF SPIRAL METALENSES

A method for designing photonic components with metasurfaces in a thin-film amorphous silicon for the visible spectrum was proposed in Reference [338]. In brief, the method can be described as follows. In diffractive optics, an optical vortex can be focused using a spiral zone plate (SZP) [359]. The transmittance of a binary SZP is given by

$$T_n(r, \varphi) = \text{sgn}\left[\cos\left(m\varphi + \frac{kr^2}{2f}\right)\right], \qquad (5.1)$$

where m is the topological charge of an optical vortex, (r, φ) are the polar coordinates, k is the wave number of light of wavelength λ, and f is the focal length of a parabolic lens.

The binary relief of such a phase SZP should has the depth

$$H = \lambda\left[2(\text{Re}\, n - 1)\right]^{-1}, \qquad (5.2)$$

where Re is the real part of the refractive index n of the SZP material. For the visible spectrum and glass material, the relief depth should be in the 300–500 nm range. If the SZP produces a near-surface focal spot, the outermost zones are comparable in size with the incident wavelength, with their depth being equal to a half-wavelength. Because of this, being nearly the same in size, the height and width of the outermost

SZP zones approximately equal $\lambda/2 \approx 300$ nm. To reduce the step depth, thus achieving an SZP of subwavelength thickness, a π-phase shift in passing the zone boundary can be realized by means of space-variant diffraction gratings. Binary subwavelength gratings in a silicon film operate as local half-wave plates, rotating the incident beam polarization vector by $\beta = 2\alpha$, where α is the angle between the grating lines and the incident wave vector. The half-wave gratings rely in their operation on the difference between the effective indices of two polarization vectors oriented along (TE-component) and across (TM-component) the grating lines [360]:

$$n_{\text{eff}}^{TE} = \left[Qn_r^2 + (1-Q)n_m^2 \right]^{1/2},$$

$$n_{\text{eff}}^{TM} = \left[Qn_r^{-2} + (1-Q)n_m^{-2} \right]^{-1/2}, \tag{5.3}$$

where Q is the fill factor (the ratio of the grating step width to its period), n_r is the refractive index of the step material, and n_m is the refractive index of the environment. For $Q=0.5$, for a π-phase shift to be obtained for TE- and TM-waves in an amorphous-silicon film with a refractive index of $n=4.35$, the depth of the grating step should be ($\lambda = 633$ nm):

$$H = \lambda \sqrt{n^2 + 1} \left[\sqrt{2}(n-1)^2 \right]^{-1} \approx 179 \text{ nm}. \tag{5.4}$$

The grating parameters were optimized following an approach described in Reference [338]. For example, Table 5.2 depicts in which way variations in the fill factor Q affect the amplitude and phase shift of the transmitted wave.

TABLE 5.2
Impact of the Fill Factor Q of a Subwavelength Grating ($H=120$ nm) on the Amplitude, Phase, and Polarization Ellipticity of the Transmitted Light ($\lambda = 633$ nm)

Fill factor, Q	Amplitude, E_x (r.u.)	Phase, E_x	Amplitude, E_y (r.u.)	Phase, E_y	Polarization	Ellipticity
0.1	1.06	0.776π	0.12	1.32π	0	0.11
0.2	0.8	0.98π	0.3	1.47π	0	0.39
0.3	0.55	1.21π	0.41	1.62π	28°	0.66
0.4	0.41	1.41π	0.41	1.74π	45°	0.55
0.5	0.36	1.56π	0.32	1.81π	42°	0.38
0.6	0.34	1.66π	0.27	1.83π	38°	0.25
0.7	0.34	1.73π	0.22	1.88π	32°	0.2
0.8	0.34	1.77π	0.17	1.8π	26°	0.2
0.9	0.35	1.78π	0.1	0π	13°	0.2

The grating steps are tilted at $\alpha = 22.5°$ with respect to the polarization vector $\mathbf{E}$ of the incident linearly polarized plane wave. The polarization vector of the transmitted wave is assumed to equal $\beta = 45°$. With the $\mathbf{E}$ vector being parallel to the x-axis, it has a single projection E_x. From Table 5.2 it is seen that the best values of Q equal 0.4 and 0.5. For further purposes, Q=0.5 is chosen thanks to a smaller ellipticity (the ratio of polarization axes).

Table 5.3 shows in which way the subwavelength grating groove depth affects the amplitude, phase, and ellipticity of the transmitted light.

From Table 5.3, an optimal depth of the grating groove is $H = 120$ nm. This value is close to the value of $H = 179$ nm which follows from Equation (5.4).

When computing a metalens, an important parameter to be taken into account is the number of space-variant gratings to perform the polarization conversion. For example, a radially polarized output beam was generated using just four space-variant subwavelength gratings that convert the output polarization by 45, 135, 225, and 315° [359]. This turned out to be sufficient for generating a near-circular subwavelength focal spot.

In this section, we study the use of 16 subwavelength gratings to generate an azimuthally polarized output beam with 16 gradations of the output polarization vectors, with the adjacent vectors making angles of 11.25°. In Reference [361] a 16-sector micrometalens was demonstrated to generate a focal spot closest to the circle when illuminated by a linearly polarized beam.

When designing subwavelength gratings for the SZPs, two rules need to be obeyed. Rule 1: The gratings should be arranged so as to produce a near-azimuthal

---

TABLE 5.3

Impact of the Height of Microrelief Features of a Subwavelength Grating (Q = 0.5) Synthesized in an Amorphous Thin-Film Silicon on the Amplitude, Phase, and Polarization Ellipticity of the Transmitted Light ($\lambda = 633$ nm). The Grating Grooves Make an Angle of $\alpha = 22.5°$ with the Incident Polarization Vector

Microrelief height, h (μm)	Amplitude, E_x (r.u.)	Phase, E_x	Amplitude. E_y (r.u.)	Phase, E_y	Polarization	Ellipticity
0.06	0.59	0.773π	0.24	1.45π	−13.5°	0.34
0.07	0.56	0.88π	0.27	1.5π	−12.6°	0.42
0.08	0.57	1.01π	0.28	1.54π	0	0.49
0.09	0.51	1.17π	0.32	1.61π	16.2°	0.58
0.1	0.46	1.32π	0.32	1.68π	27°	0.53
0.11	0.39	1.43π	0.32	1.74π	36°	0.48
0.12	0.36	1.56π	0.32	1.81π	41.4°	0.38
0.13	0.29	1.69π	0.3	1.86π	44.1°	0.26
0.14	0.28	1.8π	0.27	1.86π	43.2°	0.17
0.15	0.27	0π	0.25	0π	42.3°	0.06
0.16	0.27	0.1π	0.24	0.1π	40.5°	0.03

polarization of the output beam. Rule 2: Two adjacent gratings at the zone boundary are assumed to generate oppositely directed polarization vectors, thus producing a phase shift by π.

In this section, we designed spiral zone plates and corresponding metalenses with the topological charges $m = 1, 2,$ and 4.

Figure 5.17 depicts SZPs of the 1st-, 2nd-, and 4th-order (a), (b), and (c) and equivalent metalenses (d), (e), and (f). The calculation parameters in Figure 5.17 were as follows: the SZP focal length- $f = \lambda = 633$ nm, the metalens mask pixel is 22 nm, and

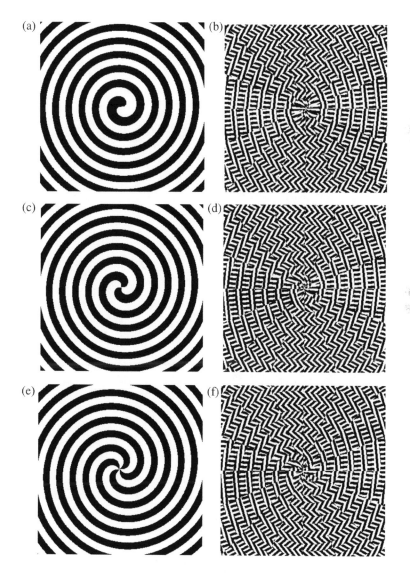

FIGURE 5.17   Binary SZPs with the topological charge m: (a) 1, (c) 2, and (e) 4 and equivalent spiral metalenses with the topological charge m: (b) 1, (d) 2, and (f) 4.

the overall size of the element 8×8 µm. The gratings in Figure 5.17 were assumed to have a period of $d = 220$ nm and a fill factor of $Q = 0.5$, meaning that the step and groove have a 110-nm width. All the gratings had a depth of $H = 120$ nm. The refractive index of the thin-film amorphous silicon was defined as $n = 4.352 + 0.486i$, with the substrate refractive index being $n = 1.5$. The film thickness was assumed to be equal to the grating depth.

The operation of the spiral metalens in Figure 5.17 can be schematically defined as a product of a matrix $\hat{R}(\varphi) = \begin{pmatrix} \cos\varphi - \sin\varphi \\ \sin\varphi \quad \cos\varphi \end{pmatrix}$ describing the rotation of the polarization vector by angle φ, the spiral plate transmittance $e^{im\varphi}$, and the spherical lens transmittance $\exp\left(-ikr^2 / (2f)\right)$. Then, depending on the incident beam polarization, such an element can produce different output beams, namely, $\begin{pmatrix} 1 \\ 0 \end{pmatrix}$, a TE-wave ($E_X$), $\begin{pmatrix} 0 \\ 1 \end{pmatrix}$, a TM-wave ($E_Y$), $\begin{pmatrix} 1 \\ i \end{pmatrix}$, right-handed circular polarization ($E_X + iE_Y$) or $\begin{pmatrix} 1 \\ -i \end{pmatrix}$, left-handed circular polarization ($E_X - iE_Y$). When illuminated by a TE-wave, the metalens with $m = 1$ produces a radially polarized output vortex:

$$\exp\left(i\varphi - i\frac{kr^2}{2f}\right)\begin{pmatrix} \cos\varphi - \sin\varphi \\ \sin\varphi \quad \cos\varphi \end{pmatrix}\begin{pmatrix} 1 \\ 0 \end{pmatrix} = \exp\left(i\varphi - i\frac{kr^2}{2f}\right)\begin{pmatrix} \cos\varphi \\ \sin\varphi \end{pmatrix} \tag{5.5}$$

converging into a doughnut intensity pattern. When illuminated by a TM-wave, the output beam is an azimuthally polarized optical vortex

$$\exp\left(i\varphi - i\frac{kr^2}{2f}\right)\begin{pmatrix} \cos\varphi - \sin\varphi \\ \sin\varphi \quad \cos\varphi \end{pmatrix}\begin{pmatrix} 0 \\ 1 \end{pmatrix} = \exp\left(i\varphi - i\frac{kr^2}{2f}\right)\begin{pmatrix} -\sin\varphi \\ \cos\varphi \end{pmatrix} \tag{5.6}$$

converging into a focal spot. When illuminated by a right-handed circularly polarized beam, the metalens (Figure 5.17(b)) produces a right-handed circularly polarized output spherical wave:

$$\exp\left(i\varphi - i\frac{kr^2}{2f}\right)\begin{pmatrix} \cos\varphi - \sin\varphi \\ \sin\varphi \quad \cos\varphi \end{pmatrix}\begin{pmatrix} 1 \\ i \end{pmatrix} = \exp\left(-i\frac{kr^2}{2f}\right)\begin{pmatrix} 1 \\ i \end{pmatrix} \tag{5.7}$$

converging into a focal spot. Finally, when illuminated by a left-handed circularly polarized beam, the metalens in Figure 5.17(b) produces a left-handed circularly polarized output optical vortex with $m = 2$:

$$\exp\left(i\varphi - i\frac{kr^2}{2f}\right)\begin{pmatrix} \cos\varphi - \sin\varphi \\ \sin\varphi \quad \cos\varphi \end{pmatrix}\begin{pmatrix} 1 \\ -i \end{pmatrix} = \exp\left(2i\varphi - i\frac{kr^2}{2f}\right)\begin{pmatrix} 1 \\ -i \end{pmatrix} \tag{5.8}$$

converging into a doughnut intensity pattern.

5.3.2 Numerical Simulation of a Spiral Metalens with $M = 1$

By way of illustration, below we numerically simulate the performance of a spiral metalens using the assumption $m = 1$. The FDTD-based simulation was implemented in the FullWAVE software. Figure 5.18 depicts the simulation results for a linearly polarized plane incident beam E_x. The images are 1.33×1.33 μm in size. Shown in Figure 5.18 are the intensity pattern (a), modules of the amplitude and phase of all transmitted field projections, namely, E_x (b), (c), E_y (d), (e), and E_z (f), (g), and the projection S_z of the Poynting vector onto the optical axis (h). The intensity pattern is seen to be in the form of bright rings with the central intensity minimum. It can be also seen that it is the axial field component (f) that gives the main contribution to the intensity (a), with its maximum intensity being approximately 10 times that of the transverse components (b) and (d), whereas the on-ring intensity (a) is 11 times that of the incident light. Thus, we can infer that when illuminated by a wave with E_x-polarization, the metalens in (b) generates a radially polarized output vector optical vortex with the topological charge $m = 1$ (g). The radial polarization stems from the fact that the intensity ring in the metalens focal plane (a) is predominantly characterized by the axial polarization (f). Note that while the Poynting vector takes the maximum value of 4.35, the maximum value of the modulus of its z-projection is 2.71 (h). Hence, after passing through the focus, only 63% of the entire energy propagates along the optical axis. It is worth noting that although compared to the axial component, the transverse E-field components are significantly smaller, it is the latter that exclusively contribute to the on-axis energy flux, which is the Poynting vector. Because of this, in (h) there is an on-axis maximum of the z-projection of the Poynting vector, with a non-zero on-axis intensity minimum found at the ring center in (a).

If the metalens of Figure 5.17(b) is illuminated by an incident plane wave whose polarization vector is rotated by 90° (E_y-polarization), a subwavelength focal spot of transverse polarization will be generated. Shown in Figure 5.19 are (a) an intensity pattern and amplitude and phase patterns for all transmitted field projections,

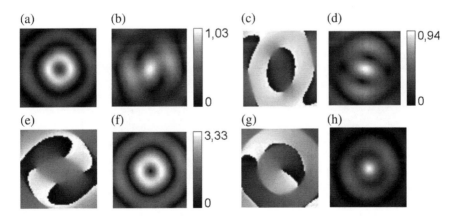

FIGURE 5.18 The metalens of Figure 5.17(a) and (b) ($m = 1$) is illuminated by a linearly polarized plane E_x-wave. Computed 600 nm from the metalens surface are: (a) the intensity pattern, the amplitude and phase patterns for the (b) and (c) E_x-component, (d) and (e) E_y-component, and (f) and (g) E_z-component; and (h) the z-projection S_z of the Poynting vector.

(b), (c) E_x, (d), (e) E_y, (f), (g) E_z, and (h) the z-projection S_z of the Poynting vector. A subwavelength near-circular focal spot of transverse polarization is seen to be generated (a). The transverse polarization stems from the fact that the intensity maxima of the transverse E-field components (b) and (d) are nearly 10 times that of the on-axis component (f). The maximum value of the Poynting vector is 2.26, whereas the maximum value of the modulus of its z-projection is 2.07 (h). Hence, almost the entire energy (90%) propagates along the optical axis.

Summing up, in this section it was shown that by illuminating a spiral metalens (Figure 5.17(b)) with a beam linearly polarized along the x-axis, a near axially polarized subwavelength optical vortex with topological charge $m=1$ can be generated. When illuminated by a plane wave linearly polarized along the y-axis, the metalens produces a subwavelength circular focal spot of predominantly transverse polarization.

5.3.3 COMPUTING THE PROJECTION OF THE POYNTING VECTOR ONTO THE FOCAL REGION OF A SPIRAL METALENS

For comparison, Figure 5.20 shows z-projections S_z of the Poynting vector onto the focal region of the metalens of (b) ($m=1$), with the incident beam linearly polarized along the x- and y-axis (a), (b)), and with a right- and left-handed circularly polarized incident beam (c), (d). As seen from Figure 5.20, there are regions in the 2D pattern where the z-projections S_z of the Poynting vector are negative, which are found (a)–(c) near and (d) on the optical axis. Hence, with a left-handed circularly polarized incident beam, a reverse energy flux occurs on the optical axis in the opposite direction to the incident light. Negative vector optical vortices with the reverse propagation of energy near the optical axis have been known in optics [143, 362–364]. Explicit relationships for the longitudinal component of the Poynting vector have been derived, making it possible to deduce the conditions for changing the sign of S_z. For instance, it has been shown [362] that a vector Bessel beam composed of two

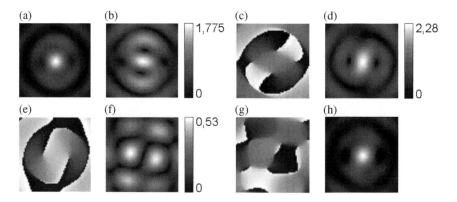

FIGURE 5.19 The metalens of Figure 5.17(a) and (b) ($m=1$) is illuminated by a linearly polarized plane E_y-wave. Computed 600 nm from the metalens surface are: (a) the intensity pattern, the amplitude and phase patterns for (b) and (c) E_x-component, (d) and (e) E_y-component, (f) and (g) E_z-component; and (h) z-projection S_z of the Poynting vector.

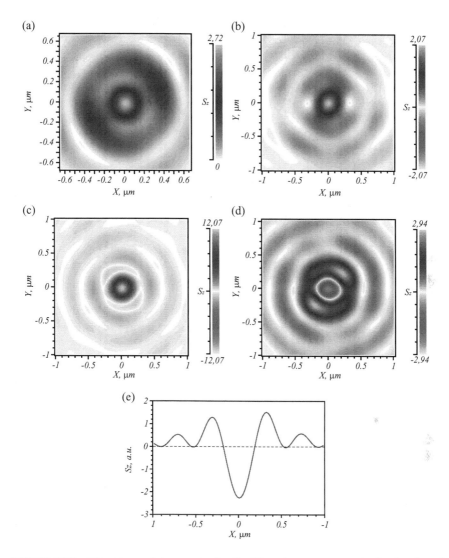

FIGURE 5.20 2D patterns of the z-projection S_z of Poynting vector onto the focal region of a metalens with $m=1$ (Figure 5.17(b)), when illuminated by a beam (a) and (b) linearly polarized along the x-axis and y-axis, (c) and (d) with right- and left-handed circular polarization; (e) profile of S_z onto the x-axis for left-handed circular polarization.

Bessel beams with radial and azimuthal polarizations has even sidelobes with the reverse energy flux on the z-axis, contrary to the positive propagation of the entire beam along the z-axis. It has also been demonstrated [363] that in the cross-section of a diffraction-free beam composed of a linear combination of two co-axial beams propagating with different phase velocities in the positive z-direction there is a small near-focus region where the energy flux is reversed. A relationship for S_z has been derived (see Equation (19) in Reference [143]) for a tightly focused linearly polarized zero-order Bessel beam, suggesting that at certain parameters, the energy flux in the outermost regions of the focal region can be reversed ($S_z < 0$). However, the reverse energy flux was not defined explicitly in Reference [143]. Below, we demonstrate that when tightly focusing the linearly polarized light, in the cross-section of the focal region, there are local regions where the energy flux is reversed ($S_z < 0$). In Reference [143], the tight focus of zero-order Bessel beams was discussed.

Below, we derive a relationship to describe the projection of the Poynting vector in the focal region for a linearly polarized nth-order Bessel beam. The projection of Poynting vector onto the optical axis is given by

$$S_Z = 0.5 \operatorname{Re} \left[\mathbf{E} \times \mathbf{H}^* \right]_z, \tag{5.9}$$

Considering that for linear polarization, $E_y = 0$, Equation (5.9) is simplified to

$$S_Z = 0.5 \operatorname{Re} \left(E_X H_Y^* \right). \tag{5.10}$$

Using Maxwell's equations for the monochromatic light

$$\operatorname{rot} \mathbf{E} = ik\mathbf{H}, \quad \mu = 1, \tag{5.11}$$

where $k = w/c$ is the wave number, w is the circular frequency, and c is the speed of light, Equation (5.10) can be replaced by

$$S_Z = \operatorname{Re} \left[\frac{i}{2k} \left\{ E_X \left(\frac{\partial E_X^*}{\partial z} - \frac{\partial E_Z^*}{\partial x} \right) \right\} \right]. \tag{5.12}$$

The amplitude E_x can be expanded into an angular spectrum of plane waves:

$$E_X(x, y, z) = \int_{-\infty}^{\infty} A(\xi, \eta) \exp \left[ik \left(x\xi + y\eta + z\sqrt{1 - \xi^2 - \eta^2} \right) \right] d\xi d\eta, \tag{5.13}$$

where $A(\xi, \eta)$ is the amplitude of a spectrum of plane waves. Based on the third Maxwell's equation, $\operatorname{div}\mathbf{E} = 0$, and considering that $E_y = 0$, the on-axis E-field component is given by

$$E_Z(x, y, z) = -\int_{-\infty}^{\infty} \frac{\xi A(\xi, \eta)}{\sqrt{1 - \xi^2 - \eta^2}} \exp \left[ik \left(x\xi + y\eta + z\sqrt{1 - \xi^2 - \eta^2} \right) \right] d\xi d\eta. \tag{5.14}$$

Then, the expression in parenthesis in Equation (5.12) takes the form:

$$\frac{\partial E_X^*}{\partial z} - \frac{\partial E_Z^*}{\partial x} = -ik \int_{-\infty}^{\infty} \frac{1 - \eta^2}{\sqrt{1 - \xi^2 - \eta^2}} A^*(\xi, \eta)$$

$$\times \exp \left[-ik \left(x\xi + y\eta + z\sqrt{1 - \xi^2 - \eta^2} \right) \right] d\xi d\eta. \tag{5.15}$$

By way of illustration, let there be an nth-order Bessel beam with its angular spectrum given by

$$A(\xi,\eta) = \exp(in\varphi)\delta(r - r_0), \quad r^2 = \xi^2 + \eta^2, \quad \tan\varphi = \eta / \xi. \tag{5.16}$$

Then, Equations (5.13) and (5.15) can be replaced by

$$E_X(\rho,\theta,z) = 2\pi r_0 i^n e^{in\theta} J_n(kr_0\rho)\exp\left(ikz\sqrt{1 - r_0^2}\right), \tag{5.17}$$

$$\frac{\partial E_X^*}{\partial z} - \frac{\partial E_Z^*}{\partial x} = \frac{2\pi kr_0(-i)^{n+1}e^{-in\theta}}{\sqrt{1 - r_0^2}}\left[\left(1 - \frac{r_0^2}{2}\right)J_n(kr_0\rho)\right.$$

$$\left. - \frac{r_0^2}{4}\left(e^{2i\theta}J_{n-2}(kr_0\rho) + e^{-2i\theta}J_{n+2}(kr_0\rho)\right)\right]\exp\left(-ikz\sqrt{1 - r_0^2}\right). \tag{5.18}$$

Substituting Equations (5.17) and (5.18) into (5.12) yields:

$$S_Z = \frac{2\pi^2 r_0^2}{\sqrt{1 - r_0^2}} J_n(kr_0\rho)\left[\left(1 - \frac{r_0^2}{2}\right)J_n(kr_0\rho) - \frac{r_0^2}{4}\cos 2\theta\left(\begin{array}{c}J_{n-2}(kr_0\rho)\\ + J_{n+2}(kr_0\rho)\end{array}\right)\right]. \tag{5.19}$$

In Equation (5.19), $kr_0 = k_r$ is the transverse projection of the wave vector. From Equation (5.19), the transverse focal region is seen to have polar-angle-dependent local regions where the projection of the Poynting vector is negative, $S_z < 0$. Considering that for tight focusing $r_0 = 1$ and assuming a small value of argument $k\rho \ll 1$ (e.g. at $\cos 2\theta = 1$), we obtain an inequality

$$2J_n(x) < J_{n-2}(x) + J_{n+2}(x), \quad x = k_r\rho. \tag{5.20}$$

Inequality (5.20) shows that in a neighborhood of the first (off-axis) zero of an nth-order Bessel function there is a horizontally oriented region where the light flux is reversed. At $n = 0$ (in the center of the focal spot), inequality (5.20) is simplified to

$$J_0(x) < J_2(x), x = k_r\rho. \tag{5.21}$$

In Figure 5.20, the maximum on-axis value of S_z equals 2.72 (a), 2.07 (b), 12.0 (c), and −2.31 (d). Hence, we can infer that when illuminating the metalens with a linearly polarized beam (a), (b), the near-axis energy flux in the positive direction approximately equals that propagating in the negative direction for the left-handed circular polarization of the incident light (c). Note that in (d), the on-axis energy flux propagates along a ring where the maximum value of S_z equals 2.94. It is also worth noting that S_z takes a zero value everywhere on the optical axis ($x = y = 0$) in (d). Surprisingly, the on-axis energy flux is maximum when the metalens is illuminated by a beam with right-handed circular polarization (c). We should also note that our analysis of Equations (5.9)–(5.19) has failed to account for the existence of the reverse flux when tightly focusing the circularly polarized light. In the next chapter, we come up with an explanation of the pattern in Figure 5.20(d).

5.3.4   The Energy Flux in the Focal Region of a Spiral Metalens with $M = 2$

For comparison, Figure 5.21 depicts 2D distributions of the projections S_z of the Poynting vector onto the optical axis in the focal region of a metalens with $m = 2$

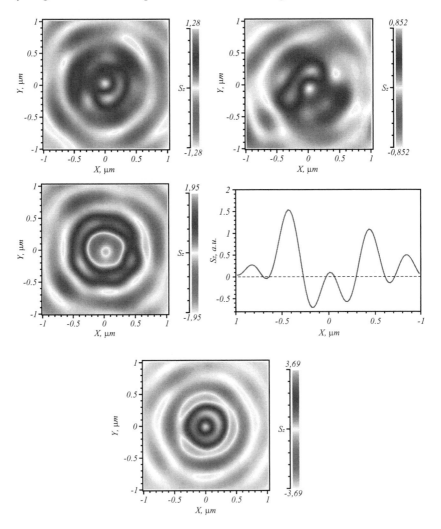

FIGURE 5.21   2D patterns of the projection S_z of the Poynting vector onto the optical axis in the focal region of a metalens with $m = 2$ (Figure 5.17(d)) for differently polarized incident beams: linearly polarized in the (a) x- and (b) y-directions, (c) right-handed, and (e) left-handed circular polarization; (d) profile of S_z onto the x-axis for right-handed circular polarization.

(Figure 5.17(d)) for differently polarized homogeneous incident beams. In Figure 5.21, the projection S_z takes the maximum value of 1.28 on an asymmetric ring in (a), 0.852 on an asymmetric ring in (b), −0.795 on the first ring and 1.94 on the second ring (c), 3.69 on the first ring and −0.47 on the second ring (e). Note that S_z takes a zero value on the optical axis in all patterns in Figure 5.21.

It is seen from Figure 5.21 that the reverse energy flux in the focus depends on the incident polarization for $m = 2$ in a similar manner to $m = 1$ (Figure 5.20), the only difference being the location of the reverse-flux regions. When illuminating the metalens in Figure 5.20(d) by a linearly polarized beam, the relatively small regions of the reverse flux are asymmetrical and characterized by low energy values (Figure 5.21(a) and (b)). Meanwhile, with the circularly polarized incident beam, the reverse-flux regions from the metalens in Figure 5.17(e) are symmetrical, account for a significant proportion of the focal spot area, and carry a significant energy comparable with that of the direct energy flux (see the near-axis region in Figure 5.21(c) – an analogous result was reported in Reference [364] for left-handed polarization, unlike right-handed polarization discussed in this study – and the first dark ring in Figure 5.21(d)).

We have shown by FDTD-simulation and theoretically, using an expansion into plane waves, that in the tight focal spot ($NA \approx 1$) generated by a spiral metalens for differently polarized homogeneous incident laser beams, there are local regions where the projection of the Poynting vector is negative, implying the negative direction of the energy flux. Above, an "optical tractor" phenomenon was mentioned, with the reverse energy flux shown to be capable of moving microparticles counter to the laser beam propagation direction. This phenomenon was found to occur in symmetrical Bessel beams [362] or result from the interference of two laser beams with different phase velocities [363]. In this section, we have for the first time shown that the reverse energy flux can also occur when tightly focusing a vortex laser beam ($m = 1, 2$) with linear ($\sigma = 0$) or circular ($\sigma = \pm 1$) polarization. Note that for left-handed polarization ($m = 2$ and $\sigma = -1$), the reverse energy flux was demonstrated in Reference [364].

6 Reverse Flux of Light Energy in Sharp Focus

6.1 FOCUSING OF OPTICAL VORTEX WITH AN INTEGER TOPOLOGICAL CHARGE *M*

When non-paraxial light fields with different polarization states propagate in space, local areas can appear in their cross-sections, where the flux of light energy has a reversed direction, i.e. the longitudinal component of the Poynting vector in such areas has a negative value. Earlier, reversed energy flux in light fields was investigated in References [6, 362–372]. It has been shown in Reference [6] that when a linearly polarized plane wave is focused by an aplanatic system, in the first dark ring of the intensity distribution in the focal plane there is an area where the flux of light energy is directed in the opposite direction to the propagation direction of the incident plane wave. For a superposition of two *m*th-order Bessel beams with tranverse electric (TE) and transverse magnetic (TM) polarizations, the possibility of a negative value of the longitudinal component of the Poynting vector on the optical axis has been shown theoretically [362]. The practically realizable case (focusing by an aplanatic system) has been considered in Reference [364] and it has been shown both theoretically and numerically that when a Laguerre–Gaussian mode of the order $(0, m) = (0, 2)$ with left circular polarization $(\sigma = -1)$ is focused, then on the optical axis in the focus there are negative values of the longitudinal component of the Poynting vector. Strongly divergent vector spherical vortex beams, which are the exact solutions of the Maxwell's equations, were considered in Reference [367]. In Reference [363], a superposition is studied of two arbitrary light fields with the different axial components of the wave vector. It has been shown that there are local areas in such light fields where the longitudinal component of the force acting on a microparticle is directed opposite to the wave vector of the light beam. In Reference [368], an optical spin torque induced by vector Bessel beams on a light-absorptive sphere of arbitrary size was calculated. It is shown that at different polarization states the sphere can rotate in opposite directions. Reverse flux on the optical axis in the focus of a vortex metalens is numerically shown in Reference [369]. The reversed propagation of energy in a vectorial Bessel beam with a fractional topological charge has been numerically shown in Reference [370]. Actually, such a light beam is a superposition of a countable number of the conventional Bessel modes. It was shown in Reference [371] that the axial component of linear momentum density flux can be negative in counterpropagating non-diffracting vortex Bessel beams. Expressions for the Poynting vector density for the vectorial X-beams were theoretically obtained in Reference [372] and the necessary conditions for the reversed energy flux were derived. In Reference [365], the reversed flux of energy was numerically shown for a non-paraxial accelerating 2D Airy beam. In Reference [366], a local wave vector is theoretically studied and conditions are considered that must be fulfilled for a light field so that it had locally backward propagation (i.e. the reversed flux of energy).

In this section, we obtain general formulae for the components of the electric and magnetic fields of an arbitrary circularly polarized optical vortex with an integer topological charge near the focus of an aplanatic system, as well as expressions for the intensity and components of the Poynting vector in the focal plane. For $m > 2$ and left circular polarization, it follows from these formulae that near the optical axis and near the focal plane there is a reverse flux of light energy that rotates along a spiral in the backward direction.

Using the Richards–Wolf (RW) formulae [6], we consider the intensity and the energy flux (components of the Poynting vector) in the plane of the sharp focus of an arbitrary circularly polarized optical vortex, focused by an aplanatic system. Expressions for the three components of the electric field strength vector of an optical vortex near the focus have been obtained in Reference [373]. Below, we derive expressions for the three components of the magnetic field strength vector, as well as for the intensity and the three components of the Poynting vector. Following [373], we suppose for the strength vector of a circularly polarized electric field $\mathbf{E} = E_x\mathbf{e}_x + i\sigma E_y\mathbf{e}_y$ ($\mathbf{e}_x$, $\mathbf{e}_y$ are unit vectors along the Cartesian coordinates) that $\sigma = 1$ means right circular polarization, whereas $\sigma = -1$ means left circular polarization. For an optical vortex with a topological charge m and with an arbitrary apodization function (real-valued function $A_m(\theta)$)

$$A_m(\theta, \varphi) = A_m(\theta)\exp(im\varphi), \tag{6.1}$$

where (θ, φ) are the angles defining a point on the converging spherical wavefront, we write the components of the electric vector $\mathbf{E}$ near the focus of an aplanatic system in the cylindrical coordinates (r, φ, z), following [373]:

$$E_x(r, \varphi, z) = -i^{m+1}e^{im\varphi}\left(I_{0,m} + \gamma_+ I_{2,m+2}e^{2i\varphi} + \gamma_- I_{2,m-2}e^{-2i\varphi}\right),$$

$$E_y(r, \varphi, z) = i^m e^{im\varphi}\left(\sigma I_{0,m} - \gamma_+ I_{2,m+2}e^{2i\varphi} + \gamma_- I_{2,m-2}e^{-2i\varphi}\right), \tag{6.2}$$

$$E_z(r, \varphi, z) = -2i^m e^{im\varphi}\left(\gamma_+ I_{1,m+1}e^{i\varphi} - \gamma_- I_{1,m-1}e^{-i\varphi}\right),$$

where

$$I_{0,m} = B\int_0^a \sin\theta\cos^{1/2}\theta A_m(\theta)e^{ikz\cos\theta}(1 + \cos\theta)J_m(x)d\theta,$$

$$I_{1,m\pm1} = B\int_0^a \sin^2\theta\cos^{1/2}\theta A_m(\theta)e^{ikz\cos\theta}J_{m\pm1}(x)d\theta, \tag{6.3}$$

$$I_{2,m\pm2} = B\int_0^a \sin\theta\cos^{1/2}\theta A_m(\theta)e^{ikz\cos\theta}(1 - \cos\theta)J_{m\pm2}(x)d\theta,$$

where $B = kf/2$, $\alpha = \arcsin(\text{NA})$, $x = kr\sin\theta$, $\gamma_{\pm} = (1 \pm \sigma)/2$, $J_{\nu}(x)$ is the Bessel function, k is the wave number of light, f is the focal length of the aplanatic system with a numerical aperture NA. Note that the theory developed in Reference [6] is valid when $kf \gg 1$. The focal intensity distributions calculated by the finite-difference time-domain (FDTD) and RW methods are shown to practically coincide at $kf > 30$ [146].

Based on Equation (6.2), an expression for the intensity in the focal plane ($z = 0$) of an arbitrary optical vortex (5.1) can be written as follows:

$$I_m(r, \varphi, z = 0) = 2\left[I_{0,m}^2 + \gamma_+\left(I_{2,m+2}^2 + 2I_{1,m+1}^2 \right) + \gamma_-\left(I_{2,m-2}^2 + 2I_{1,m-1}^2 \right) \right]. \qquad (6.4)$$

It is seen in Equation (6.4) that the intensity distribution in the focal plane is axially symmetric, since it does not depend on the azimuthal angle φ. Since the second index of the integrals $I_{p,q}$ in Equations (6.3) and (6.4) is the order of the Bessel function, it can be concluded from Equation (6.4) that the on-axis intensity ($r = 0$) equals zero for any value of $m > 2$. At $m = 1, 2$, the axial intensity of an optical vortex with left circular polarization ($\sigma = -1$) is non-zero, but in both cases ($m = 1, 2$) the light flux does not propagate along the positive direction of the optical axis: the energy flux is zero at $m = 1$ and is reversed at $m = 2$. To prove it, we write expressions for the three components of the magnetic field strength vector:

$$H_x(r, \varphi, z) = -i^m \sigma \overline{I}_{0,m} e^{im\varphi} + 2i^{m+1} \gamma_+ \sin\varphi \overline{I}_{1,m} e^{i(m+1)\varphi}$$

$$+ 2i^{m+1} \gamma_- \sin\varphi \overline{I}_{1,m} e^{i(m-1)\varphi}$$

$$+ i^m \gamma_+ e^{i(m+2)\varphi}\left(\overline{I}_{2,m+2} - \frac{2(m+1)}{kr} I_{1,m+1} \right)$$

$$- i^m \gamma_- e^{i(m-2)\varphi}\left(\overline{I}_{2,m-2} - \frac{2(m-1)}{kr} I_{1,m-1} \right),$$

$$H_y(r, \varphi, z) = -i^{m+1} \overline{I}_{0,m} e^{im\varphi} - 2i^{m+1} \gamma_+ \cos\varphi \overline{I}_{1,m} e^{i(m+1)\varphi}$$

$$- 2i^{m+1} \gamma_- \cos\varphi \overline{I}_{1,m} e^{i(m-1)\varphi} \qquad (6.5)$$

$$- i^{m+1} \gamma_+ e^{i(m+2)\varphi}\left(\overline{I}_{2,m+2} - \frac{2(m+1)}{kr} I_{1,m+1} \right)$$

$$- i^{m+1} \gamma_- e^{i(m-2)\varphi}\left(\overline{I}_{2,m-2} - \frac{2(m-1)}{kr} I_{1,m-1} \right),$$

$$H_z(r, \varphi, z) = i^{m+1} \gamma_+ e^{i(m+1)\varphi}\left(\overline{I}_{2,m+1} + \frac{2m}{kr} I_{0,m} \right)$$

$$- i^{m+1} \overline{I}_{0,m-1} e^{im\varphi} \left(\sigma\cos\varphi + i\sin\varphi \right) + i^{m+1} \gamma$$

$$- \overline{I}_{2,m-1} e^{i(m-1)\varphi},$$

where

$$\bar{I}_{0,m} = B\int_0^a \sin\theta\cos^{3/2}\theta A_m(\theta)e^{ikz\cos\theta}(1+\cos\theta)J_m(x)d\theta,$$

$$\bar{I}_{0,m-1} = B\int_0^a \sin^2\theta\cos^{1/2}\theta A_m(\theta)e^{ikz\cos\theta}(1+\cos\theta)J_{m-1}(x)d\theta,$$

$$\bar{I}_{1,m} = B\int_0^a \sin^3\theta\cos^{1/2}\theta A_m(\theta)e^{ikz\cos\theta}J_m(x)d\theta, \qquad (6.6)$$

$$\bar{I}_{2,m\pm2} = B\int_0^a \sin\theta\cos^{3/2}\theta A_m(\theta)e^{ikz\cos\theta}(1-\cos\theta)J_{m\pm2}(x)d\theta,$$

$$\bar{I}_{2,m\pm1} = B\int_0^a \sin^2\theta\cos^{1/2}\theta A_m(\theta)e^{ikz\cos\theta}(1-\cos\theta)J_{m\pm1}(x)d\theta.$$

Using Equations (6.1), (6.2), (6.5), and (6.6), we derive the longitudinal component of the Poynting vector [6] $\mathbf{S}=c\mathrm{Re}[\mathbf{E}\times\mathbf{H}^*]/(8\pi)$:

$$S_z = \frac{c}{8\pi}\mathrm{Re}\left(E_x H_y^* - E_y H_x^*\right), \qquad (6.7)$$

where c is the speed of light in vacuum and $\mathrm{Re}(\ldots)$ is the real part of a complex number in the focal plane ($z=0$):

$$S_{mz} = 2I_{0,m}\left(\bar{I}_{0,m}+\bar{I}_{1,m}\right)+2\gamma_+ I_{2,m+2}\left(\bar{I}_{2,m+2}+\bar{I}_{1,m}-\frac{2(m+1)}{kr}I_{1,m+1}\right)$$
$$+2\gamma_- I_{2,m-2}\left(\bar{I}_{2,m-2}+\bar{I}_{1,m}-\frac{2(m-1)}{kr}I_{1,m-1}\right), \qquad (6.8)$$

where the constant $c/(8\pi)$ is omitted.

It is seen in Equation (6.8) that the longitudinal flux of energy is circularly symmetric. Negative terms in Equation (6.8) show that the focal plane can contain areas with the negative longitudinal component of the Poynting vector. It is interesting to find out whether such local area can be located near the optical axis. From Equation (6.8) follows that at $m=1$ the longitudinal energy flux of the optical vortex with left circular polarization

$$S_{1z-} = 2\left[I_{0,1}\left(\bar{I}_{0,1}+\bar{I}_{1,1}\right)+I_{2,1}\left(\bar{I}_{2,1}-\bar{I}_{1,1}\right)\right] \qquad (6.9)$$

is zero on the optical axis ($r=0$), whereas at $m=2$ and left circular polarization the longitudinal flux

$$S_{2z-} = 2\left[I_{0,2}\left(\bar{I}_{0,2} + \bar{I}_{1,2}\right) + I_{2,0}\left(\bar{I}_{2,0} + \bar{I}_{1,2} - \frac{2}{kr}I_{1,1}\right)\right]$$ (6.10)

on the optical axis is reversed:

$$S_{2z-}(r = 0, z = 0) = -2B^2\left(\int_0^a \sin\theta\cos^{1/2}\theta A_2(\theta)(1 - \cos\theta)d\theta\right)^2 \leq 0.$$ (6.11)

That is, the flux of energy (6.10) near the optical axis propagates in the opposite direction with respect to the incident energy flux. At $m = 3$, instead of Equation (6.10) we get:

$$S_{3z-} = 2\left[I_{0,3}\left(\bar{I}_{0,3} + \bar{I}_{1,3}\right) + I_{2,1}\left(\bar{I}_{2,1} + \bar{I}_{1,3} - \frac{4}{kr}I_{1,2}\right)\right].$$ (6.12)

It follows from Equation (6.12) that near the optical axis ($kr \ll 1$) in the focal plane there is a reverse flux of energy, which is zero on the axis itself and increases parabolically with the radial variable r:

$$S_{3z-}(r \to 0) \approx -\frac{B^2(kr)^2}{2}\left(\int_0^a \sin^2\theta\cos^{1/2}\theta A_3(\theta)(1 - \cos\theta)d\theta\right)^2 < 0.$$ (6.13)

Comparison of Equations (6.11) and (6.13) shows that the reverse flux for $m = 3$ is approximately 3–4 times weaker than that for $m = 2$.

Now we show that near the focal plane, the transverse energy flux (both direct and reverse) of an optical vortex with left circular polarization rotates. To do this, we find the transverse components of the Poynting vector:

$$S_{x-} = -Q_m(r)\sin\varphi, \quad S_{y-} = Q_m(r)\cos\varphi,$$ (6.14)

where

$$Q_m(r) = \left(I_{0,m} + I_{2,m-2}\right)\left(\bar{I}_{2,m-1} + \bar{I}_{0,m-1}\right)$$
$$+ 2I_{1,m-1}\left(\bar{I}_{0,m} - \bar{I}_{2,m-2} + \frac{2(m-1)}{kr}I_{1,m-1}\right).$$ (6.15)

It is seen in Equation (6.14) that the energy flux in the focal plane rotates around the optical axis clockwise or counterclockwise, depending on the sign of the function $Q_m(r)$. Near the focal plane, the energy flux rotates along a spiral. It can be shown that for $m = 2, 3$, the reverse energy flux propagates along the optical axis in a spiral, rotating counterclockwise.

We performed an FDTD simulation of sharp focusing of an optical vortex by a phase zone plate. Figure 6.1 shows phase distributions of the components E_x (a, c) and E_y (b, d) of the incident light field at $m = 2$ (a, b) and at $m = 3$ (c, d), as well as

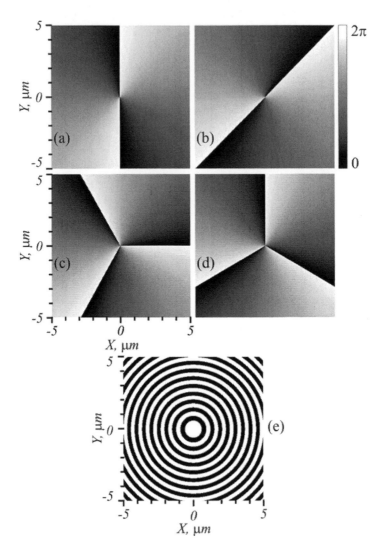

FIGURE 6.1 Phase distributions of the components E_x (a), (c) and E_y (b), (d) of the incident light field at $m=2$ (a), (b) and at $m=3$ (c), (d) respectively, and phase of the zone plate for sharp focusing the optical vortex (e).

the phase of the zone plate (PZP) (e). This PZP with a size of 10×10 μm focuses the incident field with a wavelength $\lambda=532$ nm at a distance of λ. Figure 6.2 shows distributions of the longitudinal component of the Pointing vector in the focal plane for $m=2$ (a) and $m=3$ (b), as well as the sections of the longitudinal component of the Pointing vector along the x-axis (c), and the intensity distributions for $m=2$ (d) and $m=3$ [Figure 6.2(e)]. Figures 6.2(a-c) confirm the reverse flux of light energy near the focus (or in the focus itself at $m=2$), while (d) and (e) confirm the spiral flux around the optical axis.

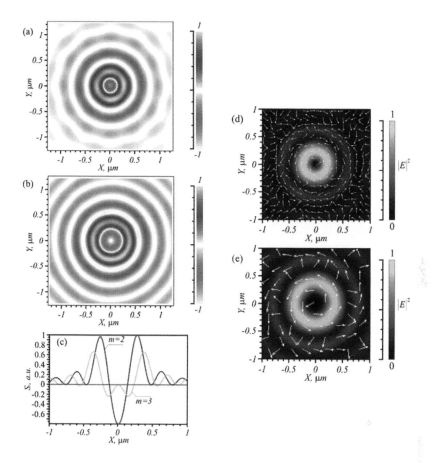

FIGURE 6.2 Distributions of the longitudinal component of the Poynting vector for $m=2$ (a) and $m=3$ (b) in the focal plane, sections of the longitudinal component of the Poynting vector along the x-axis for $m=2$ and $m=3$ (c), and intensity distributions for $m=2$ (d) and $m=3$ (e); arrows show transverse directions of the Poynting vector; sizes of areas in (d) and (e) are 2×2 µm.

In conclusion for this section, we have obtained general expressions for the intensity and for the energy flux near the focus of an arbitrary circularly polarized optical vortex with a topological charge m, focused by an ideal spherical lens (aplanatic system). Based on of these expressions, we have shown that for $m>1$ and left circular polarization ($\sigma=-1$) light energy near the optical axis propagates along a spiral in the reversed direction. Reverse flux of light energy along a spiral around the optical axis near the focus has been confirmed by the FDTD simulation of sharp focusing of an optical vortex with a phase zone plate. It is obvious that for the optical vortices with the negative topological charge $m<0$, the reverse flux of energy near the optical axis and near the focus takes place for right circular polarization ($\sigma=1$). The obtained expressions allow fast and intuitive analysis of the behavior of the intensity and of the energy flux of an arbitrary optical vortex near the sharp focus. A microparticle, trapped near the optical axis of such sharp focus, will move in the opposite direction, demonstrating the effect of the "optical tractor" [374, 375].

6.2 HELICAL REVERSE FLUX OF LIGHT OF A FOCUSED OPTICAL VORTEX WITH $M = 3$

When non-paraxial light fields with different polarization states propagate in space, local areas with a reversed direction of the light energy flux can appear in their cross-sections. The longitudinal component of the Poynting vector in such areas has a negative value. The reversed energy flux in light fields has been studied in earlier References [6, 362–366, 369, 370, 372]. For a linearly polarized plane wave focused by an aplanatic system, it has been shown [6] that in the first dark ring of the focal plane intensity distribution there is an area with the flux of light energy opposite to the propagation direction of the incident plane wave. For a superposition of two mth-order Bessel beams with TE and TM polarizations, negative value of the longitudinal component of the Poynting vector on the optical axis has been shown theoretically [362]. A feasible case (focusing by an aplanatic system) has been studied in Reference [364]. For a focused Laguerre–Gaussian mode of the order $(0, m) = (0, 2)$ with left circular polarization $(\sigma = -1)$, it has been shown both theoretically and numerically that in the focus on the optical axis there are negative values of the longitudinal component of the Poynting vector. Superpositions of two arbitrary light fields with the different axial components of the wave vector have been studied in Reference [363]. Such light fields have been shown to have local areas where the longitudinal component of force acting on a microparticle is directed opposite to the wave vector of the light beam. The reverse flux on the optical axis in the focus of a vortex metalens is numerically shown in Reference [369]. For a vectorial Bessel beam with a fractional topological charge, the reversed propagation of energy has been numerically demonstrated in Reference [370]. Such a light beam is actually a superposition of a countable number of the conventional Bessel modes. For the vectorial X-beams, expressions for the Poynting vector density were theoretically derived in Reference [372] and the necessary conditions for the reversed energy flux were obtained. In Reference [365], it has been numerically shown that a non-paraxial accelerating 2D Airy beam can have the reversed energy flux. Studying the local wave vector in Reference [366], conditions have been theoretically considered that should be fulfilled for a light field so that it would be locally travelling backward (i.e. the energy flux would be reversed).

From the works listed above, Reference [364] is the closest to our section, where the focusing of a Laguerre–Gaussian mode of the order $(0, m) = (0, 2)$ with left circular polarization $(\sigma = -1)$ by an aplanatic system has been studied, and where the negative values of the longitudinal component of the Poynting vector on the optical axis in the focus have been shown both theoretically and numerically. In contrast to [364], in this paper we show that the reverse propagation of light near the optical axis also occurs in the plane of sharp focus of an arbitrary optical vortex with the topological charge of 3 and with left circular polarization. Similar behavior of the longitudinal component of the Poynting vector (negative values near the optical axis in the focal plane) can be shown for any topological charge $m > 1$. However, the reverse flux rapidly decreases with increasing m. For $m = 2$, the reverse flux is comparable with the forward one [364]. In Reference [376], expressions are given for the components of the Poynting vector for any arbitrary integer m. However, there is no detailed analysis of the reverse energy flux phenomenon in Reference [376]. In this work, we analyze in detail the behavior of a focused optical vortex at $m=3$. At $m=3$, the reverse flux is about 2–3 times smaller than the forward flux in the focus, whereas for $m=4$ it is smaller by 8–10 times, etc. Thus, the reverse flux for optical vortices with numbers $m>3$ can be neglected.

6.2.1 LONGITUDINAL COMPONENT OF THE POYNTING VECTOR

Based on the Richards–Wolf formalism [6], we consider the focusing of an optical vortex with a topological charge $m=3$ and with left circular polarization by an aplanatic system. Following Reference [373], we assume for a circularly polarized electric field $\mathbf{E}=E_x\mathbf{e}_x+i\sigma E_y\mathbf{e}_y$ ($\mathbf{e}_x$ and $\mathbf{e}_y$ are unit vectors along the Cartesian coordinates) that $\sigma=1$ for right circular polarization and $\sigma=-1$ for left circular polarization. For an optical vortex with a topological charge m and with an arbitrary pupil apodization function (real-valued function $A_m(\theta)$)

$$A_m(\theta,\varphi)=A_m(\theta)\exp(im\varphi),\qquad(6.16)$$

where (θ,φ) are the angles defining a point on the converging spherical wavefront, we write the components of the electric vector $\mathbf{E}$ near the focus of an aplanatic system in the cylindrical coordinates (r,φ,z), following [373] $(m=3)$:

$$E_x=-I_{0,3}e^{i3\varphi}-\gamma_+I_{2,5}e^{i5\varphi}-\gamma_-I_{2,1}e^{i\varphi},$$

$$E_y=-i\sigma I_{0,3}e^{i3\varphi}+i\gamma_+I_{2,5}e^{i5\varphi}-i\gamma_-I_{2,1}e^{i\varphi},\qquad(6.17)$$

$$E_z=2i\gamma_+I_{1,4}e^{i4\varphi}-2i\gamma_-I_{1,2}e^{i2\varphi},$$

where

$$I_{0,3}=B\int_0^a\sin\theta\cos^{1/2}\theta A_3(\theta)e^{ikz\cos\theta}(1+\cos\theta)J_3(x)d\theta,$$

$$I_{2,5}=B\int_0^a\sin\theta\cos^{1/2}\theta A_3(\theta)e^{ikz\cos\theta}(1-\cos\theta)J_5(x)d\theta,$$

$$I_{2,1}=B\int_0^a\sin\theta\cos^{1/2}\theta A_3(\theta)e^{ikz\cos\theta}(1-\cos\theta)J_1(x)d\theta,\qquad(6.18)$$

$$I_{1,2}=B\int_0^a\sin^2\theta\cos^{1/2}\theta A_3(\theta)e^{ikz\cos\theta}J_2(x)d\theta,$$

$$I_{1,4}=B\int_0^a\sin^2\theta\cos^{1/2}\theta A_3(\theta)e^{ikz\cos\theta}J_4(x)d\theta.$$

where $B=kf/2$, $a=\arcsin(\mathrm{NA})$, $x=kr\sin\theta$, $\gamma_\pm=(1\pm\sigma)/2$, $J_\nu(x)$ is the Bessel function, k is the wave number of light, and f is the focal length of the aplanatic system with a numerical aperture NA.

From Equations (6.17) and (6.18), the focal plane intensity ($z=0$) reads as

$$I=|E_x|^2+|E_y|^2+|E_z|^2=2\left(I_{0,3}^2+\gamma_+I_{2,5}^2+\gamma_-I_{2,1}^2+2\gamma_+I_{1,4}^2+2\gamma_-I_{1,2}^2\right). \quad (6.19)$$

From Equation (6.19), the separate intensities for left and right circular polarization are the following:

$$I_-=2\left(I_{0,3}^2+I_{2,1}^2+2I_{1,2}^2\right),$$
$$I_+=2\left(I_{0,3}^2+I_{2,5}^2+2I_{1,4}^2\right). \quad (6.20)$$

We note that all the integrals in Equation (6.18) are real-valued functions at $z=0$. According to Equation (6.20), for a circularly polarized optical vortex with a topological charge $m=3$ the intensity distribution in the focus of an aplanatic system is axially symmetric, since it does not depend on the azimuthal angle φ. On the optical axis ($r=0$), the intensity (6.19) and (6.20) is always zero, since all the Bessel functions $J_n(x)$ in the integrals in Equation (6.19) have a positive order ($n>0$), and therefore they equal zero for zero arguments. The first light ring for I_+ is always larger than the that for I_-, since for $x\to0$ (6.20) we can write

$$I_+\approx I_{0,3}^2\approx J_3^2\approx x^6,$$
$$I_-\approx I_{2,1}^2\approx J_1^2\approx x^2 \quad (6.21)$$

Further, using the Maxwell's equation for a monochromatic field with an angular frequency ω: $i\omega\mathbf{H}=\mathrm{rot}\,\mathbf{E}$, as well as the formulae (6.17) and (6.18), we can find the transverse components of the magnetic field strength vector (the longitudinal component is not yet needed):

$$H_x=i\gamma_+\left(\frac{8}{kr}I_{1,4}-\bar{I}_{2,5}\right)e^{i5\varphi}+2\gamma_+\sin\varphi\bar{I}_{1,3}e^{i4\varphi}+i\sigma\bar{I}_{0,3}e^{i3\varphi}$$
$$+2\gamma_-\sin\varphi\bar{I}_{1,3}e^{i2\varphi}-i\gamma_-\left(\frac{4}{kr}I_{1,2}-\bar{I}_{2,1}\right)e^{i\varphi},$$
$$H_y=\gamma_+\left(\frac{8}{kr}I_{1,4}-\bar{I}_{2,5}\right)e^{i5\varphi}-2\gamma_+\cos\varphi\bar{I}_{1,3}e^{i4\varphi}-\bar{I}_{0,3}e^{i3\varphi}$$
$$-2\gamma_-\cos\varphi\bar{I}_{1,3}e^{i2\varphi}+\gamma_-\left(\frac{4}{kr}I_{1,2}-\bar{I}_{2,1}\right)e^{i\varphi},$$

(6.22)

where

$$\bar{I}_{0,3} = B\int_0^a \sin\theta\cos^{3/2}\theta A_3(\theta)e^{ikz\cos\theta}(1+\cos\theta)J_3(x)d\theta,$$

$$\bar{I}_{2,5} = B\int_0^a \sin\theta\cos^{3/2}\theta A_3(\theta)e^{ikz\cos\theta}(1-\cos\theta)J_5(x)d\theta,$$

(6.23)

$$\bar{I}_{2,1} = B\int_0^a \sin\theta\cos^{3/2}\theta A_3(\theta)e^{ikz\cos\theta}(1-\cos\theta)J_1(x)d\theta,$$

$$\bar{I}_{1,3} = B\int_0^a \sin^3\theta\cos^{1/2}\theta A_3(\theta)e^{ikz\cos\theta}J_3(x)d\theta.$$

Using Equations (6.17), (6.18), (6.22) and (6.23), we derive the axial component of the Poynting vector [6] $\mathbf{S} = c\text{Re}[\mathbf{E}\times\mathbf{H}^*]/(8\pi)$:

$$S_z = \frac{c}{8\pi}\text{Re}\left(E_xH_y^* - E_yH_x^*\right)$$

(6.24)

where c is the speed of light in vacuum and $\text{Re}(\dots)$ is the real part of a complex number in the focal plane ($z=0$):

$$S_z = 2\left[I_{0,3}\left(\bar{I}_{0,3}+\bar{I}_{1,3}\right)+\gamma_+I_{2,5}\left(\bar{I}_{2,5}+\bar{I}_{1,3}-\frac{8}{kr}I_{1,4}\right)\right.$$

$$\left. +\gamma_-I_{2,1}\left(\bar{I}_{2,1}+\bar{I}_{1,3}-\frac{4}{kr}I_{1,2}\right)\right],$$

(6.25)

where the constant $c/(8\pi)$ is omitted for brevity.

For right and left circular polarization, expressions for the axial energy fluxes can be obtained directly from Equation (6.25):

$$S_{Z+} = 2\left[I_{0,3}\left(\bar{I}_{0,3}+\bar{I}_{1,3}\right)+I_{2,5}\left(\bar{I}_{2,5}+\bar{I}_{1,3}-\frac{8}{kr}I_{1,4}\right)\right],$$

$$S_{Z-} = 2\left[I_{0,3}\left(\bar{I}_{0,3}+\bar{I}_{1,3}\right)+I_{2,1}\left(\bar{I}_{2,1}+\bar{I}_{1,3}-\frac{4}{kr}I_{1,2}\right)\right].$$

(6.26)

It is seen in Equation (6.26) that $S_{z\pm}$ is axially symmetric and there is always zero axial flux on the optical axis: $S_{z\pm}(r=0)=0$, since all the integrals $I_{\nu,\mu}$ and $\bar{I}_{\nu,\mu}$ contain the Bessel functions $J_n(x)$ of the positive orders ($n>0$), which are equal to zero for zero arguments: $J_{n>0}(0)=0$. It also follows from Equation (6.26) that despite the zero flux on the optical axis $S_{z\pm}(r=0)=0$, the flux near the axis can have backward direction, i.e. $S_{z\pm}(r\to 0)<0$. Below we show that indeed the longitudinal component of the Poynting vector near the optical axis is always negative only for left circular polarization. Near the optical axis, at small radii ($k_r \ll 1$), the Bessel functions in the integrals can be replaced by the first terms of their series expansion:

$$J_n(x) = \sum_{p=0}^{\infty} \frac{(-1)^p (x/2)^{2p+n}}{p!(p+n)!}. \tag{6.27}$$

Then, near the optical axis, Equation (6.26) for left circular polarization can be written approximately ($y = kr \ll 1$):

$$S_{z-} \approx 2\left[Ay^3 \left(By^3 + Cy^3 \right) + Dy\left(Ey + Fy^3 - \frac{4}{y}Qy^2 \right) \right]. \tag{6.28}$$

Neglecting the terms with y^3 and retaining the terms with y in Equation (6.28), instead of Equation (6.28) we get:

$$S_{z-} \approx 2Dy^2 (E - 4Q), \tag{6.29}$$

where

$$2D = B\int_0^a \sin^2 \theta \cos^{1/2} \theta A_3(\theta)(1 - \cos\theta)d\theta,$$

$$2E = B\int_0^a \sin^2 \theta \cos^{3/2} \theta A_3(\theta)(1 - \cos\theta)d\theta, \tag{6.30}$$

$$8Q = B\int_0^a \sin^4 \theta \cos^{1/2} \theta A_3(\theta)d\theta.$$

Substitution of Equation (6.30) into Equation (6.29) yields

$$S_{z-}(r \to 0) \approx -\frac{B^2(kr)^2}{2}\left(\int_0^a \sin^2 \theta \cos^{1/2} \theta A_3(\theta)(1 - \cos\theta)d\theta \right)^2. \tag{6.31}$$

As seen in Equation (6.31), for an arbitrary apodization function of the optical vortex $A_3(\theta)$ and for any numerical aperture $a = \arcsin(NA)$, the energy flux near the optical axis ($kr \ll 1$) is directed oppositely to the direction of the optical vortex propagation (i.e. $S_{z-}(r \to 0) < 0$). In particular, when $A_3(\theta) = 1$ and $a = \arcsin(NA) = \pi/2$, the integral in Equation (6.31) can be evaluated:

$$S_{z-}(r \to 0) \approx -\frac{B^2(kr)^2}{16}\sqrt{\pi}. \tag{6.32}$$

Equations (6.31) and (6.32) demonstrate the quadratic increase (in modulus) of the reverse flux with the increasing distance r.

Figure 6.3 shows the distribution of the longitudinal component of the Poynting vector, calculated by using Equation (6.26) with a constant pupil apodization function $(A_m(\theta)=1)$ for the following parameters: wavelength $\lambda=532$ nm, focal distance $f=1000\lambda$, polarization is left circular $(\sigma=-1)$, calculation area $-R\le x, y\le R$ ($R=2.5\ \lambda$ (a) and $R=5\ \lambda$ (b)).

Similarly to Equation (6.25), the energy flux can be found in the focal plane of an optical vortex with a topological charge $m=4$:

$$S_{4z} = 2\left[I_{0,4}\left(\overline{I}_{0,4}+\overline{I}_{1,4}\right)+\gamma_+ I_{2,6}\left(\overline{I}_{2,6}+\overline{I}_{1,4}-\frac{10}{kr}I_{1,5}\right)\right.$$

$$\left.+\gamma_- I_{2,2}\left(\overline{I}_{2,2}+\overline{I}_{1,4}-\frac{6}{kr}I_{1,3}\right)\right].$$

(6.33)

The integrals in Equation (6.33) can be written as:

$$\overline{I}_{0,4} = B\int_0^a \sin\theta\cos^{3/2}\theta A_4(\theta)e^{ikz\cos\theta}(1+\cos\theta)J_4(x)d\theta,$$

$$\overline{I}_{2,6} = B\int_0^a \sin\theta\cos^{3/2}\theta A_4(\theta)e^{ikz\cos\theta}(1-\cos\theta)J_6(x)d\theta,$$

(6.34)

$$\overline{I}_{2,2} = B\int_0^a \sin\theta\cos^{3/2}\theta A_4(\theta)e^{ikz\cos\theta}(1-\cos\theta)J_2(x)d\theta,$$

$$\overline{I}_{1,4} = B\int_0^a \sin^3\theta\cos^{1/2}\theta A_4(\theta)e^{ikz\cos\theta}J_4(x)d\theta.$$

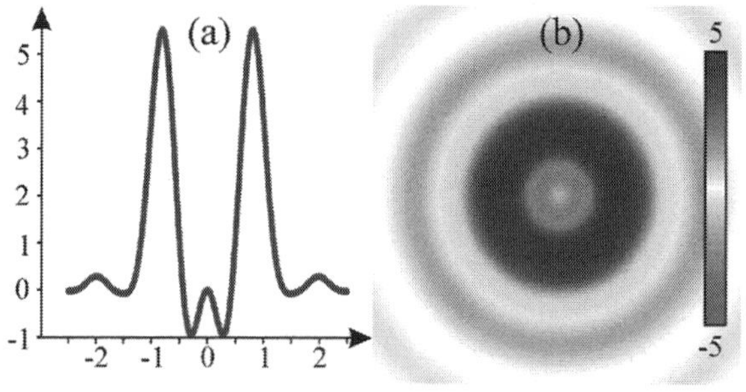

FIGURE 6.3 Transverse plot (a) and 2D distribution (b) of the longitudinal component of the Poynting vector, calculated by using Equation (6.26) with a constant pupil apodization function $(A_m(\theta)=1)$ for the following parameters: wavelength $\lambda=532$ nm, focal distance $f=1000\lambda$, polarization is left circular $(\sigma=-1)$, calculation area $-R\le x, y\le R$ ($R=2.5\lambda$ (a) and $R=5\lambda$ (b)).

Similarly to Equation (6.31), it can be shown that near the optical axis there is the reverse energy flux, which is zero on the axis itself and increases near the axis as the fourth power of the radial coordinate:

$$S_{4z-}(r \to 0) \approx -\frac{B^2 (kr)^4}{32} \left(\int_0^a \sin^3 \theta \cos^{1/2} \theta A_4(\theta)(1 - \cos \theta) d\theta \right)^2. \qquad (6.35)$$

A comparison of Equations (6.31) and (6.35) shows that the reverse flux for $m = 4$ is an order of magnitude weaker than that for $m = 3$, and therefore it can be neglected.

Note that if in Equations (6.31) and (6.35) the apodization function of an aplanatic system $\cos^{1/2}\theta$ were to be replaced by the apodization function of an ideal diffraction lens $\cos^{-3/2}\theta$ [87], then only the values of the integrals in Equations (6.31) and (6.35) would be changed, whereas the negative value of the whole expression would remain. Therefore, within the Richards–Wolf theory, it can be concluded that the phenomenon of the reverse energy flux in the focus does not depend on the apodization function of the optical system. We also note that this theory is valid when $f \gg \lambda$ with λ being the wavelength of light [6]. In the last subsection, by using rigorous vectorial simulation, we show that the reverse energy flux phenomenon also takes place even near the focusing microlens (zone plate) with the focal length $f = \lambda$.

6.2.2 SPIRAL ROTATION OF THE REVERSE FLUX NEAR THE OPTICAL AXIS

In order to find out exactly how the reverse flux (6.31) propagates near the optical axis, we derive expressions for the transverse components S_x and S_y of the Poynting vector. In its turn, it requires an expression for the longitudinal component of the magnetic field strength vector, which we have not yet given. Now we can write it:

$$H_z = \frac{-i}{k} \left(\frac{\partial E_y}{\partial x} - \frac{\partial E_x}{\partial y} \right) = \gamma_+ \left(\frac{6}{kr} I_{0,3} + \overline{I}_{2,4} \right) e^{i4\varphi}$$

$$+ \gamma_- \left(\frac{2}{kr} I_{2,1} - \overline{I}_{2,0} \right) e^{i2\varphi} \qquad (6.36)$$

$$- \overline{I}_{0,2} e^{i3\varphi} (\sigma \cos \varphi + i \sin \varphi),$$

where

$$\overline{I}_{0,2} = B \int_0^a \sin^2 \theta \cos^{1/2} \theta A_3(\theta) e^{ikz \cos \theta} (1 + \cos \theta) J_2(x) d\theta,$$

$$\overline{I}_{2,4} = B \int_0^a \sin^2 \theta \cos^{1/2} \theta A_3(\theta) e^{ikz \cos \theta} (1 - \cos \theta) J_4(x) d\theta, \qquad (6.37)$$

$$\overline{I}_{2,0} = B \int_0^a \sin^2 \theta \cos^{1/2} \theta A_3(\theta) e^{ikz \cos \theta} (1 - \cos \theta) J_0(x) d\theta.$$

From Equation (6.36), the longitudinal component of the magnetic vector of a field with left circular polarization reads as

$$H_{z-} = \left(\frac{2}{kr} I_{2,1} + \bar{I}_{0,2} - \bar{I}_{2,0} \right) e^{i2\varphi}. \tag{6.38}$$

Further, using Equations (6.38), (6.17), and (6.25), we find the transverse components of the Poynting vector only for a light field with left circular polarization (omitting the constant $c/(8\pi)$):

$$S_{x-} = -\sin \varphi \, Q(r), \quad S_{y-} = \cos \varphi \, Q(r), \tag{6.39}$$

where

$$Q(r) = \left(I_{0,3} + I_{2,1} \right) \left(\frac{2}{kr} I_{2,1} + \bar{I}_{0,2} - \bar{I}_{2,0} \right) + 2 I_{1,2} \left(\frac{4}{kr} I_{1,2} + \bar{I}_{0,3} - \bar{I}_{2,1} \right). \tag{6.40}$$

As seen in Equation (6.39), the energy flux (both the forward and the reverse flux near the optical axis) rotates counterclockwise or clockwise, depending on the sign of the function $Q(r)$. For example, near the optical axis $kr \ll 1$ and the negative terms in Equation (6.39) ($\bar{I}_{2,0}$ and $\bar{I}_{2,1}$) are compensated by the first terms in their brackets ($2I_{2,1}/(kr)$ and $4I_{1,2}/(kr)$ respectively), and the positive sign of the second factors (and the whole function $Q(r)$) is determined by the middle terms ($\bar{I}_{0,2}$ and $\bar{I}_{0,3}$). Thus, it follows from Equation (6.40) that near the optical axis ($kr \ll 1$) the reverse energy flux (6.39) propagates counterclockwise spiraling about the optical axis z.

6.2.3 FDTD Simulation

Here, we consider diffraction of a plane wave with left circular polarization ($\sigma = -1$), bounded by a circular 10-μm diameter aperture, by a spiral zone plate (SZP) with a topological charge $m = 3$ and with a focal length f equal to the wavelength $f = \lambda = 532$ nm (NA $\approx$ 1).

Transmission of the phase SZP is described by the function

$$T(r, \varphi) = \text{sgn} \left[\cos \left(m\varphi + k\sqrt{r^2 + f^2} \right) \right], \tag{6.41}$$

where $\text{sgn}(x) = \left(1, x \geq 0; -1, x < 0 \right)$ is the sign function.

The height of the relief is $h = 0.532$ μm and the refractive index of the SZP material is $n = 1.5$. The simulation was done by using the FDTD method implemented in the FullWAVE program. Figure 6.4 shows the binary phase mask of the SZP ($m = 3$), which can be used to fabricate a phase SZP. The light and dark bands in Figure 6.4 can also be treated as the phase relief of the SZP: the light bands show the ridges with the height of $h = 0.532$ μm, while the dark ones show the grooves.

Figure 6.5 shows the intensity distribution (a) and the axial energy flux (b) in the focal plane of the SZP. It is seen in Figure 6.5 that the intensity I and the flux S_{z-} do not have the circular symmetry, since the SZP itself does not have the circular

FIGURE 6.4 Binary phase mask of the SZP with $m = 3$ and with the focal length equal to the wavelength $f = \lambda = 532$ nm. Black, phase is zero, white, phase is π.

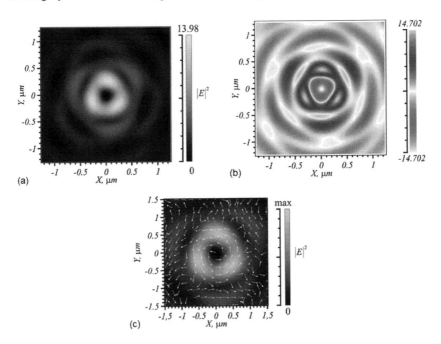

FIGURE 6.5 Distribution of (a) the intensity $I = |E_x|^2 + |E_y|^2 + |E_z|^2$, and (b) the axial component of the Poynting vector S_{z-} in the focal plane of the SZP at a distance of $z = 532$ nm; and (c) transverse orientations of the Poynting vector combined with the intensity pattern (negative, frame size is 1.5×1.5 μm).

symmetry (Figure 6.4). On the optical axis (at the center of the diffraction pattern), both the intensity and the energy flux are zero. However, near the optical axis in an area with the size nearly equal to the Airy disk (a circle in the center of Figure 6.5(b) of about 400nm in diameter), the energy flux is directed backward and the axial component of the Poynting vector is negative ($S_{z-} < 0$).

The arrows in Figure 6.5(c) show the direction of rotation of the energy flux in the focal plane. It is seen that the energy flux rotates counterclockwise in the areas where it is reversed (Figure 6.5(b)). Figure 6.6 shows the cross-sections of the 2D distributions of the intensity and of the longitudinal energy flux (Figure 6.5), drawn along the x-axis through the center. As seen in Figure 6.6, both the intensity and the energy flux are zero on the optical axis ($x = 0$), whereas near the optical axis ($-0.2\ \mu m < x < 0.2\ \mu m$) there is a reverse energy flux with its maximal value (in modulus) being a little less than 2 times smaller than the maximal value of the forward flux.

Based on the Richards–Wolf formulae describing sharp focusing by an ideal spherical lens (aplanatic system), we have obtained all three components of the Poynting vector in the focal plane of an optical vortex with an arbitrary real-valued pupil apodization function, topological charge of 3, and left circular polarization. Tending the radial coordinate of the cylindrical coordinate system to zero, we have obtained approximate expressions for these components of the Poynting vector, according to which the light energy in the focal plane near the optical axis is rotating about it along a spiral and propagates oppositely to the propagation direction of the incident light. The FDTD simulation of a plane wave with left circular polarization, diffracted by a binary spiral (vortex) zone microplate with a numerical aperture of about 1, has shown that the longitudinal

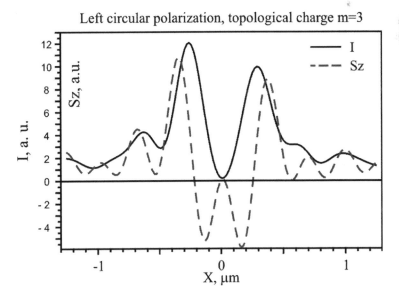

FIGURE 6.6 Intensity (solid curve) and longitudinal energy flux S_{z-} (dashed curve) along the x-axis ($y = 0$).

component of the Poynting vector near the optical axis has a negative value (although on the optical axis itself it is zero), i.e. there is a reverse flux of light energy. This effect can be used to demonstrate the "optical tractor" [374, 375], when the microparticle trapped near the focus moves in the reverse direction and rotates. Indeed, the force **F** acting on a particle of the size smaller than the wavelength is proportional to the absolute value and is directed in the same direction as the Poynting vector [363]:

$$\mathbf{F} = \frac{k}{2}\operatorname{Im}(\alpha)\operatorname{Re}\big(\mathbf{E}\times\mathbf{H}^*\big) \tag{6.42}$$

where α is the polarizability of the particle and **E** and **H** are respectively the electric and magnetic strength vectors. Referring to Equation (6.42), the longitudinal force component is proportional to the longitudinal component of the Poynting vector S_z. If some beam in its transverse distribution of S_z has negative values, then a particle illuminated by such beam should move toward the light source.

6.3 ENERGY BACKFLOW IN THE FOCUS OF AN OPTICAL VORTEX

In their classical work of 1959 [6], Richards and Wolf showed that when focusing a linearly polarized plane wave in an aplanatic system, at the bottom of the major intensity peak there is a region of energy backflow relative to that of the incident wave. In the region of interest, the on-axis projection of the Poynting vector has a small negative value approximately equal to 1% of the maximum value of the energy direct flow. An analogous result was earlier reported by V.S. Ignatovsky as far back as 1920 [5]. At the time, no significance had been attached to the said effect on account of it being too small. Meanwhile, work [6] has given an impetus to a broad range of theoretical and experimental works concerned with focusing laser beams into a tight spot [373, 377–383]. In Reference [377], attaining a sharper focus for linearly polarized optical vortices with the topological charge $m = 1$, 2, and 3 was studied. For high NA, an elliptical focal spot was shown to be extended in the direction perpendicular to that of linear incident polarization. In Reference [378], a left-hand circularly polarized optical vortex with the topological charge 1 was numerically and experimentally shown to generate a circular focal spot. Focusing circularly polarized light to a tighter spot was also reported in Reference [379]. Focusing elliptically polarized Laguerre–Gauss (LG) modes with arbitrary order numbers (n, m) was discussed in Reference [373], with ellipse-shaped intensity rings observed in the focus. In Reference [380] a sharper focus for anomalous LG modes with order numbers $(n + m/2, m)$ was obtained, with the numerically simulated focusing of radially polarized Laguerre–Gauss–Bessel beams reported in Reference [381]. In a similar way, the focusing properties of optical vortices with different topological charges and inhomogeneous (radial and azimuthal) polarization was investigated in Reference [382]. A transverse focal shift observed when tightly focusing off-axis optical vortices embedded into a Gaussian beam was numerically simulated in Reference [383].

However, none of the aforesaid papers was concerned with the study of the behavior of the Poynting vector in the focus. The first work to show theoretically that the Poynting vector has the negative axial component reported on a linear combination of two Bessel beams with TE- and TM- polarizations [362]. However, ways of generating a vortex Bessel beam with the arbitrary topological charge m, that has all three electric field components E_r, E_φ, and E_z expressed in the cylindrical coordinates (r, φ, z) were not discussed in Reference [362]. We have managed to find the only work, Reference [364], dealing with a practically feasible situation (aplanatic-system-aided focusing) where it was theoretically and numerically shown that when focusing a left-hand circularly polarized Laguerre–Gauss mode $(0, m) = (0, 2)$, the on-axis projection of the Poynting vector has negative values. Our work extends the findings of Reference [364] onto an arbitrary optical vortex with the topological charge $m = 2$. In a number of works [363, 374, 375, 384–389], a concept of an "optical tractor" was also discussed, which studied the interaction between a light field and microparticles, forcing the microparticle into motion against the illuminating beam. This force is termed as a negative optical force [363]. Putting it simply, such a situation occurs when a particle scatters more light in the forward rather than backward direction and there are no gradient forces. Hence, two different situations should be distinguished: when the local energy flow is opposite to the general energy flow in the beam versus the situation when the force of light causes the microparticle to move against the illuminating light toward the light source.

This section only deals with the first situation. By employing the Richards–Wolf theory for an aplanatic system, we show that when focusing a left-hand circularly polarized optical vortex with the topological charge $m = 2$, the on-axis projection of the Poynting vector in and near the focal plane is negative. This implies that in the region of interest the local energy flow is directed against the incident energy flow in the converging optical vortex. Previously, the presence of the energy backflow in the focus was discovered only for Laguerre–Gauss beams [364]. Below, this finding is extended onto an arbitrary optical vortex.

6.3.1 THEORETICAL BACKGROUND

By employing the Richards–Wolf theory formalism [6] we analyze below focusing a left-hand circularly polarized optical vortex with the topological charge $m = 2$ in an aplanatic system. For a circularly polarized electric field $\mathbf{E} = E_x \mathbf{e}_x + i\sigma E_y \mathbf{e}_y$, where $\mathbf{e}_x$, $\mathbf{e}_y$ are the unit Cartesian vectors, we shall assume that at $\sigma = 1$ the beam is right-hand polarized and at $\sigma = -1$ it is left-hand polarized, following [373]. In the meantime, the opposite notations were employed in Reference [364]: with left-hand polarization assumed at $\sigma = 1$ and right-hand polarization at $\sigma = -1$. Assuming an optical vortex with the topological charge m and an arbitrary pupil apodization function (a real function $A_m(\theta)$)

$$A_m(\theta, \varphi) = A_m(\theta)\exp(im\varphi),\qquad(6.43)$$

where the angles (θ, φ) define a point on the sphere in an aplanatic system, the near-focus Cartesian projections of the electric vector $\mathbf{E}$ in the cylindrical coordinates (r, φ, z) are given by (see [373])

$$E_x(r,\varphi,z) = -i^{m+1}\begin{pmatrix} I_{0,m}e^{im\varphi} + \gamma_+ I_{2,m+2}e^{i(m+2)\varphi} \\ + \gamma_- I_{2,m-2}e^{i(m-2)\varphi} \end{pmatrix},$$

$$E_y(r,\varphi,z) = i^{m}\begin{pmatrix} \sigma I_{0,m}e^{im\varphi} - \gamma_+ I_{2,m+2}e^{i(m+2)\varphi} \\ + \gamma_- I_{2,m-2}e^{i(m-2)\varphi} \end{pmatrix}, \qquad (6.44)$$

$$E_z(r,\varphi,z) = -2i^{m}\begin{pmatrix} \gamma_+ I_{1,m+1}e^{i(m+1)\varphi} \\ - \gamma_- I_{1,m-1}e^{i(m-1)\varphi} \end{pmatrix},$$

where

$$I_{0,m} = B\int_0^a \sin\theta \cos^{1/2}\theta A_m(\theta)e^{ikz\cos\theta}(1+\cos\theta)J_m(x)d\theta,$$

$$I_{1,m\pm1} = B\int_0^a \sin^2\theta \cos^{1/2}\theta A_m(\theta)e^{ikz\cos\theta}J_{m\pm1}(x)d\theta, \qquad (6.45)$$

$$I_{2,m\pm2} = B\int_0^a \sin\theta \cos^{1/2}\theta A_m(\theta)e^{ikz\cos\theta}(1-\cos\theta)J_{m\pm2}(x)d\theta,$$

where $B=kf/2$, $\alpha=\arcsin(NA)$, $x=kr\sin\theta$, $\gamma_\pm = (1\pm\sigma)/2$, $J_\nu(x)$ is a Bessel function, k is the wave number of light, f is the focal length of an aplanatic system with numerical aperture NA. For an optical vortex with the topological charge $m=2$, Equation (6.44) is rearranged to

$$E_x(r,\varphi,z) = i\left(I_{0,2}e^{i2\varphi} + \gamma_+ I_{2,4}e^{i4\varphi} + \gamma_- I_{2,0}\right),$$

$$E_y(r,\varphi,z) = -\left(\sigma I_{0,2}e^{i2\varphi} - \gamma_+ I_{2,4}e^{i4\varphi} + \gamma_- I_{2,0}\right), \qquad (6.46)$$

$$E_z(r,\varphi,z) = 2\left(\gamma_+ I_{1,3}e^{i3\varphi} - \gamma_- I_{1,1}e^{i\varphi}\right).$$

At $m=0$ (no vortex) and $\sigma=0$ (linear polarization), Equation (6.46) gives relationships for the E-field components in the focal plane (see Reference [6]). From Equation (6.46), the intensity of the E-field (power density) in the focal plane ($z=0$) is:

$$I = 2\left(I_{0,2}^2 + \gamma_+ I_{2,4}^2 + \gamma_- I_{2,0}^2 + 2\gamma_+ I_{1,3}^2 + 2\gamma_- I_{1,1}^2\right) \qquad (6.47)$$

Note that at $z=0$, all constituent integrals in Equation (6.45) are real functions. From Equation (6.47), the intensity distribution in the focus is seen to be independent of the polar angle φ, forming centrally symmetric rings. On the optical axis, at the center of the focal spot ($r=0$), the right-hand polarized vortex beam will have an intensity null $I_+ = 2\left(I_{0,2}^2 + I_{2,4}^2 + 2I_{1,3}^2\right) = 0$, while the left-hand beam will have a non-zero on-axis intensity $I_- = 2\left(I_{0,2}^2 + I_{2,0}^2 + 2I_{1,1}^2\right) \neq 0$. This follows from the fact that only the relation for the I_- intensity has a term $I_{2,0}$ that contains a non-zero on-axis Bessel function component $J_0(x=0)=1$. The remaining components of Equation (6.47) take zero values on the optical axis.

Then, using Maxwell's equations for a monochromatic field $i\omega\mathbf{H}=\mathrm{rot}\mathbf{E}$ (ω is the cyclic frequency) and Equations (6.44) and (6.45), and discarding the unneeded on-axis components, the transverse components of the H-field vector are expressed as

$$H_x = \gamma_+ e^{i4\varphi}\left(\frac{6}{kr}I_{1,3} - \overline{I}_{2,4}\right) + e^{i2\varphi}\left[\sigma\overline{I}_{0,2} - 2i\sin\varphi\left(\cos\varphi + i\sigma\sin\varphi\right)\overline{I}_{1,2}\right]$$

$$+ \gamma_-\left(\overline{I}_{2,0} - \frac{2}{kr}I_{1,1}\right),$$

$$H_y = i\gamma_+ e^{i4\varphi}\left(-\frac{6}{kr}I_{1,3} + \overline{I}_{2,4}\right) + ie^{i2\varphi}\left[\overline{I}_{0,2} + 2\cos\varphi\left(\cos\varphi + i\sigma\sin\varphi\right)\overline{I}_{1,2}\right]$$

$$+ i\gamma_-\left(\overline{I}_{2,0} - \frac{2}{kr}I_{1,1}\right),$$

$$(6.48)$$

where

$$\overline{I}_{0,2} = B\int_0^a \sin\theta\cos^{3/2}\theta A_m(\theta)e^{ikz\cos\theta}(1+\cos\theta)J_2(x)d\theta,$$

$$\overline{I}_{1,2} = B\int_0^a \sin^3\theta\cos^{1/2}\theta A_m(\theta)e^{ikz\cos\theta}J_2(x)d\theta,$$

$$(6.49)$$

$$\overline{I}_{2,4} = B\int_0^a \sin\theta\cos^{3/2}\theta A_m(\theta)e^{ikz\cos\theta}(1-\cos\theta)J_4(x)d\theta,$$

$$\overline{I}_{2,0} = B\int_0^a \sin\theta\cos^{3/2}\theta A_m(\theta)e^{ikz\cos\theta}(1-\cos\theta)J_0(x)d\theta.$$

From Equations (6.44), (6.45), (6.48), and (6.49), the longitudinal projection of the Poynting vector [6] $\mathbf{S}=c[\mathbf{E}\times\mathbf{H}^*]/8\pi$ is given by

$$S_z = \frac{c}{8\pi}\left(E_z H_y^* - E_y H_x^*\right)$$

$$(6.50)$$

up to a constant $c/8\pi$, where c is the speed of light in vacuum, and, in the focal plane, $(z=0)$ by

$$S_z = 2I_{0,2}\left(\bar{I}_{0,2} + \bar{I}_{1,2}\right) + 2\gamma_+ I_{2,4}\left(\bar{I}_{1,2} + \bar{I}_{2,4} - \frac{6}{kr}I_{1,3}\right)$$

$$+ 2\gamma_- I_{2,0}\left(\bar{I}_{1,2} + \bar{I}_{2,0} - \frac{2}{kr}I_{1,1}\right). \tag{6.51}$$

From Equation (6.51), for left-hand polarization $(\sigma = -1)$, we obtain:

$$S_z^- = 2I_{0,2}\left(\bar{I}_{0,2} + \bar{I}_{1,2}\right) + 2I_{2,0}\left(\bar{I}_{1,2} + \bar{I}_{2,0} - \frac{2}{kr}I_{1,1}\right). \tag{6.52}$$

From Equation (6.52), it follows that the energy flow $S_z^-(r, z = 0)$ in the focal plane is centrally symmetric, being independent of the polar angle φ. From Equation (6.52), the energy flow in the focal plane can be found only on the optical axis $(r=0)$. Note that in Equation (6.52) only the second term is non-zero: $4I_{2,0}\left(\bar{I}_{2,0} - 2I_{1,1}/kr\right) \neq 0$. Then, we obtain:

$$S_z^-(r = 0, z = 0) = -2B^2\left(\int_0^a \sin\theta\cos^{1/2}\theta A_2(\theta)(1 - \cos\theta)d\theta\right)^2 \leq 0. \tag{6.53}$$

Being the major result of this work, Equation (6.53) shows that an ideal spherical lens with a real apodization function $A_2(\theta)$ and arbitrary NA $(\alpha = \arcsin(NA))$ focuses a left-hand circularly polarized optical vortex with $m = 2$ into an intensity ring with an on-axis energy backflow found at its center and propagating in the opposite direction to the illuminating beam. For the simple case of a uniform illumination $(A_2(\theta) = 1)$ and maximum aperture at $a = \pi/2$ (NA = 1), the integral in Equation (6.53) can be calculated. Then, Equation (6.52) is rearranged to

$$S_z^-(r = 0, z = 0) = -2B^2\left(\frac{4}{15}\right)^2 = -1.4\left(\frac{f}{\lambda}\right)^2. \tag{6.54}$$

It can be shown that at any other positive topological charge $m \neq 2$ and any other homogeneous polarization $\sigma = 0, \pm 1$, no on-axis energy backflow will occur in the focal plane. Note that although in this case the backflow may occur elsewhere in the focal plane, it is much smaller in magnitude than the direct flow. For instance, in Reference [6] an energy backflow amounting to 1% of the maximum direct flow was shown to occur in the focal plane near the first dark ring when focusing a linearly polarized plane wave. It can be shown that a result similar to Equation (6.53) is obtained at $m = -2$ and $\sigma = 1$.

6.3.2 Numerical Simulation of Tight Focusing Using Richards–Wolf Formulae

In this subsection, using particular numerical examples, an energy backflow is shown to occur in the focus of an aplanatic system both on and near the optical axis. The energy backflow is also shown to be comparable in magnitude with the incident energy flow.

Shown in Figure 6.7 are (a) the intensity pattern and (b) the on-axis projection of the Poynting vector in the focal plane of a diffractive lens (with NA = 0.95) when focusing an optical vortex with $m = -2$ and $\sigma = 1$. The apodization function of a diffractive lens is $A_m(\theta) = \cos^{-3/2}(\theta)$. The simulation was conducted using Richards–Wolf formulae [6].

As can be seen from Figure 6.7, the energy backflow in the focal plane occurs not only on the optical axis (pattern center) but also within a circle whose radius can be estimated from Figure 6.8.

Figure 6.8 depicts two-dimensional profiles along the x-axis of the (a) intensity and (b) on-axis projection of the Poynting vector shown in Figure 6.7. From Figure 6.8, the energy backflow is seen to occur in a central "dark" intensity circle, with its radius being approximately equal to the distance from the center to the first radius on which the intensity drops twofold. The area of the circle of energy backflow is approximately equal to that of the Airy disk. From Figure 6.8, the maximum of the backflow is seen to be only twice as low as the maximum of the incident energy flow (in relative units). Figure 6.9 depicts a half-tone intensity pattern and the directions of the Poynting

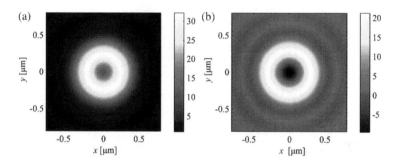

FIGURE 6.7   (a) Intensity pattern in the focal plane and (b) z-axis projection of the Poynting vector in the focal plane.

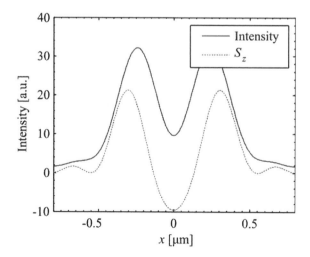

FIGURE 6.8   Profiles of the intensity (solid curve) and on-axis projection of the Poynting vector (dotted curve) along the x-axis in the focal plane (from Figure 6.28).

vector (arrows) in the focal yz-plane when focusing an optical vortex with $m=-2$ and right-hand circular polarization $\sigma=1$ in an aplanatic system with NA$=0.95$.

From Figure 6.9, the Poynting vector is seen to be directed oppositely to the incident beam on the z-axis. Figure 6.10 shows variations of the S_z projection with varying (a) topological charge m and (b) NA of an aplanatic lens.

From Figure 6.10(a), the energy backflow is seen to emerge on the optical axis only at $m=-2$ ($\sigma=1$). Figure 6.10(b) shows that the on-axis backflow becomes comparable in magnitude with the incident energy at NA>0.9 (becoming clearly noticeable).

6.3.3   FDTD-AIDED SIMULATION OF TIGHT FOCUSING

Below, results of the numerical simulation of the propagation of a left-hand circularly polarized plane wave ($\sigma=-1$) through a binary spiral zone plate SZP with $m=2$

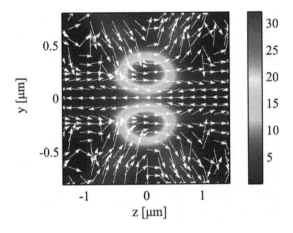

FIGURE 6.9 Intensity pattern and directions of the Poynting vector in the zy-plane when focusing an optical vortex with $m=-2$ and right-hand circular polarization $\sigma=1$ in an aplanatic system with NA$=0.95$.

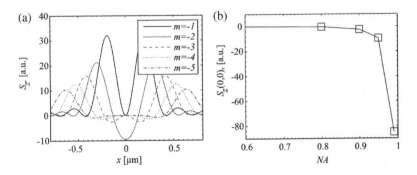

FIGURE 6.10 (a) Profile of the longitudinal Poynting vector component S_z in the focus of an aplanatic system when focusing optical vortices with different topological charges $m=-1$, $-2,...-5$ and right-hand circular polarization $\sigma=1$ (NA$=0.95$); and (b) the on-axis magnitude of S_z versus NA at $m=-2$ and $\sigma=1$.

are discussed. The incident light field ($\lambda = 532$ nm) is limited by a circular 10-μm diaphragm. The SZP with a focal length equal to the incident wavelength $f = \lambda = 532$ nm is assumed to have a refractive index of $n = 1.5$ and a relief height of 0.532 μm. The diffraction of light by the SZP was numerically simulated by solving Maxwell's equations using an FDTD technique implemented in the FullWAVE software.

From Figure 6.11(c), an energy backflow is seen to occur on and near the surface of the SZP ($S_z < 0$).

Plotted in Figure 6.12 are profiles of the intensity I and the Poynting vector projection S_z along the x-axis at a distance of $z = 532$ nm (a) and along the optical axis z (b). From Figure 6.12(a), the maximum backflow is seen to be nearly twice as high as the incident energy flow (in relative units). From Figure 6.12(b), the maximum magnitude of the on-axis energy backflow is seen to occur in the focal plane ($z = 532$ nm), quickly decaying with distance from the plane. It is worth noting that the on-axis distance at which the energy backflow is halved nearly equals the depth of focus (DOF), at which the intensity is halved, being equal to DOF $\approx 2\lambda$.

Summing up, the research findings of this work are as follows. To the best of our knowledge, we are the first to have proven, employing the Richards–Wolf formalism, that when focusing a left-hand circularly polarized optical vortex with the topological charge $m = 2$ and an arbitrary real pupil apodization function $A_m(\theta)$ into a tight focal

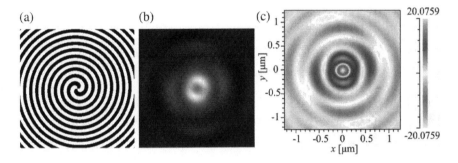

FIGURE 6.11 (a) 10×10-μm binary amplitude mask of the SZP with $m = 2$; (b) intensity pattern (a) 2×2-μm fragment) at a distance of $z = 532$ nm (numerically simulated focal spot); and (c) the gray scale to show the distribution of S_z at the same distance.

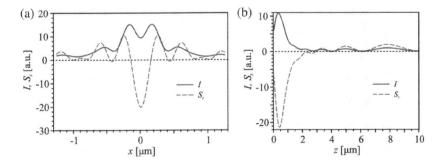

FIGURE 6.12 Profiles of the intensity I and the Poynting vector projection S_z ($y = 0$), (a) along the x-axis at a distance of $z = 532$ nm, and (b) and along the optical axis z.

spot, a backflow of light energy occurs on the optical axis, as confirmed by the negative on-axis projection of the Poynting vector. A similar effect occurs when focusing a right-hand circularly polarized optical vortex with $m = -2$. The numerical simulation based on the Richards–Wolf formulae has shown that the maximum of the on-axis energy backflow can be comparable in magnitude with the maximum of the incident energy flow. It has also been shown that the Poynting vector has negative values in the focal plane in a circle centered on the optical axis with its radius being equal to the distance from the intensity ring center to a point where the intensity is nearly halved. The diameter of the backflow circle in the focal plane may be said to be nearly equal to the Airy disk diameter. The FDTD-aided numerical simulation of the diffraction of a left-hand circularly polarized plane wave by a spiral zone microplate with $NA \approx 1$ has also shown that an on-axis energy backflow occurs at a wavelength distance (i.e. in the focal plane). The maximum value of the backflow is nearly twice as large as the incident energy flow. An on-axis length on which the energy backflow is halved is equal to the depth of focus, amounting to two wavelengths in the situation under analysis.

6.4 THE NON-VORTEX INVERSE PROPAGATION OF ENERGY IN A TIGHTLY FOCUSED HIGH-ORDER CYLINDRICAL VECTOR BEAM

Cylindrical vector beams (CVB) have been known in optics for a long time [89, 304, 390–392], and interest in the investigation of these beams is increasing [382, 393, 394]. Polarization in the cross-section of these beams changes the direction, making a complete rotation when the azimuthal angle φ changes from 0 to 2π. There have been early attempts to generalize cylindrical vector beams by considering the beams in which the direction of polarization makes several rotations. Such beams in the literature are known as high-order cylindrical vector beams. For example, the focusing of high-order radially polarized modes was investigated in Reference [395]. These modes had a form of $gl(\varphi) = \cos l\varphi i + \sin l\varphi j$ where l is the beam order, **i** and **j** are the unit vectors in the Cartesian coordinate system chosen such that the unit vector **k** coincides with the axis of symmetry of the beam. The propagation of higher-order modes with a shifted center was investigated numerically using the Fresnel transform in Reference [396]. The diffraction of higher-order modes by a gap was investigated numerically in Reference [397] also using the Fresnel transform.

Experimentally, such beams were obtained using spatial light modulators [398, 399] and elements designed as Pancharatnam-Berry phase elements [355].

In the papers cited above, only the behavior of the electric component of the light field was studied, which makes it impossible to calculate the value of the Poynting vector (PV) in the focal point. According to Reference [363], the force acting on the particle is proportional in absolute value and coincides in direction with the Poynting vector. If a beam contains a negative component of PV in its transverse distribution, then the particle illuminated by the beam should move toward the light source. The existence of areas with negative values of the longitudinal projection of the Poynting vector S_z in focal spots has been known for a long time [6]. However, only recently has it been possible to detect focal spots with negative values of PV comparable in absolute value with positive values. The propagation of light through the metalens, which simultaneously rotates the direction of polarization; the focus light was numerically simulated in Reference [369]. Using the FDTD method, it was shown that the spiral metalens illuminated by circularly polarized light forms a focal spot with negative values of the energy flow along the propagation axis S_z. The generalization of Reference [369] to the case of focusing of optical vortices with circular polarization and topological charges equal to ±1, ±2, and ±3 was made in References [376, 400]. Negative values of the longitudinal component S_z were also obtained in Reference [364], where Gauss–Laguerre beams with circular polarization were investigated, and in Reference [401], where Weber's beams were studied. While the transverse energy flow in a tightly focused azimuthally polarized beam has been discussed [402], the on-axis variations of the Poynting vector were not studied by the authors. As a consequence, the negative propagation of the energy flow was not demonstrated.

In this section, the tight focusing of high-order cylindrical vector beams was investigated. It was shown that in the focus, there are areas with the direction of the Poynting vector opposite to the direction of propagation of the beam and the negative values are comparable in absolute value with positive values. If the order of the beam is equal to two, then the region with negative values is located in the center of the focal spot. In contrast to References [369, 376, 400], where the inverse energy flow propagated along a spiral, in this section we investigate a non-vortex inverse flow with a laminar propagation of light.

6.4.1 THEORETICAL BACKGROUND

Our analysis relies upon the Richard–Wolf equations, which have been extensively utilized when studying tightly focused laser beams with due regard for the vector properties of light fields. The Richards–Wolf integral has the form [6]:

$$\mathbf{U}(\rho,\psi,z) = -\frac{if}{\lambda}\int_0^a\int_0^{2\pi} B(\theta,\varphi)T(\theta)\mathbf{P}(\theta,\varphi)$$

(6.55)

$$\times \exp\{ik[\rho\sin\theta\cos(\varphi-\psi)+z\cos\theta]\}\sin\theta\,d\theta\,d\varphi$$

where $U(\rho, \psi, z)$ is the electrical or magnetic field of focused light, $B(\theta, \varphi)$ is the incident electrical or magnetic field (θ is the polar angle and φ is the azimuthal angle), $T(\theta)$ is the apodization function, f is the focal length, $k=2\pi/\lambda$ is the wave number, λ is the wavelength, α is the maximal polar angle determined by the numerical aperture of the lens (NA=sinα), and $P(\theta, \varphi)$ is the polarization matrix for the electric and magnetic fields. The matrix $P(\theta,\varphi)$ is equal to:

$$\mathbf{P}(\theta,\varphi) = \begin{bmatrix} 1+\cos^2\varphi(\cos\theta-1) \\ \sin\varphi\cos\varphi(\cos\theta-1) \\ -\sin\theta\cos\varphi \end{bmatrix} a(\theta,\varphi) + \begin{bmatrix} \sin\varphi\cos\varphi(\cos\theta-1) \\ 1+\sin^2\varphi(\cos\theta-1) \\ -\sin\theta\sin\varphi \end{bmatrix} b(\theta,\varphi)$$

(6.56)

where $a(\theta, \varphi)$ and $b(\theta, \varphi)$ are polarization functions for the x- and y-components of the incident beam. For high-order cylindrical vector beams, the polarization functions have the form:

$$E(\theta,\phi) = \begin{pmatrix} a(\theta,\phi) \\ b(\theta,\phi) \end{pmatrix} = \begin{pmatrix} -\sin(m\phi) \\ \cos(m\phi) \end{pmatrix} \quad (6.57)$$

for the electric field, and

$$H(\theta,\phi) = \begin{pmatrix} a(\theta,\phi) \\ b(\theta,\phi) \end{pmatrix} = \begin{pmatrix} -\cos(m\phi) \\ -\sin(m\phi) \end{pmatrix} \quad (6.58)$$

for the magnetic field, where m is a positive integer number. If $m = 1$ Equations (6.57) and (6.58) describe the well-known azimuthally polarized beam.

Substituting Equations (6.57) and (6.58) into (6.55), and taking into account Equation (6.56), we can obtain all six projections of the electric and magnetic field vectors in the focal region of the aplanatic system (the apodization function has a form $T(\theta) = \cos^{1/2}\theta$ [6]):

$$
\begin{aligned}
E_x &= i^{m+1}\left[\sin m\varphi I_{0,m} + \sin(m-2)\varphi I_{2,m-2}\right], \\
E_y &= i^{m+1}\left[-\cos m\varphi I_{0,m} + \cos(m-2)\varphi I_{2,m-2}\right], \\
E_z &= -2i^{m}\sin(m-1)\varphi I_{1,m-1}, \\
H_x &= i^{m+1}\left[\cos m\varphi I_{0,m} + \cos(m-2)\varphi I_{2,m-2}\right], \\
H_y &= i^{m+1}\left[\sin m\varphi I_{0,m} - \sin(m-2)\varphi I_{2,m-2}\right], \\
H_y &= -2i^{m}\cos(m-1)\varphi I_{1,m-1},
\end{aligned}
\quad (6.59)
$$

where

$$I_{0,m} = \frac{\pi f}{\lambda}\int_0^a \sin\theta\cos^{1/2}\theta(1+\cos\theta)A_m(\theta)e^{ikz\cos\theta}J_m(x)d\theta,$$

$$I_{2,m-2} = \frac{\pi f}{\lambda}\int_0^a \sin\theta\cos^{1/2}\theta(1-\cos\theta)A_m(\theta)e^{ikz\cos\theta}J_{m-2}(x)d\theta, \quad (6.60)$$

$$I_{1,m-1} = \frac{\pi f}{\lambda}\int_0^a \sin^2\theta\cos^{1/2}\theta A_m(\theta)e^{ikz\cos\theta}J_m(x)d\theta.$$

In Equation (6.60), $x=kr\sin\theta$, $J_m(x)$ is the mth-order Bessel function, $A_m(\theta)=B(\theta, \varphi)$ is a real function describing the amplitude of the input field in the plane of the entrance pupil of the aplanatic system. It is dependent only from the angle θ and the order m of the polarization singularity.

From Equation (6.29), we obtain equations for the components of the electric and magnetic field in the cylindrical coordinate system:

$$E_r = i^{m+1}\sin(m-1)\varphi\left[I_{0,m}+I_{2,m-2}\right],$$

$$E_\varphi = i^{m+1}\cos(m-1)\varphi\left[-I_{0,m}+I_{2,m-2}\right],$$

$$E_z = -2i^m\sin(m-1)\varphi I_{1,m-1},$$

$$H_r = i^{m+1}\cos(m-1)\varphi\left[I_{0,m}+I_{2,m-2}\right], \quad\quad (6.61)$$

$$H_\varphi = i^{m+1}\sin(m-1)\varphi\left[I_{0,m}-I_{2,m-2}\right],$$

$$H_y = -2i^m\cos(m-1)\varphi I_{1,m-1}.$$

From Equation (6.61), for $m=1$ follow the well-known expressions for the components of an azimuthally field [6]

$$E_\varphi = I_{0,1} - I_{2,-1} = \frac{2\pi f}{\lambda}\int_0^a \sin\theta\cos^{1/2}\theta A_1(\theta)J_1(x)d\theta,$$

$$H_r = -\left[I_{0,1}+I_{2,-1}\right] = -\frac{2\pi f}{\lambda}\int_0^a \sin\theta\cos^{3/2}\theta A_1(\theta)J_1(x)d\theta, \quad\quad (6.62)$$

$$H_z = -2iI_{1,0} = -\frac{2i\pi f}{\lambda}\int_0^a \sin^2\theta\cos^{1/2}\theta A_1(\theta)J_0(x)d\theta.$$

In particular, from the first equation in Equation (6.62), it follows that in the focus of the azimuthally polarized beam on the optical axis ($r=0$), the intensity is equal to zero because $J_1(x)=0$ for $x=0$.

From Equation (6.59), we obtain an equation for the intensity of the electric field in the focal plane $z=0$:

$$I_m = |E_x|^2 + |E_y|^2 + |E_z|^2$$

$$= I_{0,m}^2 + I_{2,m-2}^2 - 2I_{0,m}I_{2,m-2}\cos\left[2(m-1)\varphi\right]+4\sin^2(m-1)\varphi I_{1,m-1}^2. \quad\quad (6.63)$$

From Equation (6.63) it can be seen that for $m>1$, the intensity in the focus is not radially symmetric. From Equation (6.63) it follows that only for the azimuthal polarization at $m=1$, the intensity in the focus has the shape of a symmetrical ring:

$$I_1 = (I_{0,1}-I_{2,-1})^2 \quad\quad (6.64)$$

From Equation (6.59) we can obtain a simple equation for the longitudinal projection of the Poynting vector [6]:

$$S_z = \frac{1}{2}\text{Re}\left[\left(\mathbf{E}\times\mathbf{H}^*\right)_z\right] = \frac{1}{2}\text{Re}\left(E_x H_y^* - E_y H_x^*\right) \tag{6.65}$$

In the focal plane ($z=0$), it is equal to:

$$S_{z,m} = I_{0,m}^2 - I_{2,m-2}^2 \tag{6.66}$$

From Equation (6.66), it follows that for any m the longitudinal component of PV in the focal plane has circular symmetry relative to the optical axis. This is a strange result because the intensity distribution in the focal plane (6.63) does not have circular symmetry for $m>1$. From Equation (6.66) for $m=2$, it follows that in the focal plane on the optical axis there is a reverse light energy flow (for any real amplitude in the plane of the entrance pupil $A_2(\theta)$):

$$S_{z,2}(r=0,z) = -\left(\frac{\pi f}{\lambda}\int_0^a \sin\theta\cos^{1/2}\theta(1-\cos\theta)A_2(\theta)d\theta\right)^2. \tag{6.67}$$

From Equation (6.66), it also follows that for $m=3$ on the optical axis in the focal plane, the energy flow is zero ($S_z=0$), and in the vicinity of the optical axis it increases quadratically in magnitude with the distance from the axis:

$$S_{z,3}(r\to 0,z) = -\frac{(kr)^2}{4}\left(\frac{\pi f}{\lambda}\int_0^a \sin^2\theta\cos^{1/2}\theta(1-\cos\theta)A_3(\theta)d\theta\right)^2. \tag{6.68}$$

It should be noted that if in the previous formulae, the apodization function $T(\theta)=\cos^{1/2}\theta$ is replaced by any other real function, for example, when the apodization function of the diffraction lens $T(\theta)=\cos^{-3/2}\theta$ [87], Equations (6.67) and (6.68), which prove the existence of an inverse energy flow, will not change. Only a certain value of the integrals in Equations (6.67) and (6.68) will be changed. It should also be noted that the existence of the inverse energy flow in the focus of the aplanatic system for $m=2, 3$ will be obtained for any real function $T(\theta)$.

6.4.2 Numerical Simulation Using Richards–Wolf Formulae

In our investigation, the numerical simulation was carried out using two methods: by calculating the Richards–Wolf integral (6.55) in the MATLAB environment and by solving the Maxwell equations using the FDTD method implemented in the FullWAVE software.

Figure 6.13(a) shows the direction of polarization in the investigated beam of order $m=2$. In the simulation it was assumed that the zone plate ($T(\theta)=\cos^{-3/2}\theta$ [87], NA$=0.95$) focuses the plane wave $B(\theta, \varphi)=1$. The longitudinal component of PV was calculated using (6.65), the intensity was calculated as $I=(\mathbf{EE}^*)$, where $\mathbf{E}$ is the electric field.

The results of focusing the second-order ($m=2$) cylindrical vector beam are shown in Figures 6.13(b) through 6.16. Figure 6.13(b) shows the distribution of the intensity in the focus ($z=0$).

From Figure 6.13(b), it can be seen that the intensity distribution has the shape of an asymmetrical ring. The asymmetry is explained by the redistribution of energy between components of the electric field due to tight focusing (Figure 6.14).

Figures 6.14 and 6.16 show the distribution of the longitudinal component of the Poynting vector S_z in the focal plane. Figure 6.15 shows the distribution of S_z in the transverse plane (xy), and Figure 6.16 shows the distribution in the longitudinal plane (zy) along the propagation axis of the beam (z-axis).

It is interesting that, in contrast to the focusing of a vortex beam with circular polarization in Reference [376], in this case there are no transverse components of the Poynting vector S_x and S_y in the focal plane.

Figure 6.17(a) shows the cross-section of the projection of the Poynting vector S_z for different orders of cylindrical vector beams (m equals 1, 2, 3 and 4).

From Figure 6.17(a), it can be seen that for the cylindrical vector beam ($m=1$), there are no negative values of the Poynting vector at the center of the focal spot ($S_z(x=y=0)=0$). If the order is more than two, then the distribution of the negative values of the projection of the Poynting vector has the form of rings. Only in the case

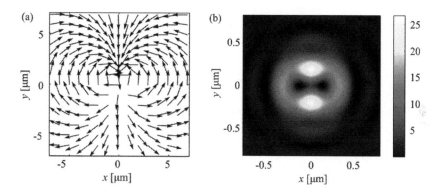

FIGURE 6.13 (a) Direction of polarization in cylindrical vector beam of the second order $m=2$; and (b) the intensity distribution in the focus ($z=0$, $m=2$): $I=I_x+I_y+I_z$.

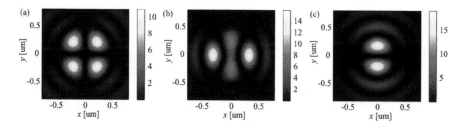

FIGURE 6.14 The distribution of the intensity components at the focus ($z=0$, $m=2$): (a) I_x, (b) I_y, and (c) I_z. The total intensity I is shown in Figure 6.34(b).

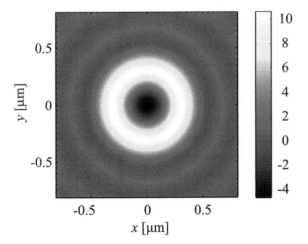

FIGURE 6.15   The distribution of the longitudinal component of the Poynting vector S_z in the focal plane ($z=0$, $m=2$)

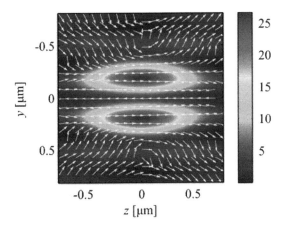

FIGURE 6.16   Intensity pattern I and directions of the Poynting vector S_z along the propagation axis of the beam (in the yz-plane).

$m=2$, the negative values of the projection of the Poynting vector onto the optical axis are observed at the center of the focal spot ($S_z(x=y=0)<0$).

Figure 6.17(b) shows the dependence of the minimum value of S_z at the center of the focal spot on the numerical aperture NA for a cylindrical vector beam of the order $m=2$. From Figure 6.17(b), it could be seen that negative values of S_z are obtained only for large numerical apertures (NA >0.8).

6.4.3   NUMERICAL SIMULATION USING THE FDTD METHOD

The FDTD method was used to verify the results obtained with the Richards–Wolf integrals. The FDTD method implemented in the FullWAVE software was used to

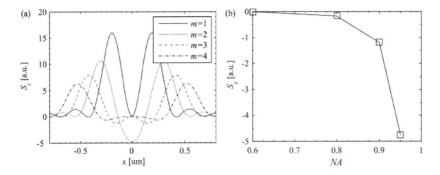

FIGURE 6.17 (a) The distribution of the longitudinal component of the Poynting vector S_z in the focus of the zone plate for the cylindrical vector beams of orders $m=1$, 2, 3, and 4; (b) the dependence of the minimum value of S_z in the center of the focal spot on the numerical aperture NA for an cylindrical vector beams of the order $m=2$.

investigate the focusing of a cylindrical vector beam of order $m=2$ multiplied by the transmission function of the Fresnel zone plate. The transmission function of the zone plate was calculated from the considerations that the focal length is equal to 532 nm (numerical aperture ~1), the focused light has a wavelength of 532 nm, the relief height is $h=0.159$ μm, and the refractive index of the material is $n=2.67$ (TiO_2). The numerical aperture of the lens is not equal to the Richards–Wolf modeled numerical aperture of the lens, because the only aim of the FDTD simulation was the qualitative comparison of the results. The grid size was equal to $\lambda/30$. The size of the calculated area was $8.6 \times 8.6 \times 1.532$ μm. Perfectly matched layers (PML) of thickness 0.5 μm were placed on the borders of the calculated region.

From Figure 6.18, it can be seen that negative values of S_z are observed in the center of the focal spot ($m=2$). Therefore, the results obtained using the FDTD method confirm the results obtained using the Richards–Wolf formulae. The asymmetry of the intensity in Figure 6.18(a) is caused by the fact that a non-radially symmetric area was calculated in the FDTD method.

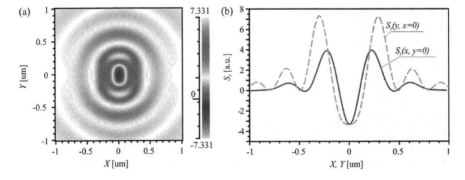

FIGURE 6.18 The distribution of the longitudinal component of the Poynting vector S_z at the focus of the zone plate ($z=532$ nm), calculated using the FDTD method.

In this section, tight focusing of high-order cylindrical vector beams was investigated. It was shown theoretically and numerically that in the focus, there are regions with Poynting vectors with directions opposite to the propagation direction of the beam, and the negative values are comparable in absolute value with positive values. If the beam order is equal to two, then the region with negative values of the projection of the Poynting vector is located in the center of the focal spot.

The possibility of obtaining focal spots with the Poynting vector opposite to the direction of beam propagation was previously shown in References [369, 376, 400]; however, in these papers, optical vortices were focused, and the energy flow has a spiral shape (spiral reverse flow). In our paper, the reverse flow has a non-vortex character. It should also be noted that a cylindrical vector beam of the second order can be obtained from a linearly polarized beam using a single element that transforms the polarization; for example, a polarizer based on subwavelength gratings [315].

Conclusion

As a conclusion, we present a brief review of modern works on optical components with a metasurface. In recent years, flat optical components of microoptics with a thickness less than the wavelength, consisting of a set of subwavelength elements (columns, slots, segments, gratings) of metal or semiconductor, which can simultaneously change the polarization, amplitude, and phase of the incident laser radiation, are studied in optics. Such photonic components are called components with a metasurface (CMPs). An overview of them can be found in References [320, 321, 403]. Below are references that will allow the reader to evaluate the progress and direction of current research in the field of metasurfaces. In particular, in a number of works [404–410], metasurfaces based on graphene are considered. In References [404, 405], metasurfaces that allow control of the phase and amplitude of the light reflected from it are implemented to control infrared radiation. Moreover, it was shown in Reference [405] that the interaction of light with a thin layer of graphene can be amplified using a subwavelength optical resonator. In Reference [407], a hyperbolic metasurface based on graphene strips for terahertz radiation and near infrared (IR) range was implemented. In Reference [408], plasmon excitation in an array of graphene microlens was studied. Plasmon excitation in a graphene taper was observed in Reference [411] by scanning near-field optical microscopy. In Reference [412], the propagation of plasmons in a graphene layer, over which an atomic force microscope (AFM) probe is located, is investigated by light with a wavelength of 11.2 μm. Filters based on graphene and dielectric layers were considered in Reference [413]. There are interesting studies of metasurfaces reacting differently to different polarizations of the light incident on them. For example, in References [414, 415], the metasurface is used to excite surface plasmons that propagate in different directions, depending on the direction of the circular polarization of the light exciting them. And in Reference [416], again, depending on the direction of rotation of the circular polarization, the layer of the metasurface acts as a collecting or diverging lens. The gradient metasurface considered in Reference [417] is capable of transforming propagating waves into surface waves with an efficiency close to 100%. And in Reference [418] the gradient metasurface is used as a lens operating in the wavelength range from 750 nm to 950 nm. In Reference [419], a one-dimensional metal lattice (the simplest metasurface) is used to control surface plasmons. It is shown numerically that the same grating with a period of 120 nm, a step width of 60 nm and a height of 80 nm, illuminated by light from different wavelengths, forms plasmons that propagate in different ways: one can observe the phenomenon of negative refraction, ordinary diffraction, and non-diverging waves. Metasurfaces based on V-shaped antennae are described in References [420, 421] In Reference [420], a two-layer metasurface operating on transmission in the visible wavelength range was considered. The metasurface under investigation consisted of a layer of V-shaped antennae of gold and a gold film with V-shaped holes that complemented the antennae in accordance with Babinet's principle. The characteristic antenna size was 150 nm, and the two gold layers were separated with a layer of 100-nm thick hydrogen-silsesquioxane. An ultrathin hologram (30 nm) based on the use of a metasurface is considered in Reference [422]. Holograms based on the metasurface, considered in Reference [423], created an image of 330×232 μm in size. A separate hologram pixel was a metal nano-column oriented in space. The spin Hall effect for photons, observed in the layer of the metasurface, was considered in Reference [424]. A metasurface that changes its optical properties under the influence of an electrical voltage applied to it is described in Reference [425] – depending on the voltage, it is possible to change the absorption in the layer of the metasurface by 30%. Metasurfaces are a special case of metamaterials. The evolution from metamaterial to metasurface and further to individual meta-devices is described in Reference [426].

References

1. M. Born and E. Wolf, *Principles of Optics: Electromagnetic Theory of Propagation, Interference and Diffraction of Light* (Elsevier, 2013).
2. E. Lommel, "Die Beugungserscheinungen einer kreisrunden Oeffnung und eines kreisrunden Schirmchens," *Abhandlungen der* Math. Cl. der königlich Bayer. Akad. der Wissenschaften **15**, 229–328 (1885).
3. H. Struve, "Ueber die allgemeine Beugungsfigur in Fernröhren," *Mem. L'Academie Imp. des Sci. Saint-petersbg.* **34**, 5 (1886).
4. P. Debye, "Das Verhalten von Lichtwellen in der Nähe eines Brennpunktes oder einer Brennlinie," *Ann. Phys.* **335**, 755–776 (1909).
5. V. S. Ignatowsky, "Diffraction by a lens having arbitrary opening," *Trans. Opt. Inst. Petrogr.* **1**, 4 (1920).
6. B. Richards and E. Wolf, "Electromagnetic diffraction in optical systems. II. Structure of the image field in an aplanatic system," *Proc. R. Soc. A Math. Phys. Eng. Sci.* **253**, 358–379 (1959).
7. L. Novotny and B. Hecht, *Principles of Nano-Optics* (Cambridge University Press, 2006).
8. J. J. Stamnes, *Waves in Focal Regions: Propagation, Diffraction and Focusing of Light, Sound and Water Waves* (CRC Press, 1986).
9. G. B. Airy, "On the diffraction of an object-glass with circular aperture," *Trans. Cambridge Philos. Soc.* **5**, 283–291 (1835).
10. E. Abbe, "Ueber einen neuen Beleuchtungsapparat am Mikroskop," *Arch. für mikroskopische Anat.* **9**, 469–480 (1873).
11. A. Taflove and M. E. Brodwin, "Numerical solution of steady-state electromagnetic scattering problems using the time-dependent Maxwell's equations," *IEEE Trans. Microw. Theory Tech.* **23**, 623–630 (1975).
12. M. Ohtsu, K. Kobayashi, T. Kawazoe, T. Yatsui, and M. Naruse, *Principles of Nanophotonics* (CRC Press, 2008).
13. S. V. Gaponenko, *Introduction to Nanophotonics* (Cambridge University Press, 2010).
14. G. de Villiers and E. R. Pike, *The Limits of Resolution* (CRC Press, 2016).
15. V. G. Veselago, "Superlens as matching device," arXiv preprint cond-mat/0501438 (2005).
16. J. B. Pendry, "Negative refraction makes a perfect lens," *Phys. Rev. Lett.* **85**, 3966–3969 (2000).
17. R. J. Blaikie and D. O. S. Melville, "Imaging through planar silver lenses in the optical near field," *J. Opt. A Pure Appl. Opt.* **7**, S176–S183 (2005).
18. D. O. S. Melville and R. J. Blaikie, "Super-resolution imaging through a planar silver layer," *Opt. Express* **13**, 2127–2134 (2005).
19. N. Fang, H. Lee, C. Sun, and X. Zhang, "Sub-diffraction-limited optical imaging with a Silver Superlens," *Science (80-.).* **308**, 534–537 (2005).
20. V. A. Podolskiy and E. E. Narimanov, "Near-sighted superlens," *Opt. Lett.* **30**, 75–77 (2005).
21. R. Merlin, "Analytical solution of the almost-perfect-lens problem," *Appl. Phys. Lett.* **84**, 1290 (2004).
22. J. B. Pendry, "Perfect cylindrical lenses," *Opt. Express* **11**, 755–760 (2003).
23. V. Podolskiy, A. Sarychev, and V. Shalaev, "Plasmon modes and negative refraction in metal nanowire composites," *Opt. Express* **11**, 735–745 (2003).
24. V. M. Shalaev, W. Cai, U. K. Chettiar, H.-K. Yuan, A. K. Sarychev, V. P. Drachev, and A. V. Kildishev, "Negative index of refraction in optical metamaterials," *Opt. Lett.* **30**, 3356–3358 (2005).

25. A. Berrier, M. Mulot, M. Swillo, M. Qiu, L. Thylén, A. Talneau, and S. Anand, "Negative refraction at infrared wavelengths in a two-dimensional photonic crystal," *Phys. Rev. Lett.* **93**, 073902 (2004).

26. A. A. Govyadinov and V. A. Podolskiy, "Metamaterial photonic funnels for subdiffraction light compression and propagation," *Phys. Rev. B* **73**, 155108 (2006).

27. R. Wangberg, J. Elser, E. E. Narimanov, and V. A. Podolskiy, "Nonmagnetic nanocomposites for optical and infrared negative-refractive-index media," *J. Opt. Soc. Am. B* **23**, 498–505 (2006).

28. Z. Jacob, L. V. Alekseyev, and E. Narimanov, "Optical hyperlens: far-field imaging beyond the diffraction limit," *Opt. Express* **14**, 8247–8256 (2006).

29. Z. Liu, H. Lee, Y. Xiong, C. Sun, and X. Zhang, "Far-field optical hyperlens magnifying sub-diffraction-limited objects," *Science (80-.).* **315**, 1686 (2007).

30. R. Merlin, "Radiationless electromagnetic interference: evanescent-field lenses and perfect focusing," *Science (80-.).* **317**, 927–929 (2007).

31. H. Liu, Shivanand, and K. J. Webb, "Subwavelength imaging opportunities with planar uniaxial anisotropic lenses," *Opt. Lett.* **33**, 2568–2570 (2008).

32. A. Husakou and J. Herrmann, "Superfocusing of light below the diffraction limit by photonic crystals with negative refraction," *Opt. Express* **12**, 6491–6497 (2004).

33. X. Wang, Z. F. Ren, and K. Kempa, "Unrestricted superlensing in a triangular two dimensional photonic crystal," *Opt. Express* **12**, 2919–2924 (2004).

34. S.-Y. Yang, C.-Y. Hong, and H.-C. Yang, "Focusing concave lens using photonic crystals with magnetic materials," *J. Opt. Soc. Am. A* **23**, 956–959 (2006).

35. T. Matsumoto, K.-S. Eom, and T. Baba, "Focusing of light by negative refraction in a photonic crystal slab superlens on silicon-on-insulator substrate," *Opt. Lett.* **31**, 2786–2788 (2006).

36. H.-T. Chien and C.-C. Chen, "Focusing of electromagnetic waves by periodic arrays of air holes with gradually varying radii," *Opt. Express* **14**, 10759–10764 (2006).

37. Q. Wu, J. M. Gibbons, and W. Park, "Graded negative index lens by photonic crystals," *Opt. Express* **16**, 16941–16949 (2008).

38. S. B. Ippolito, B. B. Goldberg, and M. S. Ünlü, "High spatial resolution subsurface microscopy," *Appl. Phys. Lett.* **78**, 4071 (2001).

39. I. Golub, "Solid immersion axicon: maximizing nondiffracting or Bessel beam resolution," *Opt. Lett.* **32**, 2161–2163 (2007).

40. D. R. Mason, M. V. Jouravlev, and K. S. Kim, "Enhanced resolution beyond the Abbe diffraction limit with wavelength-scale solid immersion lenses," *Opt. Lett.* **35**, 2007–2009 (2010).

41. Y. Lin, W. Seka, J. H. Eberly, H. Huang, and D. L. Brown, "Experimental investigation of Bessel beam characteristics," *Appl. Opt.* **31**, 2708–2713 (1992).

42. S. Ruschin and A. Leizer, "Evanescent Bessel beams," *J. Opt. Soc. Am. A* **15**, 1139–1143 (1998).

43. T. Grosjean, D. Courjon, and C. Bainier, "Smallest lithographic marks generated by optical focusing systems," *Opt. Lett.* **32**, 976–978 (2007).

44. K. Yonezawa, Y. Kozawa, and S. Sato, "Generation of a radially polarized laser beam by use of the birefringence of a c-cut Nd:YVO4 crystal," *Opt. Lett.* **31**, 2151–2153 (2006).

45. J. Li, K. Ueda, M. Musha, A. Shirakawa, and L. Zhong, "Generation of radially polarized mode in Yb fiber laser by using a dual conical prism," *Opt. Lett.* **31**, 2969–2971 (2006).

46. J. Li, K. Ueda, M. Musha, A. Shirakawa, and Z. Zhang, "Converging-axicon-based radially polarized ytterbium fiber laser and evidence on the mode profile inside the gain fiber," *Opt. Lett.* **32**, 1360–1362 (2007).

47. A. Mehta, J. D. Brown, P. Srinivasan, R. C. Rumpf, and E. G. Johnson, "Spatially polarizing autocloned elements," *Opt. Lett.* **32**, 1935–1937 (2007).

48. U. Levy, C.-H. Tsai, L. Pang, and Y. Fainman, "Engineering space-variant inhomogeneous media for polarization control," *Opt. Lett.* **29**, 1718–1720 (2004).

49. V. V. Kotlyar and S. S. Stafeev, "Sharply focusing a radially polarized laser beam using a gradient Mikaelian's microlens," *Opt. Commun.* **282**, 459–464 (2009).

50. A. Bouhelier, J. Renger, M. R. Beversluis, and L. Novotny, "Plasmon-coupled tip-enhanced near-field optical microscopy," *J. Microsc.* **210**, 220–224 (2003).

51. J. K. Kim, J. Kim, Y. Jung, W. Ha, Y. S. Jeong, S. Lee, A. Tünnermann, and K. Oh, "Compact all-fiber Bessel beam generator based on hollow optical fiber combined with a hybrid polymer fiber lens," *Opt. Lett.* **34**, 2973–2975 (2009).

52. H. Kurt, "Limited-diffraction light propagation with axicon-shape photonic crystals," *J. Opt. Soc. Am. B* **26**, 981–986 (2009).

53. W. Chen and Q. Zhan, "Realization of an evanescent Bessel beam via surface plasmon interference excited by a radially polarized beam," *Opt. Lett.* **34**, 722–724 (2009).

54. K. Watanabe, G. Terakado, and H. Kano, "Localized surface plasmon microscope with an illumination system employing a radially polarized zeroth-order Bessel beam," *Opt. Lett.* **34**, 1180–1182 (2009).

55. Y. Fu and W. Zhou, "Hybrid Au-Ag subwavelength metallic structures with variant periods for superfocusing," *J. Nanophotonics* **3**, 033504 (2009).

56. Y. Fu, R. G. Mote, Q. Wang, and W. Zhou, "Experimental study of plasmonic structures with variant periods for sub-wavelength focusing: analyses of characterization errors," *J. Mod. Opt.* **56**, 1550–1556 (2009).

57. P.-K. Wei, W.-L. Chang, K.-L. Lee, and E.-H. Lin, "Focusing subwavelength light by using nanoholes in a transparent thin film," *Opt. Lett.* **34**, 1867–1869 (2009).

58. E. Schonbrun, W. N. Ye, and K. B. Crozier, "Scanning microscopy using a short-focal-length Fresnel zone plate," *Opt. Lett.* **34**, 2228–2230 (2009).

59. P. Vahimaa, V. Kettunen, M. Kuittinen, J. Turunen, and A. T. Friberg, "Electromagnetic analysis of nonparaxial Bessel beams generated by diffractive axicons," *J. Opt. Soc. Am. A* **14**, 1817–1824 (1997).

60. Y. Kizuka, M. Yamauchi, and Y. Matsuoka, "Characteristics of a laser beam spot focused by a binary diffractive axicon," *Opt. Eng.* **47**, 053401 (2008).

61. A. P. Prudnikov, U. A. Brychkov, and O. I. Marichev, *Integrals and Series. Special Functions* (Nauka, 1983).

62. A. W. Snyder and J. Love, *Optical Waveguide Theory* (Springer Science & Business Media, 2012).

63. H. Osterberg and L. W. Smith, "Closed Solutions of Rayleigh's Diffraction Integral for Axial Points," *J. Opt. Soc. Am.* **51**, 1050–1054 (1961).

64. B. Vohnsen, S. Castillo, and D. Rativa, "Wavefront sensing with an axicon," *Opt. Lett.* **36**, 846–848 (2011).

65. S. K. Tiwari, "Generation of a variable-diameter collimated hollow laser beam using metal axicon mirrors," *Opt. Eng.* **50**, 014001 (2011).

66. D. Kuang, M. Han, H. Gao, Z. Du, and Z. Fang, "Interferometric characterization of a microaxicon with a single fringe pattern," *J. Opt.* **13**, 035501 (2011).

67. D. Kuang and Z. Fang, "Microaxicave: inverted microaxicon to generate a hollow beam," *Opt. Lett.* **35**, 2158–2160 (2010).

68. C. T. Chong, C. Sheppard, H. Wang, L. Shi, and B. Lukyanchuk, "Creation of a needle of longitudinally polarized light in vacuum using binary optics," *Nat. Photonics* **2**, 501–505 (2008).

69. K. Huang, P. Shi, X. Kang, X. Zhang, and Y. Li, "Design of DOE for generating a needle of a strong longitudinally polarized field," *Opt. Lett.* **35**, 965–967 (2010).

70. K. B. Rajesh, Z. Jaroszewicz, and P. M. Anbarasan, "Improvement of lens axicon's performance for longitudinally polarized beam generation by adding a dedicated phase transmittance," *Opt. Express* **18**, 26799–26805 (2010).

71. K. Kitamura, K. Sakai, and S. Noda, "Sub-wavelength focal spot with long depth of focus generated by radially polarized, narrow-width annular beam," *Opt. Express* **18**, 4518–4525 (2010).

72. T. Grosjean and D. Courjon, "Smallest focal spots," *Opt. Commun.* **272**, 314–319 (2007).

73. E. J. Botcherby, R. Juškaitis, and T. Wilson, "Scanning two photon fluorescence microscopy with extended depth of field," *Opt. Commun.* **268**, 253–260 (2006).

74. K. Dholakia and T. Čižmár, "Shaping the future of manipulation," *Nat. Photonics* **5**, 335–342 (2011).

75. L. Liu, C. Liu, W. C. Howe, C. J. R. Sheppard, and N. Chen, "Binary-phase spatial filter for real-time swept-source optical coherence microscopy," *Opt. Lett.* **32**, 2375–2377 (2007).

76. R. D. Romea and W. D. Kimura, "Modeling of inverse Čerenkov laser acceleration with axicon laser-beam focusing," *Phys. Rev. D* **42**, 1807–1818 (1990).

77. M. K. Bhuyan, F. Courvoisier, P. A. Lacourt, M. Jacquot, R. Salut, L. Furfaro, and J. M. Dudley, "High aspect ratio nanochannel machining using single shot femtosecond Bessel beams," *Appl. Phys. Lett.* **97**, 081102 (2010).

78. J. Fu, H. Dong, and W. Fang, "Subwavelength focusing of light by a tapered microtube," *Appl. Phys. Lett.* **97**, 041114 (2010).

79. I. Golub, B. Chebbi, D. Shaw, and D. Nowacki, "Characterization of a refractive logarithmic axicon," *Opt. Lett.* **35**, 2828–2830 (2010).

80. V. V. Kotlyar and S. S. Stafeev, "Modeling the sharp focus of a radially polarized laser mode using a conical and a binary microaxicon," *J. Opt. Soc. Am. B* **27**, 1991–1997 (2010).

81. J. W. Y. Lit and R. Tremblay, "Focal depth of a transmitting axicon," *J. Opt. Soc. Am.* **63**, 445–449 (2008).

82. M. A. Golub, S. V. Karpeev, A. M. Prokhorov, I. N. Sisakyan, and V. A. Soifer, "Focusing light into a specified volume by computer synthesized holograms," *Sov. Tech. Phys. Lett.* **7**, 264–266 (1981).

83. L. R. Staroński, J. Sochacki, A. Kołodziejczyk, and Z. Jaroszewicz, "Lateral distribution and flow of energy in uniform-intensity axicons," *J. Opt. Soc. Am. A* **9**, 2091–2094 (2008).

84. M. Abramowitz and I. A. Stegun, *Handbook of Mathematical Functions: With Formulas, Graphs, and Mathematical Tables* (Courier Corporation, 1965).

85. V. V. Kotlyar and A. A. Kovalev, "General form of hypergeometric laser beams and their particular cases," *Comput. Opt.* **31**, 29–32 (2007).

86. V. V. Kotlyar and A. A. Kovalev, "Family of hypergeometric laser beams," *J. Opt. Soc. Am. A* **25**, 262–270 (2007).

87. N. Davidson and N. Bokor, "High-numerical-aperture focusing of radially polarized doughnut beams with a parabolic mirror and a flat diffractive lens," *Opt. Lett.* **29**, 1318–1320 (2004).

88. V. P. Kalosha and I. Golub, "Toward the subdiffraction focusing limit of optical superresolution," *Opt. Lett.* **32**, 3540–3542 (2007).

89. K. S. Youngworth and T. G. Brown, "Focusing of high numerical aperture cylindrical-vector beams," *Opt. Express* **7**, 77–87 (2000).

90. Y. Fu, F. Fang, X. Zhou, Z. Xu, and Y. Liu, "Experimental investigation of superfocusing of plasmonic lens with chirped circular nanoslits," *Opt. Express* **18**, 3438–3443 (2010).

91. Y. Fu, W. Zhou, L. E. N. Lim, C. L. Du, and X. G. Luo, "Plasmonic microzone plate: superfocusing at visible regime," *Appl. Phys. Lett.* **91**, 061124 (2007).

92. L. C. Lopez, M. P. Molina, P. A. Gonzalez, S. B. Escarre, A. F. Gil, R. F. Madrigal, and A. M. Cases, "Vectorial diffraction analysis of near-field focusing of perfect black Fresnel zone plates under various polarization states," *J. Light. Technol.* **29**, 822–829 (2011).

93. H. C. Kim, H. Ko, and M. Cheng, "High efficient optical focusing of a zone plate composed of metal/dielectric multilayer," *Opt. Express* **17**, 3078–3083 (2009).

94. S. F. Yu, W. Zhou, B. K. Ng, S. P. Lau, and R. G. Mote, "Near-field focusing properties of zone plates in visible regime – New insights," *Opt. Express* **16**, 9554–9564 (2008).

95. R. G. Mote, S. F. Yu, A. Kumar, W. Zhou, and X. F. Li, "Experimental demonstration of near-field focusing of a phase micro-Fresnel zone plate (FZP) under linearly polarized illumination," *Appl. Phys. B Lasers Opt.* **102**, 95–100 (2011).

96. R. G. Mote, S. F. Yu, W. Zhou, and X. F. Li, "Subwavelength focusing behavior of high numerical-aperture phase Fresnel zone plates under various polarization states," *Appl. Phys. Lett.* **95**, 191113 (2009).

97. B. Jia, H. Shi, J. Li, Y. Fu, C. Du, and M. Gu, "Near-field visualization of focal depth modulation by step corrugated plasmonic slits," *Appl. Phys. Lett.* **94**, 151912 (2009).

98. K. R. Chen, W. H. Chu, H. C. Fang, C. P. Liu, C. H. Huang, H. C. Chui, C. H. Chuang, Y. L. Lo, C. Y. Lin, H. H. Hwung, and A. Y.-G. Fuh, "Beyond-limit light focusing in the intermediate zone," *Opt. Lett.* **36**, 4497–4499 (2011).

99. Y. Yu and H. Zappe, "Effect of lens size on the focusing performance of plasmonic lenses and suggestions for the design," *Opt. Express* **19**, 9434–9444 (2011).

100. M. Yu, F. Stief, H. Xu, N. Zhitenev, and Y. Liu, "Far-field superfocusing with an optical fiber based surface plasmonic lens made of nanoscale concentric annular slits," *Opt. Express* **19**, 20233–20243 (2011).

101. V. V. Kotlyar, S. S. Stafeev, L. O'Faolain, and V. A. Soifer, "Tight focusing with a binary microaxicon," *Opt. Lett.* **36**, 3100–3102 (2011).

102. S. S. Stafeev, L. O'Faolain, M. I. Shanina, V. V. Kotlyar, and V. A. Soifer, "Subwavelength focusing using Fresnel zone plate with focal length of 532 nm," *Comput. Opt.* **35**, 460–461 (2011).

103. J. S. Ye, G. A. Mei, X. H. Zheng, and Y. Zhang, "Long-focal-depth cylindrical microlens with flat axial intensity distributions," *J. Mod. Opt.* **59**, 90–94 (2012).

104. K. Huang and Y. Li, "Realization of a subwavelength focused spot without a longitudinal field component in a solid immersion lens-based system," *Opt. Lett.* **36**, 3536–3538 (2011).

105. G. H. Yuan, S. B. Wei, and X.-C. Yuan, "Nondiffracting transversally polarized beam," *Opt. Lett.* **36**, 3479–3481 (2011).

106. X. Li, Y. Cao, and M. Gu, "Superresolution-focal-volume induced 30 Tbytes/disk capacity by focusing a radially polarized beam," *Opt. Lett.* **36**, 2510–2512 (2011).

107. J. Lin, K. Yin, Y. Li, and J. Tan, "Achievement of longitudinally polarized focusing with long focal depth by amplitude modulation," *Opt. Lett.* **36**, 1185–1187 (2011).

108. H. Lin, B. Jia, and M. Gu, "Generation of an axially super-resolved quasi-spherical focal spot using an amplitude-modulated radially polarized beam," *Opt. Lett.* **36**, 2471–2473 (2011).

109. J. Martin, J. Proust, D. Gérard, J.-L. L. Bijeon, J. Plain, J.-L. L. Bijeon, and J. Martin, "Intense Bessel-like beams arising from pyramid-shaped microtips," *Opt. Lett.* **37**, 1274–1276 (2012).

110. F. De Angelis, F. Gentile, F. Mecarini, G. Das, M. Moretti, P. Candeloro, M. L. Coluccio, G. Cojoc, A. Accardo, C. Liberale, R. P. Zaccaria, G. Perozziello, L. Tirinato, A. Toma, G. Cuda, R. Cingolani, and E. Di Fabrizio, "Breaking the diffusion limit with super-hydrophobic delivery of molecules to plasmonic nanofocusing SERS structures," *Nat. Photonics* **5**, 682–687 (2011).

111. E. T. F. Rogers, J. Lindberg, T. Roy, S. Savo, J. E. Chad, M. R. Dennis, and N. I. Zheludev, "A super-oscillatory lens optical microscope for subwavelength imaging," *Nat. Mater.* **11**, 432–435 (2012).

112. K. A. Michalski, "Complex image method analysis of a plane wave-excited subwavelength circular aperture in a planar screen," *Prog. Electromagn. Res. B* **27**, 253–272 (2011).

113. J. H. Wu, "Modeling of near-field optical diffraction from a subwavelength aperture in a thin conducting film," *Opt. Lett.* **36**, 3440–3442 (2011).

114. V. V. Kotlyar and M. A. Lichmanov, "Electromagnetic wave diffraction on infinite circular cylinder with homogeneous layers," *Comput. Opt.* **24**, 26–32 (2002).

115. X. Li, Z. Chen, A. Taflove, and V. Backman, "Optical analysis of nanoparticles via enhanced backscattering facilitated by 3-D photonic nanojets," *Opt. Express* **13**, 526–533 (2005).

116. A. Heifetz, S.-C. Kong, A. V. Sahakian, A. Taflove, and V. Backman, "Photonic Nanojets," *J. Comput. Theor. Nanosci.* **6**, 1979–1992 (2009).

117. N. Bonod, E. Popov, P. Ferrand, M. Pianta, B. Stout, H. Rigneault, J. Wenger, and A. Devilez, "Direct imaging of photonic nanojets," *Opt. Express* **16**, 6930–6940 (2008).

118. C.-H. Chang, L. Tian, W. R. Hesse, H. Gao, H. J. Choi, J.-G. Kim, M. Siddiqui, and G. Barbastathis, "From two-dimensional colloidal self-assembly to three-dimensional nanolithography," *Nano Lett.* **11**, 2533–2537 (2011).

119. V. Astratov, "Photonic nanojets for laser surgery," *SPIE Newsroom* **1**, 2–4 (2010).

120. V. K. Valev, D. Denkova, X. Zheng, A. I. Kuznetsov, C. Reinhardt, B. N. Chichkov, G. Tsutsumanova, E. J. Osley, V. Petkov, B. De Clercq, A. V. Silhanek, Y. Jeyaram, V. Volskiy, P. A. Warburton, G. A. E. Vandenbosch, S. Russev, O. A. Aktsipetrov, M. Ameloot, V. V. Moshchalkov, and T. Verbiest, "Plasmon-enhanced sub-wavelength laser ablation: plasmonic nanojets," *Adv. Mater.* **24**, OP29–OP35 (2012).

121. F. Merola, S. Coppola, V. Vespini, S. Grilli, and P. Ferraro, "Characterization of Bessel beams generated by polymeric microaxicons," *Meas. Sci. Technol.* **23**, 065204 (2012).

122. C. Rockstuhl, M.-S. Kim, H. P. Herzig, T. Scharf, and S. Mühlig, "Engineering photonic nanojets," *Opt. Express* **19**, 10206–10220 (2011).

123. D. McCloskey, J. J. Wang, and J. F. Donegan, "Low divergence photonic nanojets from Si_3N_4 microdisks," *Opt. Express* **20**, 128–140 (2012).

124. S. S. Stafeev and V. V. Kotlyar, "Elongated photonic nanojet from truncated cylindrical zone plate," *J. At. Mol. Opt. Phys.* **2012**, 123872 (2012).

125. D. M. Palacios and G. A. Swartzlander, Jr., "The high-contrast performance of an optical vortex coronagraph," in *Current Developments in Lens Design and Optical Engineering VII*, P. Z. Mouroulis, W. J. Smith, and R. B. Johnson, eds. (2006), p. 62880B.

126. S. N. Khonina, V. V. Kotlyar, M. V. Shinkaryev, V. A. Soifer, and G. V. Uspleniev, "The Phase Rotor Filter," *J. Mod. Opt.* **39**, 1147–1154 (1992).

127. L. Allen, M. W. Beijersbergen, R. J. C. Spreeuw, and J. P. Woerdman, "Orbital angular momentum of light and the transformation of Laguerre-Gaussian laser modes," *Phys. Rev. A* **45**, 8185–8189 (1992).

128. Z. S. Sacks, D. Rozas, and G. A. Swartzlander, Jr., "Holographic formation of optical-vortex filaments," *J. Opt. Soc. Am. B* **15**, 2226–2234 (2008).

129. J. F. Nye and M. V. Berry, "Dislocations in wave trains," *Proc. R. Soc. A Math. Phys. Eng. Sci.* **336**, 165–190 (1974).

130. V. V. Kotlyar, H. Elfstrom, J. Turunen, A. A. Almazov, S. N. Khonina, and V. A. Soifer, "Generation of phase singularity through diffracting a plane or Gaussian beam by a spiral phase plate," *J. Opt. Soc. Am. A* **22**, 849–861 (2005).

131. V. V. Kotlyar, S. N. Khonina, A. A. Kovalev, V. A. Soifer, H. Elfstrom, and J. Turunen, "Diffraction of a plane, finite-radius wave by a spiral phase plate," *Opt. Lett.* **31**, 1597–1599 (2006).

132. H. Garcia-Gracia and J. C. Gutiérrez-Vega, "Diffraction of plane waves by finite-radius spiral phase plates of integer and fractional topological charge Hipolito," *J. Opt. Soc. Am. A* **26**, 794–803 (2009).

133. S. S. R. Oemrawsingh, J. A. W. van Houwelingen, E. R. Eliel, J. P. Woerdman, E. J. K. Verstegen, J. G. Kloosterboer, and G. W. 't Hooft, "Production and characterization of spiral phase plates for optical wavelengths," *Appl. Opt.* **43**, 688–694 (2004).

134. K. Sueda, G. Miyaji, N. Miyanaga, and M. Nakatsuka, "Laguerre-Gaussian beam generated with a multilevel spiral phase plate for high intensity laser pulses," *Opt. Express* **12**, 3548–3553 (2004).

135. L.-S. Zhang, W. C. Cheong, X.-C. Yuan, W. M. Lee, H. Wang, and K. Dholakia, "Direct electron-beam writing of continuous spiral phase plates in negative resist with high power efficiency for optical manipulation," *Appl. Phys. Lett.* **85**, 5784 (2004).

136. I. Moreno, N. Zhang, D. M. Cottrell, K.-J. Moh, J. Lin, J. A. Davis, and X. Yuan, "Analysis of fractional vortex beams using a vortex grating spectrum analyzer," *Appl. Opt.* **49**, 2456–2462 (2010).

137. M. W. Beijersbergen, R. P. C. Coerwinkel, M. Kristensen, and J. P. Woerdman, "Helical-wavefront laser beams produced with a spiral phaseplate," *Opt. Commun.* **112**, 321–327 (1994).

138. Y. F. Lu, L. Zhang, W. D. Song, Y. W. Zheng, and B. S. Luk'yanchuk, "Laser writing of a subwavelength structure on silicon (100) surfaces with particle-enhanced optical irradiation," *JETP Lett.* **72**, 457–459 (2000).

139. E. McLeod and C. B. Arnold, "Subwavelength direct-write nanopatterning using optically trapped microspheres," *Nat. Nanotechnol.* **3**, 413–417 (2008).

140. S. Kong, A. Taflove, and V. Backman, "Quasi one-dimensional light beam generated by a graded-index microsphere," *Opt. Express* **17**, 17343–17350 (2009).

141. A. Devilez, N. Bonod, J. Wenger, D. Gérard, B. Stout, and E. Popov, "Three-dimensional subwavelength confinement of light with dielectric microspheres," *Opt. Express* **17**, 2089–2094 (2009).

142. Z. Chen, A. Taflove, and V. Backman, "Photonic nanojet enhancement of backscattering of light by nanoparticles: a potential novel visible-light ultramicroscopy technique," *Opt. Express* **12**, 1214–1220 (2004).

143. V. V. Kotlyar, S. S. Stafeev, Y. Liu, L. O'Faolain, and A. A. Kovalev, "Analysis of the shape of a subwavelength focal spot for the linearly polarized light," *Appl. Opt.* **52**, 330–339 (2013).

144. V. V. Kotlyar, A. A. Kovalev, V. A. Shuypova, A. G. Nalimov, and V. A. Soifer, "Subwavelength confinement of light in waveguide structures," *Comput. Opt.* **34**, 169–186 (2010).

145. V. V. Kotlyar, S. S. Stafeev, and A. A. Kovalev, "Curved laser microjet in near field," *Appl. Opt.* **52**, 4131–4136 (2013).

146. S. S. Stafeev, V. V. Kotlyar, and L. O'Faolain, "Subwavelength focusing of laser light by microoptics," *J. Mod. Opt.* **60**, 1050–1059 (2013).

147. Y. E. Geints, A. A. Zemlyanov, and E. K. Panina, "Photonic nanojet calculations in layered radially inhomogeneous micrometer-sized spherical particles," *J. Opt. Soc. Am. B* **28**, 1825–1830 (2011).

148. L. Han, Y. Han, G. Gouesbet, J. Wang, and G. Gréhan, "Photonic jet generated by spheroidal particle with Gaussian-beam illumination," *J. Opt. Soc. Am. B* **31**, 1476–1483 (2014).

149. D. Grojo, N. Sandeau, L. Boarino, C. Constantinescu, N. D. Leo, M. Laus, and K. Sparnacci, "Bessel-like photonic nanojets from core-shell sub-wavelength spheres," *Opt. Lett.* **39**, 3989–3992 (2014).

150. Y. Shen, L. V. Wang, and J. Shen, "Ultralong photonic nanojet formed by a two-layer dielectric microsphere," *Opt. Lett.* **39**, 4120–4123 (2014).

151. C. Liu and L. Chang, "Optik Photonic nanojet modulation by elliptical microcylinders," *Opt. Int. J. Light Electron Opt.* **125**, 4043–4046 (2014).

152. B. B. Xu, W. X. Jiang, G. X. Yu, and T. J. Cui, "Annular focusing lens based on transformation optics," *J. Opt. Soc. Am. A* **31**, 1135–1140 (2014).

153. Y. E. Geints, A. A. Zemlyanov, and E. K. Panina, "Photonic jets from resonantly excited transparent dielectric microspheres," *J. Opt. Soc. Am. B* **29**, 758–762 (2012).

154. A. Heifetz, J. J. Simpson, S. Kong, A. Taflove, and V. Backman, "Subdiffraction optical resolution of a gold nanosphere located within the nanojet of a Mie-resonant dielectric microsphere," *Opt. Express* **15**, 17334–17342 (2007).

155. D. A. Kozlov and V. V. Kotlyar, "Resonant laser focus light by uniformity dielectric microcylinder," *Comput. Opt.* **38**, 393–396 (2014).

156. A. V. Boriskin, S. V. Boriskina, A. Rolland, R. Sauleau, and A. I. Nosich, "Test of the FDTD accuracy in the analysis of the scattering resonances associated with high-Q whispering-gallery modes of a circular cylinder," *J. Opt. Soc. Am. A* **25**, 1169–1173 (2008).

157. I. Mahariq and H. Kurt, "On- and off-optical-resonance dynamics of dielectric microcylinders under plane wave illumination," *J. Opt. Soc. Am. B* **32**, 1022–1030 (2015).

158. R. B. Vaganov and B. Z. Katsenelenbaum, *Fundamentals of the Theory of Diffraction* (1982).

159. V. B. Braginsky, M. L. Gorodetsky, and V. S. Ilchenko, "Quality-factor and nonlinear properties of optical whispering-gallery modes," *Phys. Lett. A* **137**, 393–397 (1989).

160. S. Thongrattanasiri and V. A. Podolskiy, "Hypergratings: nanophotonics in planar anisotropic metamaterials," *Opt. Lett.* **34**, 890–892 (2009).

161. K. J. Webb and M. Yang, "Subwavelength imaging with a multilayer silver film structure," *Opt. Lett.* **31**, 2130–2132 (2006).

162. H. Liu, Shivanand, and K. J. Webb, "Subwavelength imaging with nonmagnetic anisotropic bilayers," *Opt. Lett.* **34**, 2243–2245 (2009).

163. B. D. F. Casse, W. T. Lu, R. K. Banyal, Y. J. Huang, S. Selvarasah, M. R. Dokmeci, C. H. Perry, and S. Sridhar, "Imaging with subwavelength resolution by a generalized superlens at infrared wavelengths," *Opt. Lett.* **34**, 1994–1996 (2009).

164. I. Tsukerman, "Superfocusing by nanoshells," *Opt. Lett.* **34**, 1057–1059 (2009).

165. P. C. Ingrey, K. I. Hopcraft, O. French, and E. Jakeman, "Perfect lens with not so perfect boundaries," *Opt. Lett.* **34**, 1015–1017 (2009).

166. E. A. Ray, M. J. Hampton, and R. Lopez, "Simple demonstration of visible evanescent-wave enhancement with far-field detection," *Opt. Lett.* **34**, 2048–2050 (2009).

167. Z. Cao, Y. Jiang, Q. Shen, X. Dou, and Y. Chen, "Exact analytical method for planar optical waveguides with arbitrary index profile," *J. Opt. Soc. Am. A* **16**, 2209–2212 (1999).

168. M.-S. Chung and C.-M. Kim, "General eigenvalue equations for optical planar waveguides with arbitrarily graded-index profiles," *J. Light. Technol.* **18**, 878–885 (2000).

169. W. Miller Jr, *Symmetry and Separation of Variables* (Addison-Wesley Publishing Co., Inc., Reading, MA, 1977).

170. G. A. Korn and T. M. Korn, *Mathematical Handbook for Scientists and Engineers: Definitions, Theorems, and Formulas for Reference and Review* (Courier Corporation, 2000).

171. Y. R. Triandaphilov and V. V. Kotlyar, "A photonic crystal Mikaelian lens," *Opt. Mem. Neural Networks Information Opt.* **17**, 1–7 (2008).

172. J.-J. He and D. Liu, "Wavelength switchable semiconductor laser using half-wave V-coupled cavities," *Opt. Express* **16**, 3896–3911 (2008).

173. X. Lin, D. Liu, and J.-J. He, "Design and analysis of 2×2 half-wave waveguide couplers," *Appl. Opt.* **48**, F18–F23 (2009).

174. A. L. Mikaelian, "Application of stratified medium for waves focusing," *Dokl. Akad. Nauk SSSR* **81**, 569–571 (1951).

175. M. I. Kotlyar, Y. R. Triandaphilov, A. A. Kovalev, V. A. Soifer, M. V. Kotlyar, and L. O'Faolain, "Photonic crystal lens for coupling two waveguides," *Appl. Opt.* **48**, 3722–3730 (2009).

176. U. Leonhardt, "Perfect imaging without negative refraction," *New J. Phys.* **11**, 093040 (2009).

177. Y. N. Feld and L. S. Benenson, Antenna and Fiber Design, Izdatelstvo VVIA imeni Zhukovskogo, 1959, vol. 2.

178. E. G. Zelkin and R. A. Petrova, *Lens Antennas* (Sov. Radio, 1974).

179. M. S. Zhuk and U. B. Molotschkov, *The Design of Scanning and Widebandwidth Lens Antennas and Fibers* (Energia, 1973).

180. A. L. Mikaelian and A. M. Prokhorov, "Self-focusing media with variable index of refraction," in *Progress in Optics* (1980), Vol. 17, pp. 279–345.

181. A. L. Mikaelian, "General method of inhomogeneous media calculation by the given ray traces," *Dokl. Akad. Nauk SSSR* **83**, 219 (1952).

182. A. L. Mikaelian, "Method of calculation of inversion problem of geometrical optics," *Dokl. Akad. Nauk SSSR* **86**, 963 (1952).

183. A. L. Mikaelian, "Using of coordinate system for calculation of media characteristics for the given ray traces, Dokl," *Dokl. Akad. Nauk SSSR* **86**, 1101 (1952).

184. E. G. Rawson, D. R. Herriott, and J. McKenna, "Analysis of refractive index distributions in cylindrical, graded-index glass rods (GRIN Rods) used as image relays," *Appl. Opt.* **9**, 753–760 (1970).

185. C. Bao, C. Gómez-Reino, and M. V. Pérez, "Gradient-index tapered hyperbolic secant planar waveguide for focusing, collimation, and beam-size control," *J. Opt. Soc. Am. A* **14**, 1754–1759 (1997).

186. J. Liñares and C. Gómez-Reino, "Optical propagator in a graded-index medium with a hyperbolic secant refractive-index profile," *Appl. Opt.* **33**, 3427 (1994).

187. W. Streifer and C. N. Kurtz, "Scalar analysis of radially inhomogeneous guiding media," *J. Opt. Soc. Am.* **57**, 779 (1967).

188. E. T. Kornhauser and A. D. Yaghjian, "Modal solution of a point source in a strongly focusing medium," *Radio Sci.* **2**, 299–310 (1967).

189. Y. Silberberg and U. Levy, "Modal treatment of an optical fiber with a modified hyperbolic secant index distribution," *J. Opt. Soc. Am.* **69**, 960 (1979).

190. C. A. Van Duin, J. Boersma, and F. W. Sluijter, "TM-modes in a planar optical waveguide with a graded index of the symmetric Epstein type," *Wave Motion* **8**, 175–190 (1986).

191. D. W. Hewak and J. W. Y. Lit, "Solution deposited optical waveguide lens," *Appl. Opt.* **28**, 4190 (1989).

192. J. M. Rivas-Moscoso, D. Nieto, C. Gómez-Reino, and C. R. Fernández-Pousa, "Focusing of light by zone plates in Selfoc gradient-index lenses," *Opt. Lett.* **28**, 2180 (2003).

193. I. V Minin, O. V Minin, Y. R. Triandaphilov, and V. V. Kotlyar, "Subwavelength diffractive photonic crystal lens," *Prog. Electromagn. Res. B* **7**, 257–264 (2008).

194. S. Kameda, A. Mizutani, and H. Kikuta, "Numerical study on a micro prism and micro lenses with metal-dielectric multilayered structures," *J. Opt. Soc. Am. A* **27**, 749 (2010).

195. J. J. Schwartz, S. Stavrakis, and S. R. Quake, "Colloidal lenses allow high-temperature single-molecule imaging and improve fluorophore photostability," *Nat. Nanotechnol.* **5**, 127–132 (2010).

196. Y. Xu, R. K. Lee, and A. Yariv, "Adiabatic coupling between conventional dielectric waveguides and waveguides with discrete translational symmetry," *Opt. Lett.* **25**, 755 (2000).

197. A. Mekis and J. D. Joannopoulos, "Tapered couplers for efficient interfacing between dielectric and photonic crystal waveguides," *J. Light. Technol.* **19**, 861–865 (2001).

198. T. D. Happ, M. Kamp, and A. Forchel, "Photonic crystal tapers for ultracompact mode conversion," *Opt. Lett.* **26**, 1102 (2001).

199. A. Talneau, P. Lalanne, M. Agio, and C. M. Soukoulis, "Low-reflection photonic-crystal taper for efficient coupling between guide sections of arbitrary widths," *Opt. Lett.* **27**, 1522 (2002).

200. V. R. Almeida, R. R. Panepucci, and M. Lipson, "Nanotaper for compact mode conversion," *Opt. Lett.* **28**, 1302–1304 (2003).

201. P. Bienstman, S. Assefa, S. G. Johnson, J. D. Joannopoulos, G. S. Petrich, and L. A. Kolodziejski, "Taper structures for coupling into photonic crystal slab waveguides," *J. Opt. Soc. Am. B* **20**, 1817 (2003).

202. S. McNab, N. Moll, and Y. Vlasov, "Ultra-low loss photonic integrated circuit with membrane-type photonic crystal waveguides," *Opt. Express* **11**, 2927 (2003).

203. P. E. Barclay, K. Srinivasan, and O. Painter, "Design of photonic crystal waveguides for evanescent coupling to optical fiber tapers and integration with high-Q cavities," *J. Opt. Soc. Am. B* **20**, 2274–2284 (2003).

204. R. Orobtchouk, A. Layadi, H. Gualous, D. Pascal, A. Koster, and S. Laval, "High-efficiency light coupling in a submicrometric silicon-on-insulator waveguide," *Appl. Opt.* **39**, 5773 (2000).

205. S. Lardenois, D. Pascal, L. Vivien, E. Cassan, S. Laval, R. Orobtchouk, M. Heitzmann, N. Bouzaida, and L. Mollard, "Low-loss submicrometer silicon-on-insulator rib waveguides and corner mirrors," *Opt. Lett.* **28**, 1150–1152 (2003).

206. D. Taillaert, F. Van Laere, M. Ayre, W. Bogaerts, D. Van Thourhout, P. Bienstman, and R. Baets, "Grating couplers for coupling between optical fibers and nanophotonic waveguides," *Jpn. J. Appl. Phys.* **45**, 6071–6077 (2006).

207. F. Van Laere, G. Roelkens, M. Ayre, J. Schrauwen, D. Taillaert, D. Van Thourhout, T. F. Krauss, and R. Baets, "Compact and highly efficient grating couplers between optical fiber and nanophotonic waveguides," *J. Light. Technol.* **25**, 151–156 (2007).

208. B. L. Bachim, O. O. Ogunsola, and T. K. Gaylord, "Optical-fiber-to-waveguide coupling using carbon-dioxide-laser-induced long-period fiber gratings," *Opt. Lett.* **30**, 2080–2082 (2005).

209. D. W. Prather, J. Murakowski, S. Shi, S. Venkataraman, A. Sharkawy, C. Chen, and D. Pustai, "High-efficiency coupling structure for a single-line-defect photonic-crystal waveguide," *Opt. Lett.* **27**, 1601 (2002).

210. H.-J. Kim, S.-G. Lee, B.-H. O, S.-G. Park, and E. H. Lee, "High efficiency coupling technique for photonic crystal waveguides using a waveguide lens," in *Frontiers in Optics* (OSA, 2003), p. MT68.

211. J. C. W. Corbett and J. R. Allington-Smith, "Coupling starlight into single-mode photonic crystal fiber using a field lens," *Opt. Express* **13**, 6527 (2005).

212. D. Michaelis, C. Wächter, S. Burger, L. Zschiedrich, and A. Bräuer, "Micro-optically assisted high-index waveguide coupling," *Appl. Opt.* **45**, 1831–1838 (2006).

213. G.-J. Kong, J. Kim, H.-Y. Choi, J. E. Im, B.-H. Park, U.-C. Paek, and B. H. Lee, "Lensed photonic crystal fiber obtained by use of an arc discharge," *Opt. Lett.* **31**, 894–896 (2006).

214. A. L. Pokrovsky and A. L. Efros, "Lens based on the use of left-handed materials," *Appl. Opt.* **42**, 5701 (2003).

215. C. Li, J. M. Holt, and A. L. Efros, "Far-field imaging by the Veselago lens made of a photonic crystal," *J. Opt. Soc. Am. B* **23**, 490 (2006).

216. C. Y. Li, J. M. Holt, and A. L. Efros, "Imaging by the Veselago lens based upon a two-dimensional photonic crystal with a triangular lattice," *J. Opt. Soc. Am. B* **23**, 963 (2006).

217. T. Geng, T. Liu, and S. Zhuang, "All angle negative refraction with the effective phase index of-1," *Chinese Opt. Lett.* **5**, 361–363 (2007).

218. T. Asatsuma and T. Baba, "Aberration reduction and unique light focusing in a photonic crystal negative refractive lens," *Opt. Express* **16**, 8711–8719 (2008).

219. N. Fabre, L. Lalouat, B. Cluzel, X. Melique, D. Lippens, F. de Fornel, and O. Vanbesien, "Measurement of a flat lens focusing in a 2D photonic crystal at optical wavelength," in 2008 *Conference on Lasers and Electro-Optics* (IEEE, 2008), pp. 1–2.

220. P.-G. Luan and K.-D. Chang, "Photonic-crystal lens coupler using negative refraction," *PIERS Online* **3**, 91–95 (2007).

221. S. Haxha and F. AbdelMalek, "A novel design of photonic crystal lens based on negative refractive index," *PIERS Online* **4**, 296–300 (2008).

222. Z. Lu, S. Shi, C. A. Schuetz, J. A. Murakowski, and D. W. Prather, "Three-dimensional photonic crystal flat lens by full 3D negative refraction," *Opt. Express* **13**, 5592 (2005).

223. Z. Lu, S. Shi, C. A. Schuetz, and D. W. Prather, "Experimental demonstration of negative refraction imaging in both amplitude and phase," *Opt. Express* **13**, 2007 (2005).

224. J. P. Hugonin, P. Lalanne, T. P. White, and T. F. Krauss, "Coupling into slow-mode photonic crystal waveguides," *Opt. Lett.* **32**, 2638 (2007).

225. E. Pshenay-Severin, C. C. Chen, T. Pertsch, M. Augustin, A. Chipouline, and A. Tunnermann, "Photonic crystal lens for photonic crystal waveguide coupling," in *2006 Conference on Lasers and Electro-Optics and 2006 Quantum Electronics and Laser Science Conference* (2006), pp. 1–2.

226. Y. R. Triandafilov and V. V. Kotlyar, "A photonic crystal Mikaelian lens," *Comput. Opt.* **31**, 27–31 (2007).

227. A. L. Mikaelian, "Harnessing medium properties for wave focusing," *Dokl. Akad. Nauk SSSR* **81**, 2406–2415 (1951).

228. L. O'Faolain, X. Yuan, D. McIntyre, S. Thoms, H. Chong, R. M. De La Rue, and T. F. Krauss, "Low-loss propagation in photonic crystal waveguides," *Electron. Lett.* **42**, 1454 (2006).

229. D. M. Beggs, L. O'Faolain, and T. F. Krauss, "Accurate determination of the functional hole size in photonic crystal slabs using optical methods," *Photonics Nanostructures - Fundam. Appl.* **6**, 213–218 (2008).

230. J. Y. Lee, B. H. Hong, W. Y. Kim, S. K. Min, Y. Kim, M. V. Jouravlev, R. Bose, K. S. Kim, I.-C. Hwang, L. J. Kaufman, C. W. Wong, P. Kim, and K. S. Kim, "Near-field focusing and magnification through self-assembled nanoscale spherical lenses," *Nature* **460**, 498–501 (2009).

231. D. J. Goldstein, "Resolution in light microscopy studied by computer simulation," *J. Microsc.* **166**, 185–197 (1992).

232. K. Karrai, X. Lorenz, and L. Novotny, "Enhanced reflectivity contrast in confocal solid immersion lens microscopy," *Appl. Phys. Lett.* **77**, 3459 (2000).

233. F. H. Köklü, S. B. Ippolito, B. B. Goldberg, and M. S. Ünlü, "Subsurface microscopy of integrated circuits with angular spectrum and polarization control," *Opt. Lett.* **34**, 1261 (2009).

234. D. M. Karabacak, K. L. Ekinci, C. H. Gan, G. J. Gbur, M. S. Ünlü, S. B. Ippolito, B. B. Goldberg, and P. S. Carney, "Diffraction of evanescent waves and nanomechanical displacement detection," *Opt. Lett.* **32**, 1881 (2007).

235. A. Di Falco, S. C. Kehr, and U. Leonhardt, "Luneburg lens in silicon photonics," *Opt. Express* **19**, 5156 (2011).

236. T. Zentgraf, Y. Liu, M. H. Mikkelsen, J. Valentine, and X. Zhang, "Plasmonic luneburg and eaton lenses," *Nat. Nanotechnol.* **6**, 151–155 (2011).

237. H. Wu, L. Y. Jiang, W. Jia, and X. Y. Li, "Imaging properties of an annular photonic crystal slab for both TM-polarization and TE-polarization," *J. Opt.* **13**, 095103 (2011).

238. X. Zhang and Z. Liu, "Superlenses to overcome the diffraction limit," *Nat. Mater.* **7**, 435–441 (2008).

239. J. Rho, Z. Ye, Y. Xiong, X. Yin, Z. Liu, H. Choi, G. Bartal, and X. Zhang, "Spherical hyperlens for two-dimensional sub-diffractional imaging at visible frequencies," *Nat. Commun.* **1**, 143 (2010).

240. W. Chen, R. L. Nelson, and Q. Zhan, "Efficient miniature circular polarization analyzer design using hybrid spiral plasmonic lens," *Opt. Lett.* **37**, 1442 (2012).
241. S. Huang, H. Wang, K.-H. Ding, and L. Tsang, "Subwavelength imaging enhancement through a three-dimensional plasmon superlens with rough surface," *Opt. Lett.* **37**, 1295–1297 (2012).
242. F. M. Huang, N. Zheludev, Y. Chen, and F. Javier Garcia de Abajo, "Focusing of light by a nanohole array," *Appl. Phys. Lett.* **90**, 091119 (2007).
243. A. Vainrub, O. Pustovyy, and V. Vodyanoy, "Resolution of 90 nm ($\lambda/5$) in an optical transmission microscope with an annular condenser," *Opt. Lett.* **31**, 2855 (2006).
244. Y. Kuznetsova, A. Neumann, and S. R. J. Brueck, "Solid-immersion imaging interferometric nanoscopy to the limits of available frequency space," *J. Opt. Soc. Am. A* **29**, 772 (2012).
245. V. A. Zverev, *Radiooptics* (Soviet Radio, 1975).
246. V. V. Kotlyar, A. A. Kovalev, and V. A. Soifer, "Subwavelength focusing with a Mikaelian planar lens," *Opt. Mem. Neural Networks* **19**, 273–278 (2010).
247. C. J. Handmer, C. M. de Sterke, R. C. McPhedran, L. C. Botten, M. J. Steel, and A. Rahmani, "Blazing evanescent grating orders: a spectral approach to beating the Rayleigh limit," *Opt. Lett.* **35**, 2846–2848 (2010).
248. S. Thongrattanasiri, N. A. Kuhta, M. D. Escarra, A. J. Hoffman, C. F. Gmachl, and V. A. Podolskiy, "Analytical technique for subwavelength far field imaging," *Appl. Phys. Lett.* **97**, 101103 (2010).
249. K. R. Chen, "Focusing of light beyond the diffraction limit of half the wavelength," *Opt. Lett.* **35**, 3763–3765 (2010).
250. S. Ishii, A. Kildishev, V. Shalaev, K.-P. Chen, and V. P. Drachev, "Gold nanoslit lenses," in *CLEO:2011 – Laser Applications to Photonic Applications* (OSA, 2011), p. QThQ2.
251. G. Ren, C. Wang, Z. Zhao, X. Tao, and X. Luo, "Off-axis characteristic of subwavelength focusing in anisotropic metamaterials," *J. Opt. Soc. Am. B* **29**, 3103–3108 (2012).
252. V. R. Almeida, Q. Xu, C. A. Barrios, and M. Lipson, "Guiding and confining light in void nanostructure," *Opt. Lett.* **29**, 1209–1211 (2004).
253. H. Kurt and D. S. Citrin, "Graded index photonic crystals," *Opt. Express* **15**, 1240–1253 (2007).
254. Z. Cheng, X. Chen, C. Y. Wong, K. Xu, C. K. Y. Fung, Y. M. Chen, and H. K. Tsang, "Focusing subwavelength grating coupler for mid-infrared suspended membrane waveguide," *Opt. Lett.* **37**, 1217–1219 (2012).
255. G. Ren, Z. Lai, C. Wang, Q. Feng, L. Liu, K. Liu, and X. Luo, "Subwavelength focusing of light in the planar anisotropic metamaterials with zone plates," *Opt. Express* **18**, 18151–18157 (2010).
256. G. Li, J. Li, and K. W. Cheah, "Subwavelength focusing using a hyperbolic medium with a single slit," *Appl. Opt.* **50**, G27–G30 (2011).
257. V. V. Kotlyar, A. A. Kovalev, A. G. Nalimov, and S. S. Stafeev, "High resolution through graded-index microoptics," *Adv. Opt. Technol.* **2012**, 1–9 (2012).
258. V. V. Kotlyar, R. V. Skidanov, S. N. Khonina, and V. A. Soifer, "Hypergeometric modes," *Opt. Lett.* **32**, 742–744 (2007).
259. E. Karimi, G. Zito, B. Piccirillo, L. Marrucci, and E. Santamato, "Hypergeometric-Gaussian modes," *Opt. Lett.* **32**, 3053–3055 (2007).
260. V. V. Kotlyar, A. A. Kovalev, R. V. Skidanov, S. N. Khonina, and J. Turunen, "Generating hypergeometric laser beams with a diffractive optical element," *Appl. Opt.* **47**, 6124–6133 (2008).
261. J. Chen, "Production of confluent hypergeometric beam by computer-generated hologram," *Opt. Eng.* **50**, 024201 (2011).
262. B. De Lima Bernardo and F. Moraes, "Data transmission by hypergeometric modes through a hyperbolic-index medium," *Opt. Express* **19**, 11264–11270 (2011).

263. J. Li and Y. Chen, "Propagation of confluent hypergeometric beam through uniaxial crystals orthogonal to the optical axis," *Opt. Laser Technol.* **44**, 1603–1610 (2012).

264. U. Levy, M. Nezhad, H.-C. Kim, C.-H. Tsai, L. Pang, and Y. Fainman, "Implementation of a graded-index medium by use of subwavelength structures with graded fill factor," *J. Opt. Soc. Am. A* **22**, 724–733 (2005).

265. A. E. Siegman, *Lasers* (1986).

266. P. K. Tien, J. P. Gordon, and J. R. Whinnery, "Focusing of a light beam of Gaussian field distribution in continuous and periodic lens-like media," *Proc. IEEE* **53**, 129–136 (1965).

267. M. Newstein and B. Rudman, "LaGuerre-Gaussian periodically focusing beams in a quadratic index medium," *IEEE J. Quantum Electron.* **23**, 481–482 (1987).

268. J. C. Gutiérrez-Vega and M. A. Bandres, "Ince–Gaussian beams in a quadratic-index medium," *J. Opt. Soc. Am. A* **22**, 306–309 (2005).

269. M. A. Bandres and J. C. Gutiérrez-Vega, "Airy-Gauss beams and their transformation by paraxial optical systems," *Opt. Express* **15**, 16719–16728 (2007).

270. D. M. Deng, "Propagation of Airy-Gaussian beams in a quadratic-index medium," *Eur. Phys. J. D* **65**, 553–556 (2011).

271. B. E. A. Saleh and M. C. Teich, *Fundamentals of Photonics* (John Wiley & Sons, 1991).

272. P. N. Dyachenko, V. S. Pavelyev, and V. A. Soifer, "Graded photonic quasicrystals," *Opt. Lett.* **37**, 2178–2180 (2012).

273. V. V. Kotlyar, A. A. Kovalev, and A. G. Nalimov, "Planar gradient-index hyperbolic secant lens for subwavelength focusing and super resolution imaging," *Opt.* **1**, 1–10 (2012).

274. S. N. Khonina, V. V. Kotlyar, V. A. Soifer, P. Pääkkönen, J. Simonen, and J. Turunen, "An analysis of the angular momentum of a light field in terms of angular harmonics," *J. Mod. Opt.* **48**, 1543–1557 (2001).

275. D. Mendlovic and H. M. Ozaktas, "Fractional Fourier transforms and their optical implementation II," *J. Opt. Soc. Am. A* **10**, 1875–1881 (1993).

276. A. W. Lohmann, "Image rotation, Wigner rotation, and the fractional Fourier transform," *J. Opt. Soc. Am. A* **10**, 2181–2186 (1993).

277. S. N. Khonina, A. S. Striletz, A. A. Kovalev, and V. V. Kotlyar, "Propagation of laser vortex beams in a parabolic optical fiber," in *Optical Technologies for Telecommunications 2009*, V. A. Andreev, V. A. Burdin, O. G. Morozov, and A. H. Sultanov, eds. (2009), Vol. 7523, p. 75230B.

278. I. S. Gradshteyn and I. M. Ryzhik, *Table of Integrals, Series, and Products* (2014).

279. J. R. Hensler, "Method of producing a refractive index gradient in glass," (1975).

280. D. B. Keck and R. Olshansky, "Optical waveguide having optimal index gradient," (1975).

281. A. Dupuis, N. Guo, B. Gauvreau, A. Hassani, E. Pone, F. Boismenu, and M. Skorobogatiy, "Guiding in the visible with "colorful" solid-core Bragg fibers," *Opt. Lett.* **32**, 2882–2884 (2007).

282. L. Liu, G. Liu, Y. Xiong, J. Chen, W. Li, and Y. Tian, "Fabrication of X-ray imaging zone plates by e-beam and X-ray lithography," *Microsyst. Technol.* **16**, 1315–1321 (2010).

283. N. Kamijo, Y. Suzuki, H. Takano, S. Tamura, M. Yasumoto, A. Takeuchi, and M. Awaji, "Microbeam of 100 keV x ray with a sputtered-sliced Fresnel zone plate," *Rev. Sci. Instrum.* **74**, 5101–5104 (2003).

284. V. V. Kotlyar and O. K. Zalyalov, "Design of diffractive optical elements modulating polarization," *Optik (Stuttg)*. **103**, 125–130 (1996).

285. Z. Bomzon, V. Kleiner, and E. Hasman, "Pancharatnam–Berry phase in space-variant polarization-state manipulations with subwavelength gratings," *Opt. Lett.* **26**, 1424–1426 (2001).

286. Z. Bomzon, G. Biener, V. Kleiner, and E. Hasman, "Radially and azimuthally polarized beams generated by space-variant dielectric subwavelength gratings," *Opt. Lett.* **27**, 285–287 (2002).

287. A. Niv, G. Biener, V. Kleiner, and E. Hasman, "Formation of linearly polarized light with axial symmetry by use of space-variant subwavelength gratings," *Opt. Lett.* **28**, 510–512 (2003).

288. G. M. Lerman and U. Levy, "Generation of a radially polarized light beam using space-variant subwavelength gratings at 1064 nm," *Opt. Lett.* **33**, 2782–2784 (2008).

289. A. G. Nalimov, L. O'Faolain, S. S. Stafeev, M. I. Shanina, and V. V. Kotlyar, "Reflected four-zones subwavelength microoptics element for polarization conversion from linear to radial," *Comput. Opt.* **38**, 229–236 (2014).

290. S. S. Stafeev and V. V. Kotlyar, "Special aspects of subwavelength focal spot measurement using near-field optical microscope," *Comput. Opt.* **37**, 332–340 (2013).

291. R. Dorn, S. Quabis, and G. Leuchs, "Sharper focus for a radially polarized light beam," *Phys. Rev. Lett.* **91**, 233901 (2003).

292. X. Li, P. Venugopalan, H. Ren, M. Hong, and M. Gu, "Super-resolved pure-transverse focal fields with an enhanced energy density through focus of an azimuthally polarized first-order vortex beam," *Opt. Lett.* **39**, 5961–5964 (2014).

293. T. Kämpfe, P. Sixt, D. Renaud, A. Lagrange, F. Perrin, and O. Parriaux, "Segmented subwavelength silicon gratings manufactured by high productivity microelectronic technologies for linear to radial/azimuthal polarization conversion," *Opt. Eng.* **53**, 107105 (2014).

294. Z. Ghadyani, I. Vartiainen, I. Harder, W. Iff, A. Berger, N. Lindlein, and M. Kuittinen, "Concentric ring metal grating for generating radially polarized light," *Appl. Opt.* **50**, 2451–2457 (2011).

295. S. S. Stafeev, L. O'Faolain, V. V. Kotlyar, and A. G. Nalimov, "Tight focus of light using micropolarizer and microlens," Appl. Opt. **54**, 4388–4394 (2015).

296. L. . Helseth, "Optical vortices in focal regions," *Opt. Commun.* **229**, 85–91 (2004).

297. Z. Zhang, J. Pu, and X. Wang, "Tight focusing of radially and azimuthally polarized vortex beams through a uniaxial birefringent crystal," *Appl. Opt.* **47**, 1963–1967 (2008).

298. V. V. Kotlyar and A. A. Kovalev, "Nonparaxial propagation of a Gaussian optical vortex with initial radial polarization," *J. Opt. Soc. Am. A* **27**, 372–380 (2010).

299. X. Hao, C. Kuang, T. Wang, and X. Liu, "Phase encoding for sharper focus of the azimuthally polarized beam," *Opt. Lett.* **35**, 3928–3930 (2010).

300. S. Wang, X. Li, J. Zhou, and M. Gu, "Ultralong pure longitudinal magnetization needle induced by annular vortex binary optics," *Opt. Lett.* **39**, 5022–5025 (2014).

301. Z. Nie, W. Ding, D. Li, X. Zhang, Y. Wang, and Y. Song, "Spherical and sub-wavelength longitudinal magnetization generated by 4pi tightly focusing radially polarized vortex beams," *Opt. Express* **23**, 690–701 (2015).

302. Z. Chen and D. Zhao, "4Pi focusing of spatially modulated radially polarized vortex beams," *Opt. Lett.* **37**, 1286–1288 (2012).

303. S. V. Alferov, S. V. Karpeev, S. N. Khonina, and O. Y. Moiseev, "Experimental study of focusing of inhomogeneously polarized beams generated using sector polarizing plates," *Comput. Opt.* **38**, 57–64 (2014).

304. Q. Zhan, "Cylindrical vector beams: from mathematical concepts to applications," *Adv. Opt. Photonics* **1**, 1–57 (2009).

305. F. Qin, K. Huang, J. Wu, J. Jiao, X. Luo, C. Qiu, and M. Hong, "Shaping a subwavelength needle with ultra-long focal length by focusing azimuthally polarized light," *Sci. Rep.* **5**, 09977 (2015).

306. P. Suresh, C. Mariyal, K. B. Rajesh, T. V. S. Pillai, and Z. Jaroszewicz, "Generation of a strong uniform transversely polarized nondiffracting beam using a high-numerical-aperture lens axicon with a binary phase mask," *Appl. Opt.* **52**, 849–853 (2013).

307. Y. Jiang, X. Li, and M. Gu, "Generation of sub-diffraction-limited pure longitudinal magnetization by the inverse Faraday effect by tightly focusing an azimuthally polarized vortex beam," *Opt. Lett.* **38**, 2957–2960 (2013).

308. G. Therese Anita, N. Umamageswari, K. Prabakaran, T. V. S. Pillai, and K. B. Rajesh, "Effect of coma on tightly focused cylindrically polarized vortex beams," *Opt. Laser Technol.* **76**, 1–5 (2016).

309. B. Ndagano, H. Sroor, M. McLaren, C. Rosales-Guzmán, and A. Forbes, "Beam quality measure for vector beams," *Opt. Lett.* **41**, 3407–3410 (2016).

310. A. P. Porfirev, A. V. Ustinov, and S. N. Khonina, "Polarization conversion when focusing cylindrically polarized vortex beams," *Sci. Rep.* **6**, 6 (2016).

311. G. Machavariani, Y. Lumer, I. Moshe, A. Meir, and S. Jackel, "Efficient extracavity generation of radially and azimuthally polarized beams," *Opt. Lett.* **32**, 1468–70 (2007).

312. G. Machavariani, Y. Lumer, I. Moshe, A. Meir, and S. Jackel, "Spatially-variable retardation plate for efficient generation of radially- and azimuthally-polarized beams," *Opt. Commun.* **281**, 732–738 (2008).

313. R. Imai, N. Kanda, T. Higuchi, Z. Zheng, K. Konishi, and M. Kuwata-Gonokami, "Terahertz vector beam generation using segmented nonlinear optical crystals with threefold rotational symmetry," *Opt. Express* **20**, 21896–21904 (2012).

314. Z. Man, C. Min, Y. Zhang, Z. Shen, and X.-C. Yuan, "Arbitrary vector beams with selective polarization states patterned by tailored polarizing films," *Laser Phys.* **23**, 105001 (2013).

315. S. S. Stafeev, A. G. Nalimov, M. V. Kotlyar, D. Gibson, S. Song, L. O'Faolain, and V. V. Kotlyar, "Microlens-aided focusing of linearly and azimuthally polarized laser light," *Opt. Express* **24**, 29800–29813 (2016).

316. V. V. Kotlyar, S. S. Stafeev, M. V. Kotlyar, A. G. Nalimov, and L. O'Faolain, "Subwavelength micropolarizer in a gold film for visible light," *Appl. Opt.* **55**, 5025–5032 (2016).

317. T. G. Jabbour and S. M. Kuebler, "Particle-swarm optimization of axially superresolving binary-phase diffractive optical elements," *Opt. Lett.* **33**, 1533–1535 (2008).

318. S. N. Khonina and S. G. Volotovsky, "Controlling the contribution of the electric field components to the focus of a high-aperture lens using binary phase structures," *J. Opt. Soc. Am. A* **27**, 2188–2197 (2010).

319. Z. Man, C. Min, S. Zhu, and X.-C. Yuan, "Tight focusing of quasi-cylindrically polarized beams," *J. Opt. Soc. Am. A. Opt. Image Sci. Vis.* **31**, 373–8 (2014).

320. N. Yu and F. Capasso, "Flat optics with designer metasurfaces," *Nat. Mater.* **13**, 139–150 (2014).

321. A. V. Kildishev, A. Boltasseva, and V. M. Shalaev, "Planar photonics with metasurfaces," *Science (80-.).* **339**, 1232009 (2013).

322. G. M. Lerman and U. Levy, "Radial polarization interferometer," *Opt. Express* **17**, 23234–23246 (2009).

323. Z. Xie, J. He, X. Wang, S. Feng, and Y. Zhang, "Generation of terahertz vector beams with a concentric ring metal grating and photo-generated carriers," *Opt. Lett.* **40**, 359–362 (2015).

324. J. Lin, P. Genevet, M. A. Kats, N. Antoniou, and F. Capasso, "Nanostructured holograms for broadband manipulation of vector beams," *Nano Lett.* **13**, 4269–4274 (2013).

325. P. Genevet and F. Capasso, "Holographic optical metasurfaces: a review of current progress," *Reports Prog. Phys.* **78**, 024401 (2015).

326. S. S. Stafeev, M. V. Kotlyar, L. O'Faolain, A. G. Nalimov, and V. V. Kotlyar, "A four-zone transmission azimuthal micropolarizer with phase shift," *Comput. Opt.* **40**, 12–18 (2016).

327. A. B. Klemm, D. Stellinga, E. R. Martins, L. Lewis, G. Huyet, L. O'Faolain, and T. F. Krauss, "Experimental high numerical aperture focusing with high contrast gratings," *Opt. Lett.* **38**, 3410–3413 (2013).

328. Y. Yang, W. Wang, P. Moitra, I. I. Kravchenko, D. P. Briggs, and J. Valentine, "Dielectric meta-reflectarray for broadband linear polarization conversion and optical vortex generation," *Nano Lett.* **14**, 1394–1399 (2014).

329. S. Sun, K. Yang, C. Wang, T. Juan, W. T. Chen, C. Y. Liao, Q. He, S. Xiao, W. Kung, G. Guo, L. Zhou, and D. P. Tsai, "High-efficiency broadband anomalous reflection by gradient meta-surfaces," *Nano Lett.* **12**, 6223–6229 (2012).

330. L. Lan, W. Jiang, and Y. Ma, "Three dimensional subwavelength focus by a near-field plate lens," *Appl. Phys. Lett.* **102**, 231119 (2013).

331. L. Verslegers, P. B. Catrysse, Z. Yu, J. S. White, E. S. Barnard, M. L. Brongersma, and S. Fan, "Planar lenses based on nanoscale slit arrays in a metallic film," *Nano Lett.* **9**, 235–238 (2009).

332. F. Aieta, P. Genevet, M. A. Kats, N. Yu, R. Blanchard, Z. Gaburro, and F. Capasso, "Aberration-free ultrathin flat lenses and axicons at telecom wavelengths based on plasmonic metasurfaces," *Nano Lett.* **12**, 4932–4936 (2012).

333. A. Arbabi, Y. Horie, A. J. Ball, M. Bagheri, and A. Faraon, "Subwavelength-thick lenses with high numerical apertures and large efficiency based on high-contrast transmitarrays," *Nat. Commun.* **6**, 7069 (2015).

334. A. Arbabi, Y. Horie, M. Bagheri, and A. Faraon, "Dielectric metasurfaces for complete control of phase and polarization with subwavelength spatial resolution and high transmission," *Nat. Nanotechnol.* **10**, 937–943 (2015).

335. X. Ni, S. Ishii, A. V. Kildishev, and V. M. Shalaev, "Ultra-thin, planar, Babinet-inverted plasmonic metalenses," *Light Sci. Appl.* **2**, e72 (2013).

336. P. R. West, J. L. Stewart, A. V. Kildishev, V. M. Shalaev, V. V. Shkunov, F. Strohkendl, Y. A. Zakharenkov, R. K. Dodds, and R. Byren, "All-dielectric subwavelength metasurface focusing lens," *Opt. Express* **22**, 26212–26221 (2014).

337. D. Lin, P. Fan, E. Hasman, and M. L. Brongersma, "Dielectric gradient metasurface optical elements," *Science (80-.).* **345**, 298–302 (2014).

338. V. V. Kotlyar, A. G. Nalimov, S. S. Stafeev, C. Hu, L. O'Faolain, M. V. Kotlyar, D. Gibson, and S. Song, "Thin high numerical aperture metalens," *Opt. Express* **25**, 8158–8167 (2017).

339. N. M. Litchinitser, "Structured light meets structured matter," *Science (80-.).* **337**, 1054–1055 (2012).

340. Z. Zhao, J. Wang, S. Li, and A. E. Willner, "Metamaterials-based broadband generation of orbital angular momentum carrying vector beams," *Opt. Lett.* **38**, 932–934 (2013).

341. X. Yi, X. Ling, Z. Zhang, Y. Li, X. Zhou, Y. Liu, S. Chen, H. Luo, and S. Wen, "Generation of cylindrical vector vortex beams by two cascaded metasurfaces," *Opt. Express* **22**, 17207–17215 (2014).

342. Z. Zhao, M. Pu, H. Gao, J. Jin, X. Li, X. Ma, Y. Wang, P. Gao, and X. Luo, "Multispectral optical metasurfaces enabled by achromatic phase transition," *Sci. Rep.* **5**, 15781 (2015).

343. J. Sun, X. Wang, T. Xu, Z. A. Kudyshev, A. N. Cartwright, and N. M. Litchinitser, "Spinning light on the nanoscale," *Nano Lett.* **14**, 2726–2729 (2014).

344. E. Karimi, S. A. Schulz, I. De Leon, H. Qassim, J. Upham, and R. W. Boyd, "Generating optical orbital angular momentum at visible wavelengths using a plasmonic metasurface," *Light Sci. Appl.* **3**, e167–e167 (2014).

345. W. Wang, Y. Li, Z. Guo, R. Li, J. Zhang, A. Zhang, and S. Qu, "Ultra-thin optical vortex phase plate based on the metasurface and the angular momentum transformation," *J. Opt.* **17**, 045102 (2015).

346. K. E. Chong, I. Staude, A. James, J. Dominguez, S. Liu, S. Campione, G. S. Subramania, T. S. Luk, M. Decker, D. N. Neshev, I. Brener, and Y. S. Kivshar, "Polarization-independent silicon metadevices for efficient optical wavefront control," *Nano Lett.* **15**, 5369–5374 (2015).

347. M. Decker, I. Staude, M. Falkner, J. Dominguez, D. N. Neshev, I. Brener, T. Pertsch, and Y. S. Kivshar, "High-efficiency dielectric huygens' surfaces," *Adv. Opt. Mater.* **3**, 813–820 (2015).

348. H.-T. Chen, A. J. Taylor, and N. Yu, "A review of metasurfaces: physics and applications," *Reports Prog. Phys.* **79**, 076401 (2016).

349. F. Yue, D. Wen, J. Xin, B. D. Gerardot, J. Li, and X. Chen, "Vector vortex beam generation with a single plasmonic metasurface," *ACS Photonics* **3**, 1558–1563 (2016).

350. S. Mei, M. Q. Mehmood, S. Hussain, K. Huang, X. Ling, S. Y. Siew, H. Liu, J. Teng, A. Danner, and C. W. Qiu, "Flat helical nanosieves," *Adv. Funct. Mater.* **26**, 5255–5262 (2016).

351. S. Kruk, B. Hopkins, I. I. Kravchenko, A. Miroshnichenko, D. N. Neshev, and Y. S. Kivshar, "Broadband highly efficient dielectric metadevices for polarization control," *APL Photonics* **1**, 030801 (2016).

352. L. Wang, S. Kruk, H. Tang, T. Li, I. Kravchenko, D. N. Neshev, and Y. S. Kivshar, "Grayscale transparent metasurface holograms," *Optica* **3**, 1504–1505 (2016).

353. P. Genevet, F. Capasso, F. Aieta, M. Khorasaninejad, and R. Devlin, "Recent advances in planar optics: from plasmonic to dielectric metasurfaces," *Optica* **4**, 139–152 (2017).

354. L. Huang, X. Song, B. Reineke, T. Li, X. Li, J. Liu, S. Zhang, Y. Wang, and T. Zentgraf, "Volumetric generation of optical vortices with metasurfaces," *ACS Photonics* **4**, 338–346 (2017).

355. Y. Liu, Y. Ke, J. Zhou, Y. Liu, H. Luo, S. Wen, and D. Fan, "Generation of perfect vortex and vector beams based on Pancharatnam-Berry phase elements," *Sci. Rep.* **7**, 44096 (2017).

356. W. T. Chen, A. Y. Zhu, M. Khorasaninejad, Z. Shi, V. Sanjeev, and F. Capasso, "Immersion meta-lenses at visible wavelengths for nanoscale imaging," *Nano Lett.* **17**, 3188–3194 (2017).

357. R. C. Devlin, A. Ambrosio, D. Wintz, S. L. Oscurato, A. Y. Zhu, M. Khorasaninejad, J. Oh, P. Maddalena, and F. Capasso, "Spin-to-orbital angular momentum conversion in dielectric metasurfaces," *Opt. Express* **25**, 377–393 (2017).

358. R. C. Devlin, A. Ambrosio, N. A. Rubin, J. P. B. Mueller, and F. Capasso, "Arbitrary spin-to–orbital angular momentum conversion of light," *Science (80-.).* **358**, 896–901 (2017).

359. N. R. Heckenberg, R. McDuff, C. P. Smith, and A. G. White, "Generation of optical phase singularities by computer-generated holograms," *Opt. Lett.* **17**, 221–223 (1992).

360. P. Lalanne and D. Lemercier-Lalanne, "On the effective medium theory of subwavelength periodic structures," *J. Mod. Opt.* **43**, 2063–2086 (1996).

361. A. G. Nalimov and V. V. Kotlyar, "Design of a sector-variant high-numerical-aperture micrometalens," *Optik (Stuttg).* **159**, 9–13 (2018).

362. A. V. Novitsky and D. V. Novitsky, "Negative propagation of vector Bessel beams," *J. Opt. Soc. Am. A* **24**, 2844–2849 (2007).

363. S. Sukhov and A. Dogariu, "On the concept of "tractor beams," *Opt. Lett.* **35**, 3847–3849 (2010).

364. P. B. Monteiro, P. A. M. Neto, and H. M. Nussenzveig, "Angular momentum of focused beams: beyond the paraxial approximation," *Phys. Rev. A* **79**, 033830 (2009).

365. P. Vaveliuk and O. Martinez-Matos, "Negative propagation effect in nonparaxial Airy beams," *Opt. Express* **20**, 26913–26921 (2012).

366. M. V. Berry, "Quantum backflow, negative kinetic energy, and optical retro-propagation," *J. Phys. A Math. Theor.* **43**, 415302 (2010).

367. F. G. Mitri, "Vector spherical quasi-Gaussian vortex beams," *Phys. Rev. E* **89**, 023205 (2014).

368. R. Li, C. Ding, and F. G. Mitri, "Optical spin torque induced by vector Bessel (vortex) beams with selective polarizations on a light-absorptive sphere of arbitrary size," *J. Quant. Spectrosc. Radiat. Transf.* **196**, 53–68 (2017).

369. V. V. Kotlyar and A. G. Nalimov, "A vector optical vortex generated and focused using a metalens," *Comput. Opt.* **41**, 645–654 (2017).

370. F. G. Mitri, "Reverse propagation and negative angular momentum density flux of an optical nondiffracting nonparaxial fractional Bessel vortex beam of progressive waves," *J. Opt. Soc. Am. A* **33**, 1661–1667 (2016).

371. F. G. Mitri, "Counterpropagating nondiffracting vortex beams with linear and angular momenta," *Phys. Rev. A* **88**, 035804 (2013).

372. M. A. Salem and H. Bağcı, "Energy flow characteristics of vector X-Waves," *Opt. Express* **19**, 8526–8532 (2011).

373. B. Chen and J. Pu, "Tight focusing of elliptically polarized vortex beams," *Appl. Opt.* **48**, 1288–1294 (2009).

374. A. Dogariu, S. Sukhov, and J. Sáenz, "Optically induced "negative forces," *Nat. Photonics* **7**, 24–27 (2013).

375. V. Shvedov, A. R. Davoyan, C. Hnatovsky, N. Engheta, and W. Krolikowski, "A long-range polarization-controlled optical tractor beam," *Nat. Photonics* **8**, 846–850 (2014).

376. V. V. Kotlyar, A. A. Kovalev, and A. G. Nalimov, "Energy density and energy flux in the focus of an optical vortex: reverse flux of light energy," *Opt. Lett.* **43**, 2921–2924 (2018).

377. D. Ganic, X. Gan, and M. Gu, "Focusing of doughnut laser beams by a high numerical-aperture objective in free space," *Opt. Express* **11**, 2747–2752 (2003).

378. N. Bokor, Y. Iketaki, T. Watanabe, and M. Fujii, "Investigation of polarization effects for high-numerical-aperture first-order Laguerre-Gaussian beams by 2D scanning with a single fluorescent microbead," *Opt. Express* **13**, 10440–10447 (2005).

379. D. Maluenda, R. Martínez-Herrero, I. Juvells, and A. Carnicer, "Synthesis of highly focused fields with circular polarization at any transverse plane," *Opt. Express* **22**, 6859–6867 (2014).

380. M. Zhang and Y. Yang, "Tight focusing properties of anomalous vortex beams," *Optik (Stuttg).* **154**, 133–138 (2018).

381. Z. Nie, G. Shi, D. Li, X. Zhang, Y. Wang, and Y. Song, "Tight focusing of a radially polarized Laguerre–Bessel–Gaussian beam and its application to manipulation of two types of particles," *Phys. Lett. A* **379**, 857–863 (2015).

382. Z. Xiaoqiang, C. Ruishan, and W. Anting, "Focusing properties of cylindrical vector vortex beams," *Opt. Commun.* **414**, 10–15 (2018).

383. X. Zhao, X. Pang, J. Zhang, and G. Wan, "Transverse focal shift in vortex beams," *IEEE Photonics J.* 10, 6500417 (2018).

384. A. Novitsky, C.-W. Qiu, and H. Wang, "Single gradientless light beam drags particles as tractor beams," *Phys. Rev. Lett.* **107**, 203601 (2011).

385. S. Sukhov and A. Dogariu, "Negative nonconservative forces: optical "tractor beams" for arbitrary objects," *Phys. Rev. Lett.* **107**, 203602 (2011).

386. J. J. Sáenz, "Laser tractor beam," *Nat. Photonics* **5**, 514–515 (2011).

387. V. Kajorndejnukul, W. Ding, S. Sukhov, C.-W. Qiu, and A. Dogariu, "Linear momentum increase and negative optical forces at dielectric interface," *Nat. Photonics* **7**, 787–790 (2013).

388. L. Carretero, P. Acebal, C. Garcia, and S. Blaya, "Periodic trajectories obtained with an active tractor beam using azimuthal polarization: design of particle exchanger," *IEEE Photonics J.* **7**, 3400112 (2015).

389. F. G. Mitri, "Optical Bessel tractor beam on active dielectric Rayleigh prolate and oblate spheroids," *J. Opt. Soc. Am. B* **34**, 899–908 (2017).

390. M. Lax, W. H. Louisell, and W. B. McKnight, "From Maxwell to paraxial wave optics," *Phys. Rev. A* **11**, 1365–1370 (1975).

391. J. F. Nye, "Polarization effects in the diffraction of electromagnetic waves: the role of disclinations," *Proc. R. Soc. A Math. Phys. Eng. Sci.* **387**, 105–132 (1983).

392. J. V. Hajnal, "Singularities in the transverse fields of electromagnetic waves. I. Theory," *Proc. R. Soc. A Math. Phys. Eng. Sci.* **414**, 433–446 (1987).

393. Y. Han, L. Chen, Y.-G. Liu, Z. Wang, H. Zhang, K. Yang, and K. C. Chou, "Orbital angular momentum transition of light using a cylindrical vector beam," *Opt. Lett.* **43**, 2146–2149 (2018).

394. S. Matsusaka, Y. Kozawa, and S. Sato, "Micro-hole drilling by tightly focused vector beams," *Opt. Lett.* **43**, 1542–1545 (2018).

395. M. Rashid, O. M. Maragò, and P. H. Jones, "Focusing of high order cylindrical vector beams," *J. Opt. A Pure Appl. Opt.* **11**, 065204 (2009).

396. Y. Li, Z. Zhu, X. Wang, L. Gong, M. Wang, and S. Nie, "Propagation evolution of an off-axis high-order cylindrical vector beam," *J. Opt. Soc. Am. A* **31**, 2356–2361 (2014).

397. J. Qi, W. Wang, B. Pan, H. Deng, J. Yang, B. Shi, H. Shan, L. Zhang, and H. Wang, "Multiple-slit diffraction of high-polarization-order cylindrical vector beams," in *Proceedings of SPIE* (2017), Vol. 10339, p. 1033927.

398. X.-L. Wang, J. Ding, W.-J. Ni, C.-S. Guo, and H.-T. Wang, "Generation of arbitrary vector beams with a spatial light modulator and a common path interferometric arrangement," *Opt. Lett.* **32**, 3549–3551 (2007).

399. H. Chen, J. Hao, B.-F. Zhang, J. Xu, J. Ding, and H.-T. Wang, "Generation of vector beam with space-variant distribution of both polarization and phase," *Opt. Lett.* **36**, 3179–3181 (2011).

400. S. S. Stafeev and A. G. Nalimov, "Longitudinal component of the Poynting vector of the tight focused optical vortex with circular polarization," *Comput. Opt.* **42**, 190–196 (2018).

401. I. Rondón-Ojeda and F. Soto-Eguibar, "Properties of the Poynting vector for invariant beams: negative propagation in Weber beams," *Wave Motion* **78**, 176–184 (2018).

402. Z. Man, X. Li, S. Zhang, Z. Bai, Y. Lyu, J. Li, X. Ge, Y. Sun, and S. Fu, "Manipulation of the transverse energy flow of azimuthally polarized beam in tight focusing system," *Opt. Commun.* **431**, 174–180 (2019).

403. K. Yao and Y. Liu, "Plasmonic metamaterials," *Nanotechnol. Rev.* **3**, 177–210 (2014).

404. Z. Li, K. Yao, F. Xia, S. Shen, J. Tian, and Y. Liu, "Graphene plasmonic metasurfaces to steer infrared light," *Sci. Rep.* **5**, 12423 (2015).

405. F. Lu, B. Liu, and S. Shen, "Infrared wavefront control based on graphene metasurfaces," *Adv. Opt. Mater.* **2**, 794–799 (2014).

406. Q. Bao and K. P. Loh, "Graphene photonics, plasmonics, and broadband optoelectronic devices," *ACS Nano* **6**, 3677–3694 (2012).

407. J. S. Gomez-Diaz, M. Tymchenko, and A. Alù, "Hyperbolic metasurfaces: surface plasmons, light-matter interactions, and physical implementation using graphene strips," *Opt. Mater. Express* **5**, 2313 (2015).

408. L. Ju, B. Geng, J. Horng, C. Girit, M. Martin, Z. Hao, H. A. Bechtel, X. Liang, A. Zettl, Y. R. Shen, and F. Wang, "Graphene plasmonics for tunable terahertz metamaterials," *Nat. Nanotechnol.* **6**, 630–634 (2011).

409. E. H. Hwang and S. Das Sarma, "Dielectric function, screening, and plasmons in two-dimensional graphene," *Phys. Rev. B* **75**, 205418 (2007).

410. S. H. Lee, M. Choi, T. T. Kim, S. Lee, M. Liu, X. Yin, H. K. Choi, S. S. Lee, C. G. Choi, S. Y. Choi, X. Zhang, and B. Min, "Gate-controlled active graphene metamaterials at terahertz frequencies," *Tech. Dig. – 2012 17th Opto-Electronics Commun. Conf. OECC* **11**, 582–583 (2012).

411. J. Chen, M. Badioli, P. Alonso-González, S. Thongrattanasiri, F. Huth, J. Osmond, M. Spasenović, A. Centeno, A. Pesquera, P. Godignon, A. Z. Elorza, N. Camara, F. J. García de Abajo, R. Hillenbrand, and F. H. L. Koppens, "Optical nano-imaging of gate-tunable graphene plasmons," *Nature* **487**, 77–81 (2012).

412. Z. Fei, S. Rodin, G. O. Andreev, W. Bao, S. McLeod, M. Wagner, L. M. Zhang, Z. Zhao, M. Thiemens, G. Dominguez, M. M. Fogler, H. C. Neto, C. N. Lau, F. Keilmann, and D. N. Basov, "Gate-tuning of graphene plasmons revealed by infrared nano-imaging," *Nature* **487**, 82–85 (2012).

413. H. Yan, X. Li, B. Chandra, G. Tulevski, Y. Wu, M. Freitag, W. Zhu, P. Avouris, and F. Xia, "Tunable infrared plasmonic devices using graphene/insulator stacks," *Nat. Nanotechnol.* **7**, 330–334 (2012).

414. L. Huang, X. Chen, B. Bai, Q. Tan, G. Jin, T. Zentgraf, and S. Zhang, "Helicity dependent directional surface plasmon polariton excitation using a metasurface with interfacial phase discontinuity," *Light Sci. Appl.* **2**, e70 (2013).

415. J. Lin, J. P. B. Mueller, Q. Wang, G. Yuan, N. Antoniou, X.-C. Yuan, and F. Capasso, "Polarization-controlled tunable directional coupling of surface plasmon polaritons," *Science (80-.).* **340**, 331–334 (2013).

416. X. Chen, L. Huang, H. Mühlenbernd, G. Li, B. Bai, Q. Tan, G. Jin, C.-W. Qiu, S. Zhang, and T. Zentgraf, "Dual-polarity plasmonic metalens for visible light," *Nat. Commun.* **3**, 1198 (2012).

417. S. Sun, Q. He, S. Xiao, Q. Xu, X. Li, and L. Zhou, "Gradient-index meta-surfaces as a bridge linking propagating waves and surface waves," *Nat. Mater.* **11**, 426–431 (2012).

418. A. Pors, M. G. Nielsen, R. L. Eriksen, and S. I. Bozhevolnyi, "Broadband focusing flat mirrors based on plasmonic gradient metasurfaces," *Nano Lett.* **13**, 829–834 (2013).

419. Y. Liu and X. Zhang, "Metasurfaces for manipulating surface plasmons," *Appl. Phys. Lett.* **103**, 141101 (2013).

420. F. Qin, L. Ding, L. Zhang, F. Monticone, C. C. Chum, J. Deng, S. Mei, Y. Li, J. Teng, M. Hong, S. Zhang, A. Alu, C.-W. C.-W. Qiu, A. Alù, and C.-W. C.-W. Qiu, "Hybrid bilayer plasmonic metasurface efficiently manipulates visible light," *Sci. Adv.* **2**, e1501168 (2016).

421. X. Ni, N. K. Emani, A. V. Kildishev, A. Boltasseva, and V. M. Shalaev, "Broadband light bending with plasmonic nanoantennas," *Science (80-.).* **335**, 427 (2012).

422. X. Ni, A. V. Kildishev, and V. M. Shalaev, "Metasurface holograms for visible light," *Nat. Commun.* **4**, 2807 (2013).

423. L. Huang, X. Chen, H. Mühlenbernd, H. Zhang, S. Chen, B. Bai, Q. Tan, G. Jin, K.-W. Cheah, C.-W. Qiu, J. Li, T. Zentgraf, and S. Zhang, "Three-dimensional optical holography using a plasmonic metasurface," *Nat. Commun.* **4**, 2808 (2013).

424. X. Yin, Z. Ye, J. Rho, Y. Wang, and X. Zhang, "Photonic spin hall effect at metasurfaces," *Science* **339**, 1405–1407 (2013).

425. J. Lee, S. Jung, P.-Y. Chen, F. Lu, F. Demmerle, G. Boehm, M.-C. Amann, A. Alù, and M. A. Belkin, "Ultrafast electrically tunable polaritonic metasurfaces," *Adv. Opt. Mater.* **2**, 1057–1063 (2014).

426. Y. S. Kivshar, "Metamaterials, metasurfaces, and metadevices," *Aust. Phys.* **52**, 47–50 (2015).

Index